Nathaniel A. Horton

The Sewage of Worcester in its Relation to the Blackstone River

Hearings before the Joint Standing Committee on Public Health, on the matter of restraining the city of Worcester from polluting Blackstone River

Nathaniel A. Horton

The Sewage of Worcester in its Relation to the Blackstone River
Hearings before the Joint Standing Committee on Public Health, on the matter of restraining the city of Worcester from polluting Blackstone River

ISBN/EAN: 9783337377236

Printed in Europe, USA, Canada, Australia, Japan

Cover: Foto ©berggeist007 / pixelio.de

More available books at **www.hansebooks.com**

THE SEWAGE OF WORCESTER IN ITS RELATION TO THE BLACKSTONE RIVER.

HEARINGS

BEFORE THE

Joint Standing Committee on Public Health,

ON THE MATTER OF

RESTRAINING THE CITY OF WORCESTER FROM
POLLUTING BLACKSTONE RIVER.

FEBRUARY AND MARCH, 1882.

R. M. MORSE, Jun.,
GEORGE A. FLAGG,
For Petitioners.

F. P. GOULDING,
City Solicitor,
For City of Worcester.

BOSTON:
Rand, Avery, & Co., Printers to the Commonwealth.
117 Franklin Street.
1882.

JOINT STANDING COMMITTEE ON PUBLIC HEALTH.

MESSRS. NATHANIEL A. HORTON OF ESSEX,
        CHARLES Q. TIRRELL OF MIDDLESEX,
        DANIEL B. INGALLS OF WORCESTER,
                Of the Senate.

MESSRS. ARTHUR H. WILSON OF BOSTON,
        JOHN C. RAND OF MEDFORD,
        GEORGE D. CHAMBERLAIN OF CAMBRIDGE,
        JOSEPH P. HAMLIN OF BOSTON,
        JONAS C. HARRIS OF ARLINGTON,
        DAVID W. HODGKINS OF BROOKFIELD,
        JOHN B. CAMPBELL OF BOSTON,
        CHARLES SMITH OF ANDOVER,
                Of the House.

HEARING

BEFORE

THE COMMITTEE ON PUBLIC HEALTH.

STATE HOUSE, BOSTON, Feb. 21, 1882.

The CHAIRMAN (Senator HORTON). The hearing this morning is upon the whole question of Worcester sewage in its relation to the Blackstone River. Last year there was a petition from parties in the town of Millbury, asking legislation for the prevention of the pollution of the Blackstone River, by the emptying into it of the sewage of Worcester, which the Committee on Public Health considered. They viewed the locality, and, partly in view of the magnitude of the subject, and partly in view of the limited time at their disposal, they reported a recommendation that the whole subject be referred to the State Board of Health to make such recommendations for the action of the present Legislature as they might deem advisable. Although there has been, as I understand it, no formal renewal of the petition from parties in the town of Millbury, the whole subject is nevertheless referred to this Committee, under the rule, as it is presented in this report: and I desire to say to both parties in interest here (presuming that there will be two parties as there were last year), that, speaking for myself, and not for the Committee, I have not as yet read this report; and the same is true of other members of the Committee. We shall make no report to the Legislature, however, until we have carefully examined the Report of the State Board; and simply state these facts that those interested may understand the unbiassed frame of mind in which we approach the consideration of the subject, which is now open. We are ready to hear any suggestions which the parties in interest have to make.

OPENING ARGUMENT BY HON. R. M. MORSE, JUN.

Mr. Chairman and Gentlemen, — I appear in behalf of the town of Millbury, to ask the Committee, in the light of the Report of the

State Board of Health, to consider what measures should be adopted and reported to the Legislature in reference to the pollution of the Blackstone River by the sewage of the city of Worcester. It may tend to shorten this hearing, and facilitate the work of this Committee, if I occupy a little time in the opening by a statement of the history of this matter to its present condition, and of the reasons for legislation.

Prior to 1867 there was no statute under which the Blackstone River was authorized to be used for the purposes of sewerage. The river had run there from time immemorial, and it had had upon its banks a large and busy population. The city of Worcester and various towns to the north and south of it had grown up; large industries had been established, and had flourished; and there had unquestionably been, in the ordinary course of things, more or less pollution of the original purity of the water of the river. Still, no municipal action had been taken by which the character of the stream had been essentially changed; and each party who used the water of the river, whether for domestic purposes, for purposes of manufacture, or for drainage purposes, used it under the general common-law principle by which such use was subject to the equal rights of others. If any individual made any unreasonable or improper use of the water, he was liable to be restrained upon application to the Supreme Court, and was liable in damages for the injury that he did. This was the condition of things prior to 1867. In that year the Legislature passed an Act, chap. 106 of the Acts of that year, which authorized the city of Worcester, among other things, to change the channel of Mill Brook, which is a large tributary of the Blackstone River, and to use it for the purposes of a common sewer of that city; and the city, since that date, acting by authority of that Act, has adopted various orders which are stated in the case of *Butler* v. *Worcester*, reported in the 112 Massachusetts Reports, p. 541, by which it condemned the brook as a sewer. It has taken a number of years for the city to bring its water-supply and its system of sewerage to completion; and, as I am informed, it is only within a short time, comparatively, that the entire sewage of the city has been discharged by means of Mill Brook into Blackstone River. The fact, however, as I now understand it, is, that the city, acting under this special authority of the Legislature, and of course not having the power without that special Act, has discharged, and is now discharging, and proposes to discharge, the entire sewage of the city into this comparatively small stream.

Worcester is a city of something like sixty thousand inhabitants. In addition to its large number of dwelling-houses, it has very large and extensive manufactories; and the discharge from those manufactories, in addition to the discharge from the dwelling-houses, is calculated to create, and does create, very extensive and serious pollution of the waters of the river. The result has been, that within the last few years, beginning back as far as 1872, there has been a great deal of complaint, always increasing, of the injury which is caused by this sewage to the people that inhabit the banks of the Blackstone below Worcester, and who use the water of the river. A nuisance has gradually arisen on that stream. It is felt, of course, principally at Millbury, which is the first considerable town below Worcester; but it extends to a greater or less degree to all the places within the limits of the State, between Worcester and the Rhode Island line. The effect of discharging this immense amount of sewage into the river every day has been to fill up the mill-ponds, and render the water unfit for manufacturing purposes, and to create an offensive smell, which, in various ways, has affected the public health, or is calculated to affect seriously the public health; and the worst of the whole matter is, that the nuisance must necessarily increase, and that, if we leave matters in the condition in which they now are, there is practically no remedy but the absolute removal of the residences and the business of the people that occupy the Blackstone-river Valley below the city of Worcester. It will be found, I think, when this subject is investigated, that there is no place in this State, probably none in this country, where the evil of river pollution is so far reaching and so serious in its consequences as in this case of the Blackstone River. This is due to a variety of circumstances. First, because Worcester is one of the largest inland cities; secondly, because the stream is one of the smallest upon which a considerable inland city is situated; and thirdly, because there is no river in the country upon which, in so short a distance, there is crowded so large and valuable an amount of manufacturing industry as here. The Committee, therefore, can be satisfied, that, in dealing with this question, they are dealing with a matter which is most serious to those affected by it, and most important as a precedent for future legislation.

The question is a new one in the legislation of this country. What shall be done to prevent the pollution of rivers? It is an old one in other countries. In England and on the continent of Europe, the matter of the pollution of rivers is receiving very great and serious attention; and, as a result, no considerable city or town has been permitted to discharge its sewage into rivers without adopting some adequate system of purification. But in this country, where we have been inclined to help ourselves to water-supplies without any restriction, and to discharge our sewage anywhere that it is convenient, without consideration, we have gone on in a hap-hazard, negligent way, until we are being confronted with the serious evil of one city or municipality encroaching most dangerously upon the rights, upon the property, and upon the health of other communities. We need to go back, Mr. Chairman, to the old principle, always true, that one shall not use his own in such a way as to injure the rights of others; and we are going to ask you, gentlemen, in this hearing, so far to rectify and amend the legislation of 1867 as to put into it, in effect, the qualification that should have been there when it was passed, to wit, that, when the city of Worcester uses the Blackstone River for the purposes of sewage, it shall do so subject to the equal rights of the other cities and towns upon the river.

I think it is desirable, especially for the benefit of those gentlemen who were not members of this Committee last year, that I should refer, in proof of what I have been saying, to the action of the State Board of Health during the last ten years. I refer, in the first place, to the Report of that Board in 1873. In that Report, at p. 89, they say, —

"Although one of the streams which unite to form the river is extremely foul (being, in fact, dilute sewage), yet we find that the amount of impurity from this and other sources which remains in the river, by the time it reaches Blackstone, is very small, compared with the bulk of water."

That was the view taken by the State Board of Health on this question at that time. The evil had not then assumed any great magnitude, but it indicates the position which this Board early took upon the question.

In the Report made in 1874, p. 82, the condition of things is referred to, and is stated to be substantially the same; although the Board say, "An analysis might, and probably would, show the amount of impurity to be somewhat increased." At pp. 109, 110, of the same Report, they refer to the serious nuisance that will be created by this sewage. The Committee will find at p. 109 a long report, made to the State Board of Health, at their request, by Phinehas Ball, entitled, "The opportunity and possibility of utilizing sewage in the city of Worcester," Mr. Ball being a distinguished civil engineer, at one time mayor of the city of Worcester, and always, I believe, a resident thereof. This report is devoted to a consideration of the serious nature of the nuisance, and of the opportunities the city of Worcester has for abating it, and for utilizing the sewage.

In 1875 an Act was passed (chap. 192) to provide for investigating the question of the use of running streams as common sewers; and, in pursuance of that Act, an extensive investigation was made of the Blackstone River as the one first demanding attention. In the report for 1876 the State Board of Health, on p. 173, say of the Blackstone River, "It is probably more polluted than any other river in Massachusetts."

Now, Mr. Chairman, the Committee must understand that this action of the State Board of Health has been mainly of their own motion, and in pursuance of their duty to observe matters affecting the health of important communities in the State. Here is a series of reports to which I invite the careful attention of the Committee, showing that, year after year, the attention of the Board of Health was called to the necessity of doing something to remedy an evil that was sure to increase. During this time the city of Worcester has not been unobservant of what has been going on. It would be doing great injustice to the authorities and citizens of the city to suppose, for a moment, that they overlooked the fact that the discharge of the sewage into the Blackstone River was calculated to cause injury to the towns and industries below them; and I think that the Committee will find, in the inaugural addresses of several mayors of that city, and in various reports that were considered by the city governments, allusions, more or less distinct and significant, to the evil which the city of Worcester was causing to its neighbors, and to the necessity of taking some action to prevent it.

I have here the inaugural addresses of two mayors of the city of Worcester. I desire to call attention to the address of Hon. Henry Chapin, mayor *ad interim*, made on the 2d of January, 1871, in which he says, —

"The introduction of water from Lynde Brook seemed to make it necessary that some means should be devised for its disposition, in order that what was designed for a blessing might not prove to be an evil in disguise. When the system contemplated shall be consummated, and we are relieved of the danger which threatens us, another and vastly important question will present itself, which is even now extensively agitated. That question is, Cannot there be some method devised by which the sewage of the city may be utilized? Enough fertilizing power goes to waste, in the usual method of sewerage of our cities, to furnish the means of enrichment to the surrounding country. I hesitate not to prophesy that the time will come, sooner or later, when the sewage of the city of Worcester will be so utilized as to become, not only a source of income to the city, but to make many a field a garden, and many a neighborhood to blossom like the rose."

These observations were directed mainly to the use of the sewage as a matter of profit to the city, a use which we hope can be combined with the abatement of the nuisance to the inhabitants along the river below. I desire further to call the attention of the Committee to the inaugural address of his Honor, George F. Verry, made to the city of Worcester on the 1st of January, 1872. I beg the Committee to notice that this was years before the completion of the system of sewerage, and at a time when the nuisance had not reached its present proportions. He says, —

"The present sewer from what is called the Piedmont District empties into Mill Brook below Sargent's card factory; and, when the waters of this brook are diverted, there will be no means of carrying the sewage off. Deposited upon this low land, and remaining there, it would be likely to breed a pestilence in that neighborhood. It will, therefore, probably be necessary to extend that sewer through this low land, so as to connect it with the main sewer at or near Cambridge Street.

"It may also become necessary to provide a remedy for the mischief which our sewage is in danger of doing to the waters of the Blackstone River, into which it is in great part conveyed. Complaints, whether well or ill founded, are not infrequent from those who reside and do business along its valley, that the stream is greatly polluted from this cause. If these apprehensions are well founded, the business of providing a remedy will deserve, as I have no doubt it will receive, your earnest and immediate attention. No argument is necessary to enforce the performance of the duty of self-preservation, which we owe to ourselves. None should be needed to enforce the performance of that other duty 'So use your own as not to injure another,' which we owe to our neighbors.

"The subject of providing means for utilizing our sewage has been heretofore discussed, and has been recommended as a profitable enterprise in a pecuniary point of view. Of this I have no knowledge; but I am advised by the city engineer that a plan of utilizing the sewage can be adopted which is feasible, and which will at the same time relieve the Blackstone of the nuisance complained of, which plan will be submitted to your consideration if it shall be your pleasure to desire it. I would therefore recommend that an investigation of these matters be, as soon as practicable, entered upon with the view of providing a remedy, if one is required."

I do not understand that any specific action was taken upon this recommendation of Mayor Verry. The city went on increasing its discharge of sewage into the river, spending, as I was informed, last year, — and I presume that the statement was substantially correct, — a million and a half of dollars in providing a very complete system of sewage for the city; complete, that is to say, in this, — that it enabled all its citizens to empty their filth into the Blackstone River. It has also expended, as I am informed, a million and a half of dollars for its water-supply. But though the city has thus spent three millions of dollars in providing itself with an ample supply of water, and an admirable system of sewerage, it has not spent a dollar that I am aware of in the adoption of any practicable plan for diminishing the injury which it was causing to its neighbors and friends along the Blackstone River below the city. It has had warning after warning from its mayors and its city engineers. It has been told that it was causing a nuisance, and yet that the very cause of the nuisance could be utilized to give it a profit, to all of which the city gave no serious attention. In that state of things last year, a very large number of petitioners came before the Legislature asking aid from the Legislature. They came from Millbury and from the towns below. I believe that all the localities and all the industries to the south of Worcester, and within the limits of this State, were represented before this Committee. We had patient and careful attention from the Committee. The city of Worcester was represented by its chief magistrate, and was heard by its able counsel. The Committee took a view of the locality, and saw for themselves what the evil was, and what it was likely to be. We presented a bill requiring the city of Worcester to take some definite action. The Committee, however, as the chairman has stated, in view of the magnitude of the question, and of its novelty, to some extent, were unwilling peremptorily to order the city of Worcester to abate this nuisance; and therefore they came to the conclusion, in their wisdom, that the proper thing to do was to send this whole matter to the State Board of Health, an impartial, intelligent, and competent tribunal, and ask them to investigate it, and to report to this Legislature whether a practicable plan for the removal of this nuisance could be adopted.

The State Board of Health, in obedience to that resolve, have taken up the subject.

Mr. GOULDING. You mean the State Board of Health, Lunacy, and Charity?

Mr. MORSE. Yes, sir; but the lunacy and charity divisions of the Board were not specially concerned with this subject.

Mr. GOULDING. Millbury was not investigated.

Mr. MORSE. The Board referred this subject to a commission of three experts, whose expenses were authorized to be paid by the Governor and Council. I mention these facts because, in considering the weight to be given to their report, it should be borne distinctly in mind that the town of Millbury, the petitioners in this case, and the other persons aggrieved by the nuisance, had nothing to do with the selection of the commission. The Committee will find the report of the Board of Health on this subject in the general Report of the Board for the present year. The first reference to the matter is on p. lxv : —

"As the consideration of this report . . . will bring more directly to public attention than ever before the rapidly increasing pollution of streams not used as sources of water-supply for domestic uses (but which, as in the case of the Blackstone at Millbury, are becoming too foul even for manufacturing purposes, and as objectionable to residents on their banks as open sewers would be), it is time to ask whether the State must not take one step more, and protect rivers not used for domestic water-supply in the interests of the residents upon their banks, and of the manufacturers themselves. A comparison of the chemical analyses of waters of the Blackstone River made in 1881 with a large number made by the State Board of Health in 1875 reveals a very serious increase in the percentages of polluting matter."

I will read the resolve of last year upon which the Board acted : —

"*Resolved*, That the State Board of Health, Lunacy, and Charity is hereby authorized and directed to examine and consider the question of the disposition of the sewage of the city of Worcester, especially with a view to prevent the pollution of the Blackstone River and its tributaries, and report its conclusions in print to the next Legislature, with recommendations as to a definite plan for the prevention of such pollution. For this purpose the Board may employ such assistants, and incur such engineering or other expenses, as shall be approved by the Governor and Council." *Approved* May 12, 1882.

After quoting this resolve, the Report of the Board (p. 66) says, —

"The Board at once entered upon the investigation of this question, and, after due notice given to all the parties in interest, spent parts of two days, early in July, in Worcester, when a hearing was had, at which appeared the city of Worcester and the town of Millbury, represented by city and town officers, or committees duly appointed. One result of this hearing was, that the Board voted to request the city of Worcester and the town of Millbury to submit in writing such evidence of experts, as to methods of disposal of the Worcester sewage, as each municipality should deem proper, especially with a view to prevent the pollution of the Blackstone River and its tributaries. Another was, that Dr. C. F. Folsom of the National Board of Health, J. P. Davis, C.E., of New York, and Dr. H. P. Walcott, health officer of this Board, were appointed a committee to consider the matter of the disposal of Worcester sewage, and report their conclusions to this Board. They have presented their report with plans and estimates of expense, which, together with the documents furnished by the town of Millbury, will be found in the special Sanitary Appendix. Our recommendations as to a definite plan for preventing the further pollution of the Blackstone River are given in Part Fifth."

I may add, in this connection, that, of the gentlemen appointed, Dr. Walcott is of course known to this Committee as the health officer of the Board. Dr. Folsom is a distinguished expert upon this class of subjects ; and Mr. Davis was formerly city engineer of Boston, one of the most distinguished ever in its employ, and the one who laid out the extensive system of sewerage which is now in course of construction in that city.

A further reference to this subject is to be found on p. ccviii, under the title, " The disposal of the sewage of the city of Worcester."

I will not read that in detail, because I propose to refer in another connection to the plans that are recommended. I will say, however, that the Board there state, as the result of the investigation of this commission, that they recommend the system of " intermittent downward filtration," supplemented, if necessary, by broad irrigation, as best adapted to the existing state of things, and the best method of disposing of the sewage of the city of Worcester. The Committee will also find in the Appendix to the Report of the State Board of Health, on p. 117, the detailed report of the experts, which states most if not all of the facts that the Committee will desire to have before them in determining the question now under consideration. That report occupies a good many pages. It shows the extent to which the river is polluted by sewage, the amount of the ordinary flow of the river, of the sewage, and the reasons for recommending the specific plan before mentioned. On p. 134 in the Appendix will be found a report from the town of Millbury, a communication from a committee of that town to the secretary of the State Board of Health, Lunacy, and Charity, transmitting a report prepared for that town by George E. Waring, jun., a distinguished expert upon sanitary matters, which is printed on p. 137 of the Appendix. I may state that Col. Waring was employed by the town of Millbury at their own expense to consider this subject, and to make a recommendation of a practicable method for disposal of the sewage. Appended to that on p. 145 is an estimate of the cost of the plan recommended by Col. Waring, signed by two civil engineers of reputation, Mr. Ball and Mr. Heald, and by Amos Pike, contractor. I may state here for the information of the Committee, that, as I understand the two plans that are recommended, — one by Col. Waring, and the other by the experts selected by the State Board of Health, — they agree in this, that it is essential that the sewage of the city should be kept separate from the ordinary flow of the stream. As we now look at it, that principle should have been understood and stated when this system of sewage was first entered upon by the city of Worcester. It would have saved a good deal of cost and trouble, if at that time the whole matter had been carefully investigated ; and I assume that the city of Worcester would then have been more ready than it is to-day to adopt such a system. But the city came here and got its legislation, and then adopted its system of sewerage without consideration, in fact, of other localities. It looked to what would be the best for its own interests. Whereas, upon consideration, I believe the city would say that it ought to have taken into account the interests and rights of others. The city did, however, adopt the system without regard to its effect upon the people below ; and under that system they have used one channel, the old channel of Mill Brook, with such alterations of it as they have found it convenient to make for the disposal, not only of the ordinary flow of the stream, but of the entire sewage.

Now, both Col. Waring and the commission of experts say that that is a wrong principle, and that it is essential that the sewage of the city should be kept distinct from the ordinary flow of the stream. The main difference between the two plans recommended, as I understand it, is this: that the commission of experts report that it is desirable and necessary to construct two lateral sewers, one on either side of Mill Brook, into which the sewage shall flow, and that the sewage kept separate in that way from the ordinary flow of the stream shall be carried down to a point below the city where it can be pumped up, and then allowed to flow upon a large tract of cheap land to be specially prepared for the purpose, the flowing to be conducted under a system of intermittent filtration, — that is to say, using a portion of the land one week, and another portion another. In this way the sewage will be purified; and the water, freed from the offensive matter, will then go, by a course shown on the plans, back into the river. This plan allows the ordinary flow of Mill Brook to go directly into the Blackstone River. It takes the sewage of the city in separate sewers to localities where its contents can be purified and the water returned to the river.

That is the plan recommended by this commission. It requires a large outlay, but the expense is not excessive as compared with the amounts which the city of Worcester has already invested in its water-supply and sewage system, nor is it beyond the reasonable ability of the city to make such an expenditure. It is a proper part of the cost of its water and sewerage systems. The cost of this plan is estimated by the commission at $408,490; this estimate including the separate system of sewers, pumping-station, land, and land damages, the preparation of the land, and all other items. To this the commission add, that, if a system of utilizing the sewage should be entered upon, it will involve a further outlay of one hundred thousand dollars. That is, of course, a matter which concerns the city of Worcester more than it does anybody else. For the purpose merely of removing the nuisance, an expenditure of four hundred thousand dollars in the first place is unquestionably needed.

Col. Waring, on the other hand, reports, that in his opinion it is practicable to divide the channel of the present sewer, Mill Brook, into three channels, and to make the lateral sewers that I have spoken of, inside of the present large sewer. In other words, he avoids the expense of constructing new sewers, and believes that the existing sewer would be sufficient, if properly divided. The Board of Health do not agree with him in that opinion. The expense of carrying out the plans proposed by Col. Waring is $206,500. I may say that he proposes to discharge the sewage, and to purify it at a different point, and in a somewhat different manner from those recommended by the plan of the experts' commission; but that is not very essential to the point under consideration.

The Committee, then, will have before them two possible practicable methods for purifying this sewage. Each of them undoubtedly involves considerable expense. To remedy an evil of this magnitude must cost a large sum; but it is an expense which, sooner or later, must be met, and it is less now than it will be hereafter. I say it must be met, because the Committee are brought face to face with this alternative; and every Committee of the Legislature will be, until the question is decided, either that the city of Worcester, in consequence of taking this brook for a sewer, and discharging its sewage into it, is thereby entitled to ruin the rest of the valley of the Blackstone, to drive out its inhabitants, and destroy its industries, or else it must provide in some way for taking care of its sewage. You cannot by any process reason away the proposition that a city of sixty thousand people, constantly increasing as we believe and hope it will increase, cannot go on discharging its enormous mass of filth, year after year, into this little river without finally polluting it to such an extent that not only its water cannot be used for any purpose whatever, but that its banks must become uninhabitable.

Mr. Chairman, I have referred to the report of the State Board of Health, not because I consider that it is necessary that the petitioners here should show the Committee or the city of Worcester how and in what way it can remedy this evil, but because I know, that, in view of the friendly relations of my clients to their good neighbors and friends in the city of Worcester, they ought not to complain of what the city has done under an Act of the Legislature, unless they can show that the injury might have been prevented or may now be stopped.

Before I close, however, and come to the specific legislation which we ask the Committee to recommend, I wish to say a few words on the question of the legal liability of the city of Worcester. I do it because I think there has been considerable misapprehension in the public mind on that subject. I admit the possibility of a difference of opinion upon questions of this sort. I may be mistaken; but I desire to present my own view, and then the reasons why, notwithstanding that opinion, I still ask for legislation.

The general principle was stated by Mayor Verry, — who was not only a good mayor, but is a good lawyer; and my friends here will agree to it. It is, that one shall not use his own so as to injure the rights of his neighbor. That principle has been applied time out of mind to the use of water in a running stream. Under it our courts have held that the owner of a mill privilege must return the water to the stream, subject only "to those slight and substantially immaterial obstructions and retardations which necessarily result from exercising the right of a mill privilege above" (7 Gray, 348).

So, again, the court has held that one may not pollute a running stream. Repeatedly injunctions have been applied for, and obtained, by an owner or dweller upon a stream, against one who fouled the water above him. The court has always held, that while a certain inappreciable amount of injury may be caused to a running stream, which the court will not take account of, yet, that wherever that injury is of such a grave character as to be excessive, as to involve an unreasonable use of the water, the court will restrain it, or the person offending will be liable for damages.

Now, apply that principle to the case of the city of Worcester. I take it that there is no question but that, prior to the Act of 1867, any individuals who discharged their sewage into the Blackstone River, whether from the city of Worcester or from any town above or below that city, would be liable in damages to any person who could prove that he was thereby injured. When the Act of 1867 was passed, however, which authorized the city of Worcester to take that brook for sewage purposes, a new class of liabilities arose. In that Act was a provision that a person injured by the taking of the brook, or by the taking of land, or by any other proceeding under the Act, was entitled to file his petition and to recover his damages. And the courts have decided, in reference to that portion of the Act, that so far as the taking of the brook for sewage purposes involved, as a natural and necessary consequence, an injury to an individual or his property, his remedy was to be sought through a petition under that Act. That was decided, as my learned friends here know very well, first in the case of *Merrifield* v. *Worcester*, 110 Mass., p. 216. But I desire to call the attention of the Committee to the language of the Supreme Court in making that decision. It is an important case, and will be referred to, very likely, as much upon the other side as upon this. It was a suit brought by the owner of a lot of land on both sides of Mill Brook in Worcester, who alleged that he had a right to have the water of the brook flow pure and uncorrupted, and that the defendant, that is, the city of Worcester, had deposited in said Mill Brook and the waters thereof, at points in the channel, above and higher than his works, great quantities of filth, dirt, gravel, etc. In other words, he alleged that the city of Worcester had turned its sewage into Mill Brook. He brought an action against the city to recover damages; and the court, in its decision, said, that "so far as he has suffered damage from any proper exercise of the power and rights conferred" by the Act of 1867, authorizing the taking of Mill Brook as a sewer, he had no right of action; his only right was to petition for damages under the statute. But, at the conclusion of its opinion, the court says, "Whether the damage which the plaintiff has suffered is attributable in any degree to the improper construction or unreasonable use of the sewers, or to the negligence or other fault of the defendant in the care and management of them, is a question which does not appear by the record to have been tried." In other words, the court in that opinion asserted the principle, which is referred to in other cases, that, although the right to damages of the person injured by the taking of Mill Brook as a sewer, so far as such injury is the natural and necessary consequence of that taking, is limited to a petition under the Act, yet, if the city, by the improper construction or unreasonable use of the sewers, was negligent, there would be a liability which could be enforced by a suit at law. I refer also to the case of the *Washburn & Moen Manufacturing Company* v. *The City of Worcester*, in the 116th volume of Massachusetts Reports, p. 458. That was a suit in equity brought by the Washburn & Moen Manufacturing Company to restrain the city of Worcester from causing a nuisance upon its premises by reason of the discharge of its sewage into Mill Brook. The court refers again to the principle I have stated, as laid down in the case of *Merrifield* v. *The City of Worcester*, that the liability of the city for damages occasioned as the natural and necessary consequence of the taking of the brook as a sewer, could be recovered only by way of petition; and then it goes on to say, that "the Bill does not allege any negligence of the city, either in the manner in which the sewage was discharged from the mouth of the sewer, or in *omitting to have proper precautions to purify it.*" Here, again, is the assertion of the principle, by implication, that, if the city of Worcester omits to take proper precautions to purify the sewage discharged into Mill Brook, it is liable to an action at law for damages; it is liable to restraint in a suit in equity; and, in case the nuisance which it occasions is a public nuisance, it is liable to indictment for maintaining the nuisance.

Still further, as illustrating this principle, I wish to call attention to the case of *Badger* v. *the City of Boston*, which will be found in the 130th volume of Massachusetts Reports, p. 170. I would not ordinarily occupy the time of the Committee in citing legal authorities, but I think that in the course of this hearing it may be important to have these questions considered. This was a suit by Erastus B. Badger against the city of Boston, which arose in this way: In 1876 the Legislature passed an Act authorizing the city of Boston to construct urinals in the public streets, and to take land for that purpose. The Act further provided, that any person who was injured in his property, by reason of the construction of a urinal, might apply for an assessment of his damages in the same way as for land taken for highway purposes. Mr. Badger thereupon brought a petition for damages, in which he alleged that a urinal had been constructed near his place, and that it was a nuisance to him, and an injury to his property. At the trial he undertook to show that the urinal was offensive and a nuisance. The court, however, declined to hear the testimony, upon the ground, that, in a petition for the assessment of damages under that Act, the fact that the urinal proved to be a nuisance was not a subject for consideration, nor for the allowance of damages. In effect, the principle is laid down that the Legislature is never to be presumed as authorizing a municipality or an individual to create a nuisance. It gives it specific powers; it authorizes it to take land; it authorizes it to do certain things; and it says to individuals who are injured, "So far as your injury is the natural and necessary consequence of what is done under that Act, you can come in and petition for a jury, in the way in which damages are assessed for the taking of land for highways; but after that land has been taken, and that work has been constructed, if, as a consequence of the negligent way in which the work is constructed, or the negligent way in which the public work is carried on, the injury is occasioned, that is a new and independent claim, a new source of liability, on which an action at law may be maintained, or a bill in equity to obtain an injunction may be sustained." Mr. Justice Endicott, who gives the opinion in that case, says, " If this urinal, by reason of its management or use, becomes a public nuisance, the city may be liable to indictment for thus maintaining it, or be subject to an action of tort by the person who suffers special damage thereby;" but then proceeds to say, that in this petition the court could not consider it. To the same general propositions I cite the important cases of *Haskell* v. *New Bedford*, 108 Mass. 208; *Brayton* v. *Fall River*, 113 Mass. 218; *Boston Rolling Mills* v. *Cambridge*, 117 Mass. 396.

Now, if I make myself clear to the Committee, the application of the principles laid down in these cases, and in others which I have not referred to, is this: The Act of 1867 authorized the city of Worcester to take Mill Brook for the purposes of sewage, to construct new channels, and to take land therefor; to alter the boundaries of the brook, and to do a great many things which are necessary to the use of that stream for sewerage purposes. So far as any individual was injured, as the natural and necessary consequence of the taking of Mill Brook as a sewer, he could recover damages on his petition the same as if the city had taken his land, or injured his property, in the construction of a highway, or in the alteration of the grade of a road. But if the city, having appropriated this brook for the purposes of a sewer, so constructed the sewer as that it caused injury to the people who had a right in that river, or so managed the discharge of its sewage as that it was not properly purified and freed from polluting substances, then from the time that that injury began, and so long as it continues, it is liable to suits for damages on the part of every individual who receives special injury in his person or his estate. It is liable to a suit in equity, on the application of persons who are specially affected by its wrongful acts to compel an abatement thereof; and if, as we fear, this evil shall become of such magnitude as to constitute a public nuisance, the city is liable to indictment by the grand jury of the county. That I believe to be the law affecting the city of Worcester.

Now, Mr. Chairman, my friends on the other side may say, — my friend, Mr. Goulding, did say it last year, — "If you have a claim for damages in suits at law, why not bring your suits at law? Why not bring your Bill in Equity? Why not get an indictment? Why not involve the city of Worcester in a great litigation?" That may be the result, Mr. Chairman. The people on the Blackstone River below the city of Worcester must protect their rights; but they have not thought it the part of good neighbors, in view of the fact that the evil has come gradually, and, I am bound to concede, almost imperceptibly, to the apprehension of everybody on that stream, at once to go to law about it. They have had an inclination, and have shown it year after year, to avoid litigation; and they will await the decision of this Legislature, after it shall have been put in possession of all the facts, as to whether or not a proper modification shall be made of the Act of 1867.

I cannot state the views of the people who ask for legislation here better than by adopting the language which our distinguished senator, Mr. Hoar, addressed to his fellow-townsmen of Worcester on the eve of the last election, and after this question of a remedy for the pollution caused by the sewage of the city of Worcester had become a prominent issue in that election.

"Mr. Hoar said that he did not desire that Worcester should send representatives or a senator to the Legislature, *to get her off*, as if she had done or were doing some wrong, for which she was to be indicted as for a nuisance. Worcester cannot afford to put forth her strength to do a wrong to her neighbors and friends, the towns below her in the Blackstone Valley. Our representatives and senator ought in this matter to represent the justice of the whole State, and not a mere local interest. They should do just what they ought to do if they represented Berkshire or Essex, or one of the towns interested. While he was not one of the largest taxpayers, he was so situated that the burden of taxes pressed upon him most heavily; but he would rather Worcester should pay one million dollars than do a wrong to one of these towns. It is a great and serious thing to poison the air, to pollute the streams, or destroy the health of the homes of a town like Millbury or Sutton or Northbridge or Uxbridge or Blackstone. Worcester must call to her aid all the resources of science, all the experience of other cities and countries, all the ingenuity of mechanic art, to avoid such a result, whatever may be the cost. For one, he desired his representatives in the Legislature to meet the question in this spirit."

Nothing can be said better, stronger, clearer, nobler, than that statement from Mr. Hoar. He recognizes the injury that has been done, and is being done. He recognizes that it will involve a great cost to remove it. He recognizes that the natural, selfish instinct of most people will be to say, " Well, so long as we get rid of our sewage, we won't trouble ourselves until we are compelled to do it, in regard to its effect upon other people ; " and then he says, whatever may be the cost, the city of Worcester must summon to its aid all the resources of science and all experience to devise and adopt some plan for the removal of this nuisance. Mr. Chairman, while there are such men in Worcester, and such a spirit shown there, it would certainly be in the highest degree unfair and unneighborly for the people injured by this evil to take any legal action ; and that is why they have forborne to do it.

I will also call attention to an editorial from " The Worcester Gazette " upon the same subject. It appeared Tuesday evening, Nov. 8, 1881.

"Senator Hoar takes a just view of the sewage controversy, and one not essentially different from that already advanced by us. The ancient fashion, when a difference of opinion arose between cities or individuals, was to break heads with cudgels, or send out opposing armies to destroy each other. The progress of civilization has modified these customs. Hard words, and writs, injunctions, and attachments are the weapons most in use at the present day; but human nature is still belligerent, and men fight more readily than they compromise. The peaceful settlement of the Alabama Claims by arbitration was looked upon in the light almost of a new invention for the prevention of wars,— at least the beginning of a new era of enlightenment in the world's history. We do not at present advise either arbitration or compromise in the issue between this city and the towns in the valley below; but if the citizens of Worcester will approach the subject in a spirit somewhat broader and more liberal than heretofore, and will recognize, that, while they have their rights and necessities, they have not after all been behaving in a very neighborly manner, we think a great deal of money, useless wear and tear of mind, and waste of energy can be saved. We still hope that some practical and profitable way of utilizing the sewage may be discovered. No system could well be more wantonly wasteful and expensive than that now in common use here and in the cities of the world. This is Yankee land, and in the taking out of patents there is no end. A portion of all this ingenuity might very well be turned toward this question, and it would be a sagacious step on the part of the city government to offer a handsome reward for such a discovery."

I had the honor of saying to this Committee, a year ago, Mr. Chairman, that I believed that the inventive ability of our engineers and scientific men would discover a practical method, adapted to our climate and our situation, of freeing sewage from improper matters, precisely the same as it has made useful and valuable improvements in all other directions. The fact is, that only within the last few years has the attention of our people been directed to sanitary questions. Now they have become, as this Committee are specially aware, of the greatest importance. Almost all the cities have their water-supply; most of them have systems of sewerage: but, in securing those great luxuries which are only the necessities of modern civilization, they have not given sufficient attention to the effect which the use of these conveniences may have upon their neighbors, and also, I may say, to the waste which they themselves are making of what they have.

The object of this hearing is to ask this Committee to require the city of Worcester to take some steps in the matter. It has been, up to this time, a matter of investigation, a matter of consideration. Now we ask the Committee to require the city of Worcester to take some action. I have prepared a Bill, which is intended to be as fair and reasonable as it is possible for a Bill to be framed, upon the basis of the obligation of the city of Worcester to take care of its sewerage. It does not require that the city shall adopt the specific plan recommended either by the State Board of Health or by Col. Waring. It does not give to the State Board of Health, even, the power to determine what the city shall do. It does not require the city to do any thing at present but investigate: but it does require it within four months to adopt some system for abating the nuisance it causes; and it provides, that, after the expiration of four months, the sewage shall not be discharged into Blackstone River, until it has been properly purified.

I may say that the Bill is framed largely upon the Act reported by this Committee, and passed by the Legislature of last year, requiring the city of Boston to purify the water flowing through the sewer into lower Mystic Pond. That Bill not only received the approval of this Committee and of the Legislature, but it has been sustained by the Supreme Court in a suit in equity brought to compel the city of Boston to carry out its provisions. The question of the constitutionality of the Act was raised in the hearing before a single justice, who held that the Act was constitutional. From this decision the city appealed; but, before the case was heard by the full court, the city consented that a decree might be entered for the petitioners, and for a perpetual injunction, thereby assenting to the validity of the Act. I may say, however, that that Act was open to a good many objections which would not apply to this one; because, in that case, as the Committee will remember, the sewer which the city of Boston was required to purify was merely an artificial sewer, and not an ancient water-course.

The further provisions of the Act which I submit to this Committee are the ordinary ones, allowing the city to take land and construct works, and authorizing proper appropriations for that purpose. Then, there is at the end a provision similar to the section in the Act of last year in reference to Mystic River, by which the selectmen of any of the towns upon the Blackstone River are permitted to apply to the Supreme Court sitting in equity for an injunction, or for other appropriate action in case of a violation of the Act.

Let me restate our case before this Committee. The city of Worcester is polluting, in an unexampled way, the water of this river, creating a great nuisance there. We do not say to the city, "You must stop that nuisance at once." We appreciate the difficulty of stopping it at once. We simply say, "Take, in addition to all the years that you have had to give to this subject, four months more to consider it. Employ such experts as you please, devise any scheme, — the least expensive that will answer the purpose is, of course, the most desirable for you, and it is just as satisfactory for us, — but, at the end of the time named, do something definite to prevent the continuance of this nuisance." After the city has determined what it will do, we ask only the further provision, that, if the selectmen of any of the towns affected can satisfy the Supreme Court that the city has not done enough, the court may, by injunction or some other process, require something further to be done.

I may add, that we are prepared with evidence upon the existence and extent of the nuisance. We are prepared to show by the testimony of those who live upon the river below the city of Worcester, that this evil is of very great magnitude; that it is affecting very seriously the value of property, and, what is more important, the safety of life. We are also prepared, by the presence of Dr. Walcott, to give to the Committee the results, in perhaps a more satisfactory shape than even the printed report of the investigations made by the State Board of Health. We shall be prepared at another meeting, in case the Committee desire to go further in the matter, to present Dr. Folsom, and also Col. Waring, in order to assist the minds of the Committee, as far as it is in our power, in determining the important practical question whether the city of Worcester can do any thing. We are willing to go as far as the Committee may desire, and we propose to submit ourselves very largely to their direction as to the order of the hearing; but of course we shall be influenced very much by the position which our friends from Worcester may take. On the one hand, we have no desire to burden the Committee with long hearings upon questions of fact. On the other hand, we do not desire the matter to be disposed of without a realizing sense, on the part of every gentleman on this Committee, of the seriousness and importance of the question.

Mr. GOULDING. We came here without knowing what the proposition of the other side was, and without knowing what course they would pursue in this hearing. When they have finished their case, we shall be ready to present ours; and we are indifferent, of course, as to what method is pursued here, except that we desire as short a hearing as possible. I desire to say that I shall not, at any stage of this hearing, consume any considerable time in any opening statement. I shall controvert most of the positions of law, and many of the positions of fact, which the counsel takes upon the other side, from the same evidence which he has himself adduced. I shall take occasion carefully to analyze these reports of the State Board of Heath, as far as they bear upon this question; and I should like to have the privilege of doing that in the closing argument, even if it should take more time, rather than to make two arguments upon the matter.

The CHAIRMAN. Is there any likelihood that the city of Worcester will accept this or any similar Bill?

Mr. GOULDING. Oh, no, sir! not at all. We shall show, we think, that there is no occasion for any thing of the sort.

The CHAIRMAN. How far does the city of Worcester admit the fact of a nuisance here?

Mr. GOULDING. We deny that there is any nuisance. We shall offer to show, from evidence that has already been put in, that there is no nuisance, as a necessary implication. Of course, it depends upon what you mean by nuisance; but we speak in the legal sense of a nuisance.

Mr. MORSE. I understand, then, that we are to proceed in the ordinary way, with the introduction of evidence?

The CHAIRMAN. Yes, sir.

EVIDENCE FOR THE PETITIONERS.
TESTIMONY OF NATHAN H. GREENWOOD.

Q. (By Mr. FLAGG.) — You are one of the selectmen of Millbury? *A.* — I am.

Q. — State where you live in Millbury. *A.* — About one mile north of the centre of the town, near Burling Mills, so called.

Q. — How near the Blackstone River? *A.* — Within five or six rods.

Q. — How long have you resided there? *A.* — About thirty-nine years, with the exception of six years.

Q. — State what you remember of the condition of Blackstone River fifteen years ago, as compared with its present condition. *A.* — It was comparatively pure. The waters were a great deal clearer. For instance, there is a bridge crossing it near my house; there is quite a deep place on one side, where the water is twelve or fifteen feet deep, perhaps more; then, in a bright, sunny day in the summer, you could see the bottom quite distinctly. I have seen it many times as a boy. Now, on the other side, where the water is not more than a foot deep, you cannot see the bottom in the brightest day in summer.

Q. — What have you noticed as to smell? *A.* — There is, in the summer more particularly, quite an offensive smell. Of course, at this time of year the smell is not so bad; but in the summer-time it is very offensive.

Q. — How far from the river have you noticed that smell? *A.* — Well, I can't say exactly how far, but quite a considerable distance; quite a number of rods I have noticed it.

Q. — Will cattle drink of the stream at your farm? *A.* — Well, very seldom, without they are very thirsty.

Q. — How wide is Blackstone River at your place of residence? *A.* — Well, right directly opposite my house, there is quite a projection runs out in coves; but a short distance above there, I should say, or below, I should think it wasn't more than twenty-five feet in width; that is, in the summer, with the natural flow of the water.

Q. — About how deep at that place? *A.* — It is quite shallow all through there, except this hole that I speak of, just below the bridge, which is merely a place of three, four, or five rods.

Q. — Do you use the stream for bathing now? *A.* — No, sir, we do not.

Q. — Did you in former years? *A.* — It was customary, when I was a boy, until within a dozen or fifteen years, perhaps ten years, for the boys to use it.

Q. (By the CHAIRMAN.) — You mean they do not use it as a bathing-place on account of the impurities? *A.* — Yes, sir: they would get more dirt on than they would get off by bathing there now.

Q. (By Mr. FLAGG.) — Can you state what action was taken by the town of Millbury in regard to this matter?

Mr. GOULDING. I suppose the action of the town of Millbury would be shown by its records, if you mean votes.

Q. — In general, whether or not the town of Millbury has taken any action? *A.* — It has taken the action of appointing a committee to endeavor to find some remedy, or have it remedied.

Q. — Now, as to the health of your family the last few years? *A.* — Well, the older members of it, my mother and aunt, have been unwell. Last fall my aunt was very sick. The physician we had said it was malarial disease, — malarial fever. That it was caused by the river, of course I cannot say.

Q. — Did you attribute this ill health in your family to pollution of the river? *A.* — I did.

Q. (By Mr. MORSE.) — Did the physician? *A.* — I don't know as he said so to me.

Cross-Examination.

Q. (By Mr. GOULDING.) — Your residence is exactly where, sir? *A.* — It is near the Burling Mills, just across the river.

Q. — On what road or street? *A.* — It is on the old road leading from Worcester to Millbury.

Q. — Then it is on the west side of the Blackstone River? *A.* — The west side of the Blackstone River.

Q. — You go over the new Millbury road, crossing the river north of Burling Mills, to get to your place? *A.* — Yes, sir.

Q. — You are about five or six rods from the Blackstone River? *A.* — Yes, sir.

Q. — At that point, has the Blackstone River two channels, or one? Does the water run in two channels, or in one? *A.* — Except at high water, it runs in one channel; at high water it runs over and through another channel.

Q. — The channel it runs in is the channel that leads it to the Burling Mills? *A.* — No, sir.

Q. — Is there not a channel that takes the bulk of the water of the river to Burling Mills? *A.* — There is.

Q. — Is that the natural channel of the Blackstone River? *A.* — It is not: it is the old canal.

Q. — Do you live opposite to that channel? *A.* — No, sir.

Q. — You live above it? *A.* — I live opposite to it, on the west side, and further across the river from there.

Q. — And opposite where you live, there is this artificial channel, which takes the water of the river to Burling Mills? *A.* — Yes, sir.

Q. — Is that the old Blackstone canal? *A.* — It is a large portion of it.

Q. — Now, does not, except in times of freshets, all, or nearly all, the water that runs down the river, go through that channel? *A.* — When the mill is running, the larger part of it, in dry weather, goes that way in the daytime; in the nights it does not.

Q. — There is a roll-way on the west side of the channel, above the mill, to lead off the water when it is not wanted for the mill? *A.* — There is.

Q. — Is the water turned into that channel at the north end of it? *A.* — All of it, when the water doesn't run over their dam. They have a dam there. When the water does not run over it, all, except what leaks through, goes down that channel.

Q. — Now, when you speak of the channel of the river near your place, do you mean this natural channel of the river, which runs parallel with this canal? *A.* — I do.

Q. — No considerable part of the water goes there, except at times of high water? *A.* — It goes there when the Burling Mill isn't running.

Q. — The mill is generally running, is it not? *A.* — In the daytime: it is not in the habit of running nights.

Q. — You are not out there to see it much, nights? *A.* — I can smell it, though, if I can't see it. It is not necessary to *see* the Blackstone River to know where it is.

Q. — Perhaps, Mr. Witness, you understood me as asking whether you could smell it or not. I asked you whether you were out there nights to see it. *A.* — I answered it: I said I was not.

Q. — Now, what does your family consist of, Mr. Greenwood? *A.* — I have a wife and three children.

Q. — Your mother, you say, lives with you? *A.* — My mother lives with me.

Q. — Have you always lived in the same house? *A.* — Always, with the exception of six or seven years that I was away.

Q. — Is your own health pretty good? *A.* — Comfortable.

Q. — What is your business? *A.* — I am a farmer at present.

Q. — How old are you? *A.* — Thirty-nine years.

Q. — Ever been sick at all? *A.* — Yes, sir.

Q. — Well, when? *A.* — I have not been sick abed for some time, with the exception of once, a year ago last fall.

Q. — What were you sick with then? *A.* — I don't know.

Q. — Any idea what it was? *A.* — Neuralgia pain, I guess: I believe the doctor called it that. That was not at home: I was away from home at the time.

Q. — Where were you? A. — I was in Northborough.

Q. — Any rivers down there? A. — I don't attribute that to the sewer, Mr. Goulding, if you please. I don't want you to understand that I attribute that to the sewer.

Q. — When were you sick at any other time besides that? A. — I haven't said that I was sick at all.

Q. — I haven't said you did; but you said you had been sick in your life. Now, I ask you if you have ever been sick, except that time when you were at Northborough? A. — Why, yes.

Q. — When? A. — I can't remember the dates.

Q. — What diseases have you had? A. — Nothing particular, as I know of.

Q. — Have you ever had any malaria? A. — I never have had any malaria: no, sir.

Q. — How long have you been married? A. — Six years.

Q. — Wife in good health? A. — Yes, sir.

Q. — How old are your children? A. — One is five years, and the youngest one is a year and a half.

Q. — Children in pretty good health? A. — Yes, sir.

Q. — How old a lady is your mother? A. — Seventy-six.

Q. — When was it you said she was sick? A. — She hasn't been well for a great many years.

Q. — What has been the matter with her? A. — I don't know.

Q. — You don't know? A. — No.

Q. — Have you any idea what it is? A. — It is a sort of debility. No, I don't know what it is: I don't know what to call it.

Q. — A kind of debility? A. — Yes.

Q. — Accustomed to have a doctor? A. — Not very often; sometimes.

Q. — Ever asked the doctor what was the reason of your mother's debility? A. — I don't know that I ever did.

Q. — Have you the slightest idea what the cause of your mother's debility is? A. — She has been so ever since I can remember.

Q. — Then, for thirty-nine years she has been in a debilitated condition? A. — I don't remember for thirty-nine years.

Q. — Say thirty-five, or thirty-three, or thirty-four years? A. — Thirty years.

Q. — Has she lived there all the time? A. — Yes, sir.

Q. — And been in about the same condition of debility all that time? A. — Well, with the exception of growing more so.

Q. — The infirmities of age affect her somewhat? A. — I presume so.

Q. — When was it she was sick? Last fall, did you say? *A.* — No, sir.

Q. — Hasn't your mother been sick, except this general debility? *A.* — No, sir. I said my aunt had been sick.

Q. — Your aunt was sick, then? *A.* — Yes, sir, she was.

Q. — How old is your aunt? *A.* — She is about seventy.

Q. How long has she lived there? *A.* — She has lived there all her life.

Q. — Was she ever sick before this occasion, that you remember of? *A.* — I don't remember any serious sickness that she had, except once before.

Q. — When was that? *A.* — I can't tell when it was. It was when I was away; it must have been some eight or nine years ago, — eight years, perhaps.

Q. — You were not there? *A.* — I was not there at the time.

Q. — Do you know what was the matter with her then? *A.* — I do not.

Q. — Do you remember what season of the year it was? *A.* — It was in the summer.

Q. — You don't know what the matter was? never inquired? *A.* — I presume I did know; but I have forgotten what the trouble was.

Q. — How long was she sick? *A.* — I can't tell you now exactly.

Q. — With the exception of those instances that you have related, you know of no other sickness? *A.* — No, sir.

Q. — How long was she sick this last time? *A.* — She isn't over it yet; that is, she isn't so well as she was before she was sick. The doctor called her dangerously sick for about a week.

Q. — The doctor never told you what the cause of it was, and you never inquired of him? *A.* — He said he didn't know what the cause of it was, if I remember correctly; he didn't want to say; he couldn't say what the cause of it was, or didn't want to say.

Q. (By Mr. MORSE.) — Did the doctor come from Worcester? *A.* — No, sir.

Q. (By Mr. GOULDING.) — Where did he come from? *A.* — From Millbury.

Q. — He didn't want to tell you what the matter was? *A.* — He didn't tell me: I don't know that I asked him.

Q. (By the CHAIRMAN.) — Do you appear as representing the selectmen officially, or as a citizen? *A.* — I appear at the request of the committee of the town.

Q. — A committee chosen in town-meeting? *A.* — Yes, sir.

TESTIMONY OF GEORGE D. CHASE.

Q. (By Mr. FLAGG.) — What is your business, Mr. Chase? *A.* — I work in the sash and blind shop in Millbury.

Q. — You live in Millbury, and are one of the selectmen? *A.* — Yes, sir.

Q. — Live near the river? *A.* — Perhaps a matter of fifteen or twenty rods from it.

Q. — You work there near the river? *A.* — Within two or three rods.

Q. — How long have you known the river? *A.* — Some twenty years or more.

Q. — Will you state its condition, as you remember it, fifteen years ago, and as you know it to-day? *A.* — Fifteen years ago the river was clear. It was used a good deal at that time for bathing; that is, amongst the boys, I amongst the rest of them. I remember it more particularly on that account. I used to be round the river a great deal; and the water was very clear, more particularly at the upper dam, where we used to go in bathing, — at Mr. Morse's shop. It was an excellent place for bathing; had a nice bank, and the water was clear. We used to go in there very often.

Q. — Was that the sash and blind shop where you now work? *A.* — That was the sash and blind shop where I work now. But no bathing has been done in the river for some time: it is not fit for the purpose. The river at the present time, or rather in the summer season, is very offensive; much more so within a year, or a year and a half, than ever before.

Q. — Where do you notice that? *A.* — I notice that more particularly near the place of David Harrington than anywhere else, although I notice it at the shop.

Q. — Where is David Harrington's place situated on the river? *A.* — It is perhaps an eighth of a mile north-west, or north, of the sash and blind shop.

Q. — It is an eighth of a mile from the river? *A.* — Oh, no, sir! It is right on the river. The buildings are within thirty feet.

Q. — Do you notice this smell at your house? *A.* — I do sometimes during the summer season: yes, sir.

Q. — Now, you speak of its offensiveness. What do you mean by offensiveness? What do you notice? *A.* — A very strong smell. I can't define it in any other way. It is very offensive, I know. I notice it more particularly at the shop, because, in falling over the dam, of course it stirs it up.

Q. — The dam at the sash and blind shop? *A.* — Yes, sir.

Q. — Do you take any precautions to keep this offensive odor from you while at work? *A.* — Yes, sir: we shut the windows.

Q. — Is that customary among the workmen there at the shop? *A.* — It has been done a great many times.

Q. — Has there been an increase in this bad condition of things during the last few years? *A.* — Yes, sir, very marked.

Q. — Do you notice it is growing worse from year to year? *A.* — Yes, sir.

Q. — At your house do you notice these odors? *A.* — Yes, sir.

Q. — What has been the effect upon your family? *A.* — It has not been good. I have not been in good health for the last two or three years. The doctor didn't seem to know what it was at first, but he has finally laid it to that; and, in fact, my boy has been sick; and that is the only way that I could account for it.

Q. — Was your attention called last summer to some dead fish in the stream? *A.* — Yes, sir.

Q. — Will you tell what you saw? *A.* — Notice came to me one Sunday afternoon — I forget the date now — that the river had the appearance of cotton balls, as the man expressed it, below the dam at Mr. Morse's shop. He said the authorities had got to do something about it, or ought to do something about it; and I went up and saw Mr. Greenwood, and, with Mr. Greenwood and Mr. Whitney, I went out to the river, and we found it in a very bad shape. The river there, for an eighth of a mile, perhaps, was covered with fish; and we hunted round for some men and a boat to take them out. We got one boat out with two men, and went after some more; and, by the time we got back with the second lot, the first boat-load had given up. I asked them what the trouble was, and they said they couldn't go that. We finally got another boat out; they worked a while, and they gave up; and finally Mr. Greenwood ran across a couple of men, and they went out and finished the job. Since then, in settling with the parties, we have had to pay quite a considerable sum for cleaning the river, they claiming that they had received a great deal of injury by taking the fish out; that they had been sick, and had lost a good deal of time. The doctor said, in the case of one man in particular, that, if he had not had an excellent constitution, it would have hurt him permanently: he has been sick, laid up, some time since then.

Q. — What was the condition of the river at that time as to depth of water? *A.* — Well, at that time, if I remember right, I think it was low.

Q. — What time of year was it? *A.* — It was in summer: I can't say now just what time.

Q. — Those fish appeared just below the dam of the sash and blind factory? *A.* — Yes, sir.

Q. — How far down below? *A.* — It was down below the Providence Railroad bridge: the river there is very crooked.

Q. — Was it, in fact, between the two dams? *A.* — Yes, sir, it was between the two dams.

Q. — Those fish that the men took out, I don't understand were decayed fish? *A.* — Some of them were; that is, they were dead fish.

Q. — How long had they been dead? *A.* — They made their first appearance that day, that morning, or that afternoon: that was the first that we had heard of it.

Q. — How large a space did those dead fish cover? *A.* — I should think perhaps we worked over some six hundred feet; that is, we cleaned the river.

Q. — In this distance of six hundred feet, how many baskets full did they take out? *A.* — There were some fifteen or twenty bushels: I can't say just how many. There was over fifteen and under twenty.

Q. — What kind of fish were they? *A.* — Suckers.

Q. — What other kind of fish are there in the river? *A.* — Most all pouts; that is, there have been pouts, pickerel, perch, and the other common fish found in country streams: but for the last few years I see very few other fish than pouts and suckers; that is, I have not noticed any.

Q. — What is the condition of the pond at the sash and blind factory, as to its appearance now, compared with what it was fifteen years ago? *A.* — It is very different. Fifteen years ago there was quite a large pond; at the present time it is filled up very much. At one point I notice, particularly above the cemetery, the water has changed to land, and is changing very fast: it is filling up very fast indeed. The pond has changed in size a great deal within the last four or five years.

Q. — What sort of growth appears in the pond? *A.* — It is a kind of weed: I don't know what it is, — coarse water-grass and weeds.

Q. — Are there any ice-houses by the pond? There were, some years ago, but they have been abandoned.

Q. — Would you buy ice for your family taken from that pond? *A.* — No, sir.

Q. — Formerly the village was supplied with ice from that pond, was it not? *A.* — Yes, sir: there were two ice-houses within gun-shot that used to supply the town, but they have been abandoned.

Cross-Examination.

Q. (By Mr. GOULDING.) — Mr. Chase, how long have you worked in that sash-factory? *A.* — I have worked there now nine years, or it will be nine years next month, continuously. Before that, I worked there.

Q. — It was burned down within a few months, wasn't it? *A.* — Yes, sir.

Q. — And has been rebuilt since? *A.* — Yes, sir.

Q. — That sash-factory is about how far below Quinsigamond Village? *A.* — I should think perhaps a matter of three and a half miles, or four miles, — somewhere there.

Q. — What do you do in the factory? *A.* — Work on window-frames.

Q. — Which story do you work in? *A.* — Since the fire, I have worked in the first story; before that, I worked in the second story.

Q. — About how many have been employed there since you worked there? *A.* — The average number of men is about fifty, — from fifty to fifty-five.

Q. — Do you keep up the same number throughout the year? *A.* — About the same: they come and go some.

Q. — I noticed you said, in answer to a question whether it was the custom to shut the windows, that it had been done a great many times. Do you mean to say it is the custom to keep the windows shut in summer at that factory? *A.* — It was the custom last summer. We didn't keep them shut all the time, but we did a great deal of the time.

Q. — A great deal of the time it was the custom last summer, — was it prior to last summer? *A.* — No, sir.

Q. — Was the water unusually low last summer? *A.* — Sometimes in the summer it was low. I don't know as it was much lower than it has been.

Q. — Were considerable areas of that pond exposed during the summer? *A.* — Not so much as what has been, — not any more.

Q. — Not any more than has been? *A.* — No, sir.

Q. — But were there not considerable areas exposed during the summer? *A.* — No, sir.

Q. — Pond full all the time? *A.* — The usual depth of water there.

Q. — Well, this thing you noticed last summer? *A.* — Yes, sir.

Q. — How large is that pond? *A.* — I cannot say: I should think it was somewhere about thirty acres, — somewhere along there.

Q. — Is the health of the hands at that mill as good as at other places generally? *A.* — No, sir.

Q. — What facts have you noticed about that? *A.* — I have noticed that the hands have been complaining, and they all cuss the river.

Q. — Have they cussed the river always, ever since you have been there? *A.* — No, sir.

Q. — When did they begin to cuss it? *A.* — They commenced cussing it a year or two ago, but more so last summer.

Q. — Their cursing was more terrible last summer than before? *A.* — Yes, sir.

Q. — Although they cursed, they didn't leave the mill, they kept at work there? *A.* — It has been a matter of consideration with a great many whether they should not leave.

Q. — I didn't ask you about their consideration, but whether they did actually leave. *A.* — Some, I think, have left.

Q. — Who has left? *A.* — I can't remember who has left: some of them have.

Q. — Do you know anybody who has left on account of the scent there, — that you can testify left because the smell was bad? *A.* — I can't think of any one now who left for that special reason.

Q. — Well, it is in summer, I suppose, that it is the worst? *A.* — Yes, sir: it is natural that it should be.

Q. — You have lived there a long time, you say, in that vicinity, and known the river? *A.* — I have known the river for twenty years.

Q. — How old are you? *A.* — Thirty years. I have known the river thirty years. I have lived near to it, now, nearly twenty years, — fifteen years or so.

Q. (By the CHAIRMAN.) — Do you mean the Committee to understand that any who did leave, left on account of the smell? *A.* — No, sir: I won't say that they left on account of the smell.

Q. (By Mr. GOULDING.) — You don't know of any? *A.* — No, sir, not now.

Q. — Has this cursing increased since the last hearing before the Legislature? *A.* — No, sir.

Q. — They cursed about as much before? *A.* — I should think so.

Q. — Mr. Morse is your employer, and he is one of the leading gentlemen who are making this agitation? *A.* — Mr. Morse is on the committee: yes, sir.

Q. — Isn't he one of the leading gentlemen who are agitating this subject? *A.* — He is one of the men; but he is not alone, by any means.

Q. — He is one of the leading men? *A.* — Yes, sir.

Q. — You spoke about a miraculous draught of dead fish there: when was that exactly? *A.* — I could not tell you, sir, exactly.

Q. — No; but as near as you can tell, when was it? *A.* — I cannot tell what month it was in: it was in the summer, some time, I know.

Q. — Last summer? *A.* — Last summer.

Q. — Your attention was called to the fact that a large number of fish were floating in the water, below C. D. Morse's mill, as I understand it? *A.* — Yes, sir.

Q. — And covered a large area? *A.* — Yes, sir.

Q. — How many of you went to work getting them out? *A.* — I think there were six or eight different men on the river at different times.

Q. — How long a period did this work cover of getting them out? *A.* — It covered a good share of a Sunday afternoon.

Q. — What did you do with the fish? *A.* — Buried them.

Q. — Did you ever see any such phenomenon as that before? *A.* — No, sir.

Q. — Have you since? *A.* — No, sir.

Q. — Have you ever heard any cause for it mentioned by anybody? *A.* — No, sir. I don't know the reason for it; never heard of any.

Q. — Never heard that poison was put into the river for the purpose of killing the fish, — never heard a rumor of that kind? *A.* — I have heard a rumor of that kind.

Q. — What was the rumor you heard? *A.* — Well, that they were poisoned: that is all I know about it.

Q. — That is as definite as any thing you have heard about it? *A.* — Yes, sir.

Q. — You have heard that the fishes were poisoned by somebody? *A.* — By somebody, yes, sir.

Q. — You don't know whether it is true or not? *A.* — No, sir.

Q. — But you do know that you never saw such a thing before, and never have seen it since? *A.* — Yes, sir.

Q. — There are a great many suckers and eels in the river now, are there not? *A.* — I don't know: I haven't had any occasion to find out.

Q. — Any musquash or mink there? *A.* — No, sir. Years ago musk-rats used to build there a great deal. In and around the ring-meadow about Mr. Morse's pond, where their houses used to be counted by the hundred, I was going to say, now you cannot count a dozen.

Q. — Can you count ten or eleven? *A.* — I don't think you can.

Q. — There are some there? *A.* — There may be a few there.

Q. — You will admit a few rats? *A.* — Well, a few.

Q. — Any mink? *A.* — No, sir. You may see one once in a while, but not many.

Q. — Have you seen any of late there? *A.* — Mr. Greenwood, I believe, caught one not a great while ago.

Q. — Now, you spoke of Mr. Harrington's house; and that, you say, is very near the pond of Mr. Morse? *A.* — Yes, sir.

Q. — How old a man is Mr. Harrington? *A.* — He is a man, I should think, seventy-five years old.

Q. — He has pretty good health? *A.* — Not first class; you couldn't expect it.

Q. — Good as men average at that age? *A.* — Yes, sir: I should think so.

Q. — Is he a brother of Mr. Stephen Harrington of Worcester, do you know? *A.* — I can't say.

Q. — You say he has always lived there? *A.* — No, sir, I don't think he has. He has lived there as long as I can remember.

Q. — You can remember twenty-five years back, perhaps? *A.* — Yes, sir.

Q. — Do you know of any death in his family? *A.* — Yes, sir.

Q. — Who has died? *A.* — An old lady — I don't know how she was related to him — died this last winter.

Q. — At how early an age was she cut off? *A.* — She was along in the latter part of life. I can't say how old she was: somewhere about eighty years old, I should think.

Q. — Any other deaths in his family? *A.* — Not that I can state. I have not kept the track of his family.

Q. — Don't you know that this lady had really reached the age of ninety-two years? *A.* — I can't say how old she was.

Q. — Didn't you hear her spoken of as one of the oldest people in the town? *A.* — I suppose I heard how old she was; but I did not lay it up against her because she had lived to be ninety years old.

Q. — Now you think of it, you think she may have been ninety years old? *A.* — She may have been.

Q. — Your own residence is how far from this pond? *A.* — I should think twenty or twenty-five rods.

Q. — Which side of the pond are you? *A.* — The north-east side.

Q. — Any house between you and the pond? *A.* — Yes, sir.

Q. — What does your own family consist of? *A.* — Wife and boy and grandmother.

Q. — How old is your grandmother? *A.* — She was ninety last June.

Q. — How long has she lived there? *A.* — She has lived there about twenty years.

Q. — Is her health pretty good, considering her age? *A.* — It is, sir.

Q. — How old is your wife, — about your age? *A.* — Yes, sir.

Q. — She is in pretty good health? *A.* — Fair.

Q. — You say your boy has had a little ailing? When was he sick? *A.* — He was taken two years ago this winter.

Q. — In the winter? *A.* — Yes, sir.

Q. — What was the matter? *A.* — I didn't know at that time: some kind of fever.

Q. — Did the doctor tell you what the cause of it was? *A.* — He couldn't tell of any thing, unless it was malarial fever.

Q. — Did he say it was malarial fever? *A.* — He couldn't tell what it was. He said he couldn't think of any thing else.

Q. — How long was he sick? *A.* — Six weeks; that is, he was up, and then taken down again.

Q. — Six weeks each time, or does that include both times? *A.* — Each time.

Q. — How old is the boy? *A.* — Seven in June.

Q. — Go to school? *A.* — No, sir.

Q. (By the CHAIRMAN.) — Do I understand that it was generally considered in Millbury that the death of those fishes was caused by the pollution of that stream? *A.* — I don't know the reason for it. I don't know of any reason, if it was not that. What the cause was of their death, I don't know.

Q. — Did they pass through the wheel-way? *A.* — They were below the tail-race of Mr. Morse's mill; but I don't think they got over there.

Q. — Any possibility of any schools of fish being killed on the passage through, or any thing of that kind? *A.* — No, sir.

Q. — What I wanted to get at was, whether it was the general opinion of people in that vicinity that the fish were killed by the impurity of the water, or not? *A.* — Yes, sir.

Q. — You stated that you had heard a rumor that the fish were poisoned? *A.* — I said I heard a rumor that the fish were poisoned; but the general opinion was that they were killed by the condition of the river.

Q. — How do you account for the fact that there has been no such phenomenon since or before? *A.* — I don't know, unless it was caused by the condition of the river at that time. The river was somewhat low; and whether the water got too warm for them, or the water was too rich, and that, with the heat together, cooked them, or what it was, I don't know; but evidently they died, and we had to take care of them. The river was in very bad shape, at any rate. The water was low at that time.

Q. (By Mr. GOULDING.) — There are one or two questions suggested by this last examination, and one is, where were these fish with reference to Singletary Brook? Were any of them below Singletary Brook? *A.* — Yes, sir.

Q. — Singletary Brook is a brook that comes into Blackstone River from the west, is it not, and comes down from Bramanville, that way? *A.* — Yes, sir.

Q. — Were some of these fish in that Singletary Brook? *A.* — I don't know of any being in there.

Q. — But some of them were in the river below Singletary Brook? *A.* — Yes, sir.

Q. — There was no dam or any thing to prevent the fish running up Singletary Brook, if they wanted to run out of the impurities of the Blackstone River? *A.* — No, sir.

Q. — You do not undertake yourself to account for this singular appearance of this large number of dead fish on that day, and never before or after, do you? *A.* — I don't know what the reason was.

Q. — Didn't a good many people think that they were poisoned, for a purpose, by somebody? *A.* — I never heard but one person speak of it.

Q. — Who was that? *A.* — That was Mr. L. L. Whitney.

Q. (By Mr. FLAGG.) — On that day, do you remember whether there was any current at that place, or not? *A.* — No, sir.

Q. — What was the color of the water? *A.* — The color of the water was very muddy and yellow, if I remember right; in fact, the water last summer was different from what it ever was before. I don't know the cause of it. It was not any different that day from what it had been for days before.

TESTIMONY OF LEVI L. WHITNEY.

Q. (By Mr. FLAGG.) — You live in Millbury? *A.* — Yes, sir.

Q. — You are one of the selectmen of Millbury? *A.* — Yes, sir.

Q. — You have been on the board in previous years? *A.* — I was on the board the year before last, and for two years before that.

Q. — Have you been a member of the Legislature? *A.* — I was a member of the Legislature last session.

Q. — State where you live in Millbury? *A.* — I live in what is known as Bramanville, up about three-quarters of a mile from the Blackstone River, — half to three-quarters of a mile west from the river.

Q. — You are familiar with the Blackstone River? *A.* — I pass it nearly every day.

Q. — How long have you been familiar with it? *A.* — I have lived there now for ten years.

Q. — Will you state its appearance ten years ago, as compared with what it is to-day? *A.* — Well, there has been a gradual change in the appearance of the river every year: it has grown gradually more impure every season since I have been there. It was not a pure stream, by any means, ten years ago; but it was in a better condition than it is at the present time. Ten years ago, the sewage of the city of Worcester (that part that entered the river) was not probably more than one-third what it is at the present time. The river was in as bad a condition, I think, as it ever has been in my remembrance, a year ago last summer: that was a very dry season, and the river was very low. This last season was not quite so dry a season, and there was more water running; and we did not notice it quite as perceptibly as we did two years ago.

Q. — State what you have noticed about the river, as to its color, its odor, and appearance? *A.* — The general appearance of the river is like that of any filthy, polluted stream. The water is black; the banks are black and filthy. The filth that comes down the river accumulates on the banks; and the ice, as it forms on the river, has a yellowish appearance. Instead of being clear, pure ice, it is dark, discolored ice.

Q. — Would you think of using that ice for domestic purposes? *A.* — No, I don't think I should: it is not suitable for any purpose whatever.

Q. — As to the color of the water, and the smell of the water? *A.* — The color is black, like any polluted stream; and there is more or less offensive odor that comes from it all the time as I pass by it. I notice it in going on the highway from Millbury to Worcester. As I drive up the Millbury road, I notice the odor that comes from the river.

Q. — How long a distance do you notice that? *A.* — Well, the whole distance from Millbury to Quinsigamond, where the sewer enters the Blackstone River. The river runs perhaps thirty rods from the road, almost parallel with it.

Q. — Now, is this odor which you speak of that you notice something about which there is no mistake, or is it imagination? *A.* — Well, there cannot be any question what it is: it arises from the river. We notice it as we pass over the river, and as we are going towards the river; and, when we get within a few rods of it, we notice it there; and, as we pass over it, it becomes stronger; and then, as we ride up the Worcester road, and all the way from the village to Quinsigamond Village. There is'nt much imagination about it: it becomes a reality to those who have to endure it.

Q. — Can you describe to the Committee the nature of the odor? *A.* — I don't know as I can.

Q. — You can give them some idea, can you not? *A.* — It does not differ very much from any other odor which would be made of the same material of which this is made. I don't know as I can describe it any better: I don't know that it is any different from that.

Q. (By Mr. MORSE.) — How far from the river can you smell that odor? *A.* — At times for a quarter of a mile. Some families that live a quarter of a mile distant say that they are very much troubled with it just at night, in summer, in their houses.

Q. (By Mr. FLAGG.) — You were familiar with those dead fish that were in the river last year, that have been talked about? *A.* — The thing was brought to my attention.

Q. — Has Mr. Chase described it as it was? *A.* — He has described it to you as I saw it when I got to the river.

Q. — Does Singletary Brook come in there? *A.* — Yes, sir.

Q. — Do not the waters of the Blackstone back up until they reach the first privilege on Singletary Brook, John R. Rhodes's? *A.* — The waters of the Blackstone River set back up, I might say, right to the dam of John Rhodes.

Q. — Will you tell as to the management of the case for the town, whether the committee is composed of individual manufacturers and mill-owners, or whether they act by vote of the town? *A.* — There was a committee appointed by the town to take charge of the matter; and that committee had instructions from the town, by vote, to make an effort to prevent the pollution of the river, or to remedy the evil. That committee is made up of yourself [Mr. Flagg], Mr. C. D. Morse, and Mr. Waters. I believe you have the honor of being chairman of that committee.

Q — How many inhabitants has Millbury? *A.* — The last census, I think, gives us 4,700, — something very near 4,700.

Q. — What is the general business carried on there? *A.* — The principal business is manufacturing. The cotton and woollen manufacture is the principal business located on the stream.

Q. — Will you enumerate those mills? *A.* — The first mill on the Blackstone River, after it enters Millbury, is what is known as the Burling Mills. That is a woollen-mill.

Q. — About how many people are dependent upon the Burling Mills? *A.* — It is a seven-set mill. I should say they employ 125 to 140 hands in the mill; and, judging from other mills that I know more about, I should say the number dependent upon them would be about 250 or 300. The next is Mr. Morse's mill. That Mr. Chase is more familiar with than I am. He says that there are from 50 to 55 employed there. That is a sash and blind shop; and nearly all the employés, I might say all, are men. They would represent from 30 to 45 families. The next mill below that is the Atlanta Mill, which is a woollen-mill, and a four-set mill, I think it is, which would employ about 100 hands, I should think. The next mill is the Millbury Cotton-Mill, manufacturing print cloths, with about 150 looms, employing 100 to 125 hands. The next mill below that is what is known as the Cordis Mills, one of the largest corporations we have in the place, manufacturing tickings. I should think they employed somewhere from 175 to 200. The next mill below that is the Messrs. Simpson Satinet-Mills, perhaps a little larger than the Atlanta Mill, employing perhaps 150 hands, more or less.

Q. — What is the distance between the first you mention, the Burling Mills and the Morse Mill? *A.* — I should say it was a mile: it may not be quite that.

Q. — Within how short a distance are all those other mills situated?

A. — It is about the same distance from Mr. Morse's Mill to the Atlanta Mills; then the Atlanta Mills and the Millbury Cotton-Mills are within a hundred feet of each other; the Cordis Mills, perhaps, half a mile below; and Mr. Simpson's, perhaps a little short of half a mile from the Cordis Mill. Below Simpson's is the print-cloth mill. They have about a hundred and fifty looms, — a hundred and seventy-five perhaps. I don't know exactly the size of it: but it is the only mill in Sutton on the stream; and from that it goes to Saundersville, in Grafton, and so on, down to Farnumsville, and the cotton-mills along there.

Q. — Whether or not from the Burling Mills down to Blackstone there is any place where the river runs naturally like a river, — whether the field has not been, in fact all, taken up by these mills, — whether there is any thing but a succession of ponds? *A.* — The pond of one flows back, usually to the wheel of the other. They are all about as thick as they can be, and get the flow that they are entitled to; not much of any space between them. One pond flows back to the wheel of the other.

Q. — How long have you lived in Millbury? *A.* — Ten years.

Q. — You were born there? *A.* — No, sir: I was born in Princeton.

Q. — You lived in Millbury a good many years ago? *A.* — I lived there from 1854 to 1859.

Q. — Were all those dams in existence then? *A.* — They were: yes, sir.

Cross-Examination.

Q. (By Mr. GOULDING.) — You live out in Bramanville? *A.* — Yes, sir.

Q. — How many mills are there on Singletary Brook? *A.* — Seven.

Q. — What kind of mills are they? *A.* — There are three woollen-mills and four cotton-mills.

Q. — Which is nearest to Blackstone River? *A.* — The first mill is Mr. Rhodes's, — a cotton-mill.

Q. — All these mills sewer into the river, I suppose. Their privies are all over the stream, are they not? *A.* — No, sir.

Q. — None of them? *A.* — Some of them are; not all of them.

Q. — What ones are not? *A.* — The upper mill is not; and two of the others, I think, are not.

Q. — Who owns them? *A.* — Mr. Walling owns one, and Crane and Waters the other.

Q. — Do they use the water to wash the wool in the woollen-mills? *A.* — Yes, sir: they use the water.

Q. — And for dyeing purposes? *A.* — Yes, sir.

Q. — How far is the first privilege from the junction of Singletary Brook with the Blackstone River? *A.* — About half a mile: it may not be quite that. I should say it was about half a mile.

Q. — You say that you have observed this smell as you were driving to Worcester? *A.* — Yes, sir.

Q. — Any particular place where you discover it more than at others? *A.* — I don't know as there is any particular place, unless it be some point where you come nearer the river. Around the turn where Mr. Chase lives, you come near the river; and, as you pass by Mr. Morse's pond, you come as near the river as at any place.

Q. — You would observe it as bad as anywhere where Mr. Chase lives? *A.* — I should think that it would be as strong there as anywhere on the road.

Q. — You mean Mr. George F. Chase's? *A.* — Yes, sir.

Q. — Is there any odor of dye-stuffs perceptible? *A.* — No, sir.

Q. — None whatever? *A.* — No, sir.

Q. — How far did you ever smell that river? I mean, how far away from the river were you, at the farthest, when you have smelt it? *A.* — Well, perhaps twenty rods.

Q. — When was that, if you can tell the time, and under what circumstances? *A.* — No: I made no memorandum of it.

Q. — Well, under what circumstances have you smelt it twenty rods from the river? *A.* — No extraordinary circumstances: the usual time in the summer, when the stream is low. Almost any time in the summer you can smell it. I have no doubt, if you ever rode down the Millbury road in summer, you have smelt the same smell that I have a good many times.

Q. — You don't recall any particular time when you smelt it? *A.* — No, sir, no particular time: it is a general complaint.

Q. — How long ago did you ever smell it twenty rods from the river? *A.* — Any season, the last four years.

Q. — Prior to that, did you ever notice it? *A.* — Not as bad as it has been since.

Q. — Did you ever notice it prior to that time? *A.* — I think it has been noticeable any time the past twelve years.

Q. — Have you noticed it? *A.* — I don't know as I have, particularly, myself. I don't live on the banks of the river.

Q. — Are you on any committee of the town connected with this matter? *A.* — No, sir.

Q. — Do you own any of these factories, or have any interest in them, on the Blackstone River? *A.* — No, sir.

Q. — Where are your factories? *A.* — I live in Bramanville, upon what is known as Singletary Brook.

Q. — Do you own more than one mill up there? *A.* — I have one which I am operating now, and perhaps a small interest in another.

Q. — Which mills are you interested in? *A.* — One is a cotton-mill, and one (Crane & Waters) is a woollen-mill. It is the second mill on Singletary Brook from the Blackstone River.

Q. — In your woollen-mill this water is used for dyeing your wools? *A.* — We use the water for all the purposes that it is usually used for in a woollen-mill.

Q. — And discharge it in the way that woollen-mills ordinarily discharge their water? *A.* — Yes, sir.

Q. — Who called your attention to those dead fish first? *A.* — Word was sent to me, during Sunday, that the river was full of dead fish. I don't recollect, now, who first called my attention to it. I know I hitched up and drove down there.

Q. — How long were you down there? *A.* — I was there all the afternoon.

Q. — Do you remember when it was exactly? *A.* — It was some time in July.

Q. — You don't remember the Sunday? *A.* — I don't remember the Sunday: no, sir.

Q. — Have you any way of fixing that date? *A.* — It was, I should say, somewhere the last part of July. Mr. Hull thinks it was some Sunday in August. I got the impression that it was earlier than that.

Q. — You have the impression that it was in July; but the other gentleman, who speaks to you, thinks it was in August? *A.* — Yes, sir. I had the impression that it was in July; but it was in warm weather.

Q. — Did you help get some of the fish out? *A.* — No, sir: I didn't help get them out.

Q. — You had nothing to do with it? *A.* — No, sir.

Q. — How long did you observe the operation? *A.* — I was there during the afternoon. We employed some men to get them out.

Q. — Did you ever see any thing of the kind before? *A.* — I don't know that I ever have: no, sir.

Q. — And never since? *A.* — No, sir.

Q. — It was a singular thing? *A.* — I thought so at the time.

Q. — Did you ever hear that those fish were probably poisoned by somebody who put poison in for the purpose of killing them? *A.* — You have got about three questions into one. I can answer one of the questions. You can cut the question up.

Q. — You may divide it to suit yourself, and answer it in detail. *A.* — The first question is, if I ever heard they were poisoned. It came to my notice some time within a week, or shortly after, that they might have been poisoned.

Q. — You say it came to your notice: what do you mean by that?

A. — Some one, who did not know any more about it than I did, suggested that they might have been poisoned.

Q. — Is that any answer to my question? *A.* — That is all the answer I have to make to that.

Q. — Had you ever heard any thing about it before that? *A.* — No, sir.

Q. — Have you any means of accounting for this singular appearance of these fish on that Sunday afternoon, and never before or after? *A.* — No, sir.

Q. — Have you ever undertaken to account for it in your own mind? *A.* — No, sir, I have not.

Q. — Can you now give a reason which you think is satisfactory at all? *A.* — I don't know how to account for it, or any thing about it. I have never taken any pains to investigate the matter, or to find out about it. As one of the selectmen, it was my duty to remove them as soon as possible; and I should think it was the wisest course to do to take them out of the river.

Q. — I understand you to have answered that you have no opinion about it? *A.* — No, sir, I have not any idea.

Re-direct.

Q. (By Mr. FLAGG.) — Have you ever heard anybody say that they thought, or had any reason to believe, that those fish had been poisoned, or any thing more than a casual remark by somebody, that they might have been? *A.* — No, sir, none whatever.

Q. — That is all that that poisoning story amounts to? *A.* — At the time when I learned of it, I thought it came from an unreliable source; that it was not any thing that it was worth while to investigate, and look into the matter at all.

Q. — Was there any story about it? Was there any thing more than a mere surmise? *A.* — That is all. I don't think that any one knew any thing about it, — only surmised that they might have been poisoned.

Q. (By Mr. GOULDING.) — Was there not a circumstantial story about it, whether true or false? *A.* — Not to my knowledge.

Q. (By Mr. FLAGG.) — Was there any story as to their being poisoned? *A.* — Not that I ever knew of; not that I ever had any foundation for.

The CHAIRMAN. — Mr. Whitney, as far as your observation extends, how far do you think that these movements in the town of Millbury are based upon apprehensions concerning the public health, and how far upon apprehensions concerning the destruction of mill-power, and of the industries upon which the prosperity of the place depends? *A.* — There is a general feeling about the town — I think it comes from our physicians there — that the people living on the banks of the Blackstone River are more or less affected from living where they do. I think there are a number of families where there has been sickness, which the physicians attribute to that cause. And as far as the damage to property is concerned, take the Burling Mills, for instance: they have not been running very much for the last six or eight months, — not to their full capacity, — and I think that the manufacturers attribute their stoppage, and the damage to the business, as much to the water which they had to use in connection with making their goods, or more, than to any other one thing. They manufacture a great many goods, and have to send them to market in a damaged condition, from the effects of the water that they use to cleanse their goods. As to the injury to the water-power, I should not, perhaps, be a competent person to judge; but I think it is a damage to every one who is on the Blackstone from twenty-five to forty per cent. I think there is a depreciation of the property. If it was to be put into the market to-day, I think it would not bring as much by from twenty-five to forty per cent as it would if the stream was what it was fifteen years ago.

Q. — I merely wanted to know how far this is a health question, and how far it is a property question. *A.* — I think it enters into both largely.

Q. (By Mr. GOULDING.) — You spoke about the Burling Mills. Mr. Harrington had a large interest in that, had he not? *A.* — He did.

Q. — He was here last winter, and told the Committee about the effect on his woollens? *A.* — I think he testified before the Committee.

Q. — He testified, did he not, that in his judgment the effect on the water, which prevented his washing his wool with it, was produced by the manufacturers above, and not by the sewage? *A.* — I don't recollect that he testified that way.

Q. — Don't you know that they have got artesian wells there that supply them with all the water they need to wash their wools, and that they use the water from those wells exclusively? *A.* — I think they have, within the last few months. I only know it from hearsay.

Q. — Have you been to the Burling Mills to examine, or have you examined the product of the mills, so as to have any personal knowledge about it? *A.* — No, sir.

TESTIMONY OF SAMUEL E. HULL.

Q. (By Mr. FLAGG.) — You live in Millbury? *A.* — I do.
Q. — You have been one of the selectmen? *A.* — Yes, sir.

Q. — State how far from the river you live? *A.* — I should think one-eighth of a mile, more or less. I don't know exactly.

Q. — How far from the river do you work? *A.* — About ten feet from it.

Q. — How long have you been familiar with the river? *A.* — Well, thirty years or more.

Q. — State to the Committee where you used to live, and where you live now? *A.* — I formerly lived in the northerly part of Millbury. Part of the farm where I was born borders on the banks of the river. The house, perhaps, is a quarter of a mile from the river, — maybe more.

Q. — How far from Quinsigamond? *A.* — About two miles.

Q. — So that for thirty years you were familiar with the river; you were two miles from Quinsigamond, and now you are familiar with it at Millbury? *A.* — I was familiar with it at Millbury at the same time.

Q. — State its condition as you remember it when you lived two miles from Quinsigamond. *A.* — The stream was perfectly clear and pure, and no one at that time seemed to think it was otherwise. The boys and older persons used to bathe in it whenever they wished; but since that time the river has changed very materially.

Q. — You have heard the testimony of the preceding witnesses? *A.* — I have.

Q. — You agree with them in what they said about the condition of the river in former years, and as to its present condition? *A.* — I do: yes, sir.

Q. — You work now, you say, ten feet from the river? *A.* — The end of the room in which I work is about ten feet from the river. There is a narrow driveway between the end of the building and the river.

Q. — State what you notice about the river from the place where you work? *A.* — I have noticed, in the summer-time particularly, a very offensive odor; and in the morning, when I go and open the windows, it is extremely offensive, and I have been obliged to shut them.

Q. — Is this odor from the dye-stuffs used in the mills on Singletary Brook in Bramanville? *A.* — No, sir.

Q. What is the nature of the odor? *A.* — Well, perhaps I could not describe it any better than Mr. Whitney did, — that it is a very offensive odor. The most I can say of it is, that it stinks.

Q. — Well, is it of the nature of privy odor? *A.* — Yes, sir.

Q. — Do the men where you work take any precautions to shut it out in summer there? *A.* — They do.

Q. — What has been the effect of it on your health, do you think? *A.* — I think it has not been favorable.

Q. — What have you noticed about it? *A.* — A year ago last summer I was able to work nearly all the time, but didn't feel well. In the winter I felt better. Last summer I was obliged to be away from my work for five or six weeks : I was sick, and I attributed it to the bad state of the river.

Q. — What was the nature of your sickness? *A.* — Well, I don't know as I can describe it: it was a sort of exhaustion.

Q. — Am I not right in saying that it was of the nature of dysentery? *A.* — Yes, sir: it came to that at last. And mornings, when I would go to the shop and open the windows, the smell would fairly gag me, as you might say: it was very offensive. I presume that part of the shop in which I work is in reality the worst room in the shop, where you get the odor more perceptibly than in any other room.

Q. — Did this affect your stomach at the shop? *A.* — It has, yes, sir, — caused vomiting.

Q. — What has been the general opinion of the workmen in the shop there as to the effect of the river? *A.* — The general opinion is, that it is not conducive to health.

Q. — What has been the general opinion throughout the town, among the people, as to the effect of the odors? *A.* — That is the general opinion of the town, so far as I have been able to learn. If it would be proper for me to say, one gentleman here asked if any one had left the shop on account of the smell? I know one young man who left the shop, and his father told me he left because he thought his health would be much better than it would be if he remained there. He went to Worcester, and is working in Worcester now.

Q. (By the CHAIRMAN.) — On account of the effluvia? *A.* — Yes, sir.

Q. — Not on account of business? *A.* — No, sir, not on account of business. He is working in the same business at Worcester.

Q. — Does he get the same or more pay? *A.* — I think he does not get so much pay.

Q. (By a member of the Committee.) — Did he leave Mr. Morse's mill? *A.* — Yes, sir.

Q. (By Mr. FLAGG.) — Is it a matter of general talk among those who work there, that they had better work in some healthier place? *A.* — Yes, sir.

Q. — Wasn't the place healthy enough ten years ago? *A.* — It had been always considered as healthy as any place of the kind, as far as I know.

Cross-Examination.

Q. (By Mr. GOULDING.) — Where do you work, sir? *A.* — I work for C. D. Morse & Co.

Q. — And you work in the room nearest the pond? *A.* — Yes, sir: nearer the river than any of the other rooms. The race-way runs in front of the mill; and I work in the rear part of the mill, down by the river.

Q. — You work down in the basement? *A.* — No, sir: it is not a basement, but it extends back of the main building, — an L.

Q. — Is it over the flume? *A.* — No, sir.

Q. — Anywhere near the flume? *A.* — Well, it is below the flume.

Q. — How long have you worked there? *A.* — Ten years ago last January.

Q. — For ten years you have worked in the same room? *A.* — No, sir, not in the same room that I work in now. The building has been built within five or six years.

Q. — Have you always worked equally near to the river? *A.* — Not quite so near: no, sir.

Q. — Nearly the same? *A.* — Yes, sir: the difference is simply the length of the building in which I am now.

Q. — As a general thing you have the windows open in summer, when you work, don't you? *A.* — As a general thing, I want to have them open if possible, and usually do.

Q. — As a matter of fact you have them open most of the time, but sometimes shut them? *A.* — Yes, sir, very frequently.

Q. — Have you always been in pretty good health, or is your health somewhat uncertain? *A.* — I have always considered myself healthy. I have not been sick but once, excepting last summer, to call a doctor.

Q. — When were you sick before? *A.* — Eight years ago last fall, I think.

Q. — With the exception of that, you have always been healthy until last summer? *A.* — No, sir, I didn't say healthy. I said I never had been obliged to call a physician. A year ago last summer I was quite ill; and, if I had felt able, I should not have continued at my business.

Q. — But you did work all through the summer? *A.* — Yes, sir.

Q. — Had you been healthy up to that time, always? *A.* — Yes, sir, perfectly healthy; always considered myself well.

Q. — Last summer you had the dysentery? *A.* — Yes, sir.

Q. — When did you have that trouble? *A.* — Well, I did not feel well, say the fore part of July; and it kept growing upon me, this bad feeling, and finally the dysentery came on about the last of July or first of August.

Q. — How long were you sick with the dysentery? *A.* — It was two or three weeks before that was checked. I was away from the shop for five or six weeks.

Q. — Who was your doctor? *A.* — Dr. Webber.

Q. — Of Millbury? *A.* — Yes, sir.

Q. — How near to the river do you live? *A.* — I should say perhaps between an eighth and a quarter of a mile. I don't know exactly, but an eighth, certainly: it may be a quarter.

Q. — Which way from the river? *A.* — It would depend upon where you place the river. The river runs all around me, as you might say.

Q. — Whereabouts, exactly, is your residence? How far from the shop? Can't you locate your residence any nearer, in reference to the river? *A.* — I should say that the river at the shop is as near to my house as at any point.

Q. — In what direction are you from the shop? *A.* — I am in a south-easterly direction.

Q. — Who was this young man whose father told you that he left the shop, and went to Worcester on account of his health? *A.* — His name is Wood.

Q. — What is his father's name? *A.* — Zebedee.

Q. — And the young man's? *A.* — I think his name is Zebedee, jun. I am not positive as to the young man's name; but we have always called him "Zeb." I don't know whether that is his name or not.

Q. — Do you know for whom he works in Worcester? *A.* — I do not: I can ascertain.

Q. — When did he go to Worcester? *A.* — Well, it was in the summer, I think: I can't tell you when.

Q. — Last summer? *A.* — Yes: I should say it was some time previous to Mr. Morse's fire; I should say in July or August, perhaps.

Q. — The fire was last October? *A.* — Yes, sir.

Mayor STODDARD (of Worcester). At some time during these proceedings, I desire to have the Committee examine this locality, from Worcester to Millbury. I know there are some members of the Committee who have seen it; but I think it would be well for all the members of the Committee to visit it. I should hope that, before these hearings close, the gentlemen of the Committee would come to Worcester, and examine our system of sewerage.

Adjourned to Wednesday at 10 A.M.

SECOND HEARING.

WEDNESDAY, Feb. 22, 1882.

THE hearing was resumed at 10 A.M.

TESTIMONY OF THOMAS WHEELOCK.

Q. (By Mr. FLAGG.) — Mr. Wheelock, you reside in Millbury? *A.* — Yes, sir.

Q. — How long have you resided there? *A.* — A little over ten years.

Q. — Where do you do business? *A.* — Worcester.

Q. — Where is your house situated in relation to the Blackstone River? *A.* — It is about forty rods, perhaps, in a north-east direction, in a direct line from the river.

Q. — State what you have noticed at your house, of the condition of the river. *A.* — Well, I have noticed at times in summer, when the water is low, a very disagreeable smell there.

Q. — Are you accustomed to drive back and forward between Millbury and Worcester? *A.* — Not very often, sir: I go usually on the cars; but I have driven up.

Q. — What have you noticed when you have been driving back and forward? *A.* — The same disagreeable odor.

Q. — You are a member of the school committee, and have been for some years? *A.* — Eight years, I think.

Q. — Will you state to the Committee what you have noticed in visiting the schools, taking the school nearest Worcester first, and then going on down the river? *A.* — The school at Burling Mills is up towards Worcester.

Q. — How far from the mouth of the sewer is the Burling Mills? *A.* — Perhaps three miles, sir. The schoolhouse stands upon a hill, about, I should say, two hundred feet from the river; and in the summer-time, when I have been there, and the windows were open, I could see or feel that same smell that I always smell when I come down the road.

Q. (By the CHAIRMAN.) — Excuse me: do you pay taxes on property in Worcester? *A.* — I only pay a tax on personal property: I have a store there.

Q. (By Mr. FLAGG.) — How large a school is this one at Burling Mills? *A.* — The average attendance is about forty scholars throughout the year.

Q. — In visiting that school, what have you noticed there as to the river? *A.* — I have noticed the same disagreeable smell there.

Q. — What is the next school on the river, or near the river? *A.* — The next school on the river, or near the river, is the school that is called the Union Building.

Q. (By Mr. GOULDING.) — Where is the Union Building? *A.* — That is down near the depot, and perhaps two hundred and fifty or three hundred feet from the river.

Q. (By Mr. FLAGG.) — Is that in the centre of the town of Millbury? *A.* — Yes, sir; it is the schoolhouse that is called the Union Building.

Q. — How many schools are there there? *A.* — There are four schools in one building, common schools; and then, a few rods on, not many rods from that building, is the high-school building, in which there is a high school. The high school will average in its attendance about sixty scholars through the year. The grammar school in the Union Building will average about thirty-five; the next lower grade, about forty; the primary and sub-primary, so called, will average in their attendance, I should say, sixty in the sub-primary, and fifty in the primary. Those four schools are in the Union Building. There are more scholars there; but I am speaking of the average attendance.

Q. — How many scholars are there in all the five schools? *A.* — Thirty-five in the grammar, say about forty in the intermediate, fifty in the primary, and sixty in the sub-primary.

Q. — What has been called to your attention in this building as to the river? *A.* — Well, some of the teachers, two or three of them I have in mind who have been there, who have occasionally called my attention to this same disagreeable smell.

Q. — The next school on the river is situated how far from this one? *A.* — I should say three-quarters of a mile.

Q. — How many schools in that school-building? *A.* — Four.

Q. (By Mr. GOULDING.) — That is down below Millbury? *A.* — Yes, sir.

Q. (By Mr. FLAGG.) — It is in the town of Millbury? *A.* — Yes, sir; in the lower part of the town.

Q. — About how many scholars in the four schools? *A.* — I should say about two hundred.

Q. — How far is this school-building from the banks of the river? *A.* — It is a matter that has not been called to my attention at all; but I should think it would be about two hundred feet.

Q. — What has been called to your attention as to the river? *A.* — The same thing has been called to my attention by the teachers in the summer; not so much in the winter.

Q.— You say you have been ten years in Millbury: during that time you have become acquainted with the people and their feelings? *A.* — Somewhat.

Q. — In your opinion, is the agitation upon this subject due to a fear of the loss of water-power alone, or to a general fear as to its effect upon health? *A.* — Well, I had supposed that the people regarded their health as of more importance than any other subject that was before them.

Q. — Are the people there, other than the mill-owners, interested in this matter? *A.* — I think they are, sir, decidedly.

Q. (By the CHAIRMAN.) — You mean that they regard their health as more important as applied to this particular matter, not on general principles? *A.* — Yes, sir.

Q. (By Dr. WILSON.) — How long have you noticed this smell? *A.* — Well, I have noticed it for the last four or five years.

Q. (By Mr. MORSE.) — Whether, during that time, it has increased or diminished? *A.* — It certainly, in my opinion, has not decreased; and I think that I have smelt as much or more of it: but I do not drive up and down the road from my house to Millbury now as much as I did four or five years ago. I go almost altogether in the cars.

Q. (By Mr. FLAGG.) — State whether you noticed this smell from your house as badly five years ago as it is now. *A.* — I do not think we smelt it as badly five years ago as we did last summer.

Cross-Examination.

Q. (By Mr. GOULDING.) — What is your business? *A.* — I have a boot and shoe store in Worcester.

Q. — Where is your boot and shoe store? *A.* — 38 Front Street, Worcester.

Q. — Where did you live before you lived at Millbury? *A.* — Before I lived in Millbury, I lived in Elizabeth, N.J.

Q. — And you came to Millbury just ten years ago? *A.* — Well, it is ten years ago last May, I think, sir, since I came there, and about ten years ago last September since I bought the place which I now occupy.

Q. — Were you connected with Millbury before? *A.* — No, sir.

Q. — Never were there? had no relatives there? *A.* — Well, my wife had some relatives there.

Q. — Is your wife a Millbury woman? *A.* — She is a Sutton woman, sir.

Q. — And your residence is forty rods from the river in which direction? *A.* — About south-west, I should think the river was, from my house.

Q. — Are you above Morse's factory, up towards Worcester? *A.* — Yes, sir,

Q. — How far above Morse's factory towards Worcester are you? *A.* — I should think three-quarters of a mile, or a little less.

Q. — Which side of the main road to Worcester are you? *A.* — I am on perhaps it would be called the east side of the main road.

Q. — You are on the main road, are you? *A.* — No, sir: my house is not on the main road. It is on what is called Park Hill Road. My land comes right down to the main road.

Q. — Is it on the old road to Worcester? *A.* — I presume it is; but I don't know.

Q. — At times, in summer, when the water is low, you have noticed this smell? *A.* — Yes, sir.

Q. — What does your family consist of? *A.* — My family consists of a wife and three children now, sir.

Q. — How old are your children? *A* — My children are from twenty to twenty-nine years of age.

Q. — They live at home? *A.* — Two of them are at home, and the other one is close by there.

Q. — The other lives close by? *A.* — Yes, sir.

Q. — How near to the river? *A.* — About the same distance that my house is, sir.

Q. — You say you are three-quarters of a mile above Mr. Morse's factory. Are you opposite Morse's pond? Does Morse's pond flow back, in other words, to a point opposite your house? *A.* — No, sir: I think it is below.

Q. — How much below the head of Morse's pond? *A.* — I really cannot tell you: I have not examined that locality at all.

Q. — Does not Morse's pond flow up to a point nearly opposite your house? *A.* — I have not looked at the locality: I am not able to tell you certainly.

Q. — You are not able to tell from memory? *A.* — No, sir: it is so seldom that I go that way, that I pay but little attention to the water.

Q. — In going from your house to the Millbury station, don't you go by Mr. Morse's pond? *A.* — Yes, sir: in going down I would go through the village; but I do not go by his shop. The pond comes up towards my house; but to state the number of feet or rods, I cannot.

Q. — Cannot you give an idea of how near to your house the head of the pond comes? *A.* — I really do not know where the head of it is.

Q. — What is the prevailing direction of the winds in summer-time? Have you observed that? *A.* — Well, I should suppose they were west and south, — west winds largely.

Q. — In summer? *A.* — Yes, sir.

Q. — Have you ever observed the direction of the wind at the times when you have noticed these smells? *A.*—Well, I think I have noticed the smell more when the wind comes from the north and west, or north-west. I won't be positive about that: it is a matter to which my attention has not been called.

Q.—It may be south-west, may it not, as far as you remember? *A.*—I should be rather inclined to think that we get it stronger when it comes from the north-west than when it comes from the south-west.

Q.—Do you mean to testify, that when you have a good, clear, bracing breeze from the north-west, you get the smell more than you do when the wind is from the south-west? *A.*—I am not so clear about "a clear, bracing wind;" but, when the wind comes from that direction, I think I have smelt it stronger there than I have when it comes from the south. I am not positive about that: it is a matter that my attention has never been called to. I know I have smelt it; but I did not go out and look at the direction of the wind.

Q.—Then, it is true that at times you have smelt it, but you have not noticed the direction of the wind? *A.*—I have noticed in both of those two schools, the first two I have spoken of, when the windows and doors were open, and the wind blowing though the schoolhouse in summer, the same smell. I should not smell it so strong if it came from the east or from the south.

Q.—When the wind is in that direction, would you smell it at all? *A.*—Very likely I might some: I don't say that I should not.

Q.—It is when the water is low that you have observed this smell? *A.*—It is when the water is low I have observed it more than when it is high, of course.

Q.—When was this schoolhouse that you have visited at Burling Mills built? *A.*—I am not able to tell you.

Q.—Is it an old building, or one recently built? Was it built before you came to town? *A.*—Built before I came there, sir.

Q.—What times do your schools have their summer vacation? *A.*—The summer vacation commences usually the first week in July.

Q.—And continues until when? *A.*—We usually commence about the last week in August.

Q.—Have you noticed this smell before the vacation, or after? *A.*—Well, I think I have noticed it both before and after.

Q.—Where do those scholars that go to this school at Burling Mills mostly live? *A.*—Well, they mostly live, perhaps, in the village near Burling Mills.

Q.—What do you mean, nearer to the river than the schoolhouse mostly? *A.*—Some of them would be nearer, and some of them would be just about the same distance.

Q. (By a member of the Committee.)—You say you live east or west of this main road? *A.*—It is called east; a little north-east.

Q. (By Mr. GOULDING.) — These things you have noticed when you have been visiting the school in summer? *A.* — Yes, sir.

Q. — Now, this building that you call the Union Building is just north of the road that goes to Bramanville, is it not? *A.* — Yes, sir.

Q. — And the Providence Depot is near the schoolhouse? *A.* — About two hundred feet right east, I should say, of the depot, or north-east.

Q. — And the river is the other side of the depot? *A.* — Yes, sir.

Q. — Does the river, after it passes the depot, get nearer to the school than it is at the depot? *A.* — I shouldn't think it was much nearer.

Q. — Now, is that bridge that crosses the road that goes to Bramanville called Gowan's Bridge sometimes? *A.* — I have heard them speak of Gowan's Bridge; but I could not really swear whether that is the bridge or not.

Q. — Does the water from the next dam below set up as far as the depot, or as far as that school? *A.* — I cannot tell, sir.

Q. — Occasionally, you say, they have called your attention to the offensive smell there? *A.* — Yes, sir.

Q. — This school that is three-quarters of a mile below the Union Building is below the Cordis Mills, is it not? *A.* — I don't know which mill they call the Cordis Mill: I know where the Union School Building is.

Q. — You have devoted yourself mostly to the schools? *A.* — I am either in the schools or in Worcester all the time, sir.

Q. — If you do not know where Gowan's Bridge is, or where the Cordis Mill is, or any of those localities, you really do not know much about the sentiment of Millbury from any inquiry you have made yourself. Have you been around through the town to inquire what they think of this matter in particular? *A.* — No, sir, I don't go round the town inquiring particularly what they say about it. The people of Millbury are occasionally in my store, — quite frequently, perhaps, come in there; and I hear them express their feelings about it, and I see them occasionally in meetings. I do not go around particularly to make inquiry, but simply see them as you and I would see our neighbors anywhere and everywhere.

Q. — Have you any knowledge from any source, — I do not ask you as intimating that you have not, but I want to get your idea about it, — have you any knowledge from any source so that you can testify to this Committee that you know what the public sentiment of Millbury is on this question? *A.* — I think I have.

Q. — What is the business of Mr. C. D. Morse? *A.* — He is a manufacturer of sashes, blinds, and doors.

Q. — What is the business of Mr. Flagg, besides being a good lawyer? *A.* — Mr. Flagg has been interested in manufacturing there.

Q. — What is the business of Mr. Waters, the other gentleman on the Committee? *A.* — I think he is engaged in manufacturing.

Q. — Is there anybody that you know of that is agitating this subject, as a member of any committee, that is not a manufacturer? *A.* — I don't know as I know of anybody in Millbury that, when you talk about it with them, but what are agitating this subject. The people are talking about it. It is not simply the manufacturers: it is the men that we come in contact with every day.

Q. — I did not ask you that question. I asked you, do you know of anybody in the town of Millbury, that is a member of any committee agitating this subject, that is not a manufacturer? *A.* — Now, you have got the word "committee" so many times that I really don't comprehend you, Mr. Goulding.

Q. — I will endeavor to state it slowly, so that you can comprehend it. Do you know of anybody in Millbury who is engaged in this subject, or doing any thing about it, as a member of any committee, representing anybody, who is not also a manufacturer? *A.* — I don't know that I do, sir.

Q. (By the CHAIRMAN.) — Are those manufacturers the people who are usually selected on committees in general town affairs, aside from this? *A.* — I think they have been on other committees. I am not certain about that. I take but little interest in the political affairs of the town.

Q. — I merely wanted to know whether they are the class of men whom the people of the town are in the habit of selecting for service on committees? *A.* — I think they are.

Q. — On other questions? *A.* — On other questions.

Q. (By Dr. WILSON.) — Will you tell me whether your town is in the habit of selecting for committees for any duty the men as much or more interested in the matter to be considered by the Committee than any other people in town? How is it about that? Are they in the habit of taking men who are themselves pretty actively interested? *A.* — I think they do.

Q. (By Mr. FLAGG.) — I forgot to ask you one question, Mr. Wheelock. During the last two or three years, has there been any noticeable sickness among the scholars in these schools? *A.* — Two years ago this winter, our schools were very badly broken into on account of a disease which was in town; and last year, and even this winter, our average attendance has been very much lower than it was the previous three or four years, before I was a member of the board.

Q. (By Dr. WILSON.) — Any epidemic in the schools? *A.* — The diseases last year, and the year before, were called by the physicians scarlet fever and canker-rash. Our schools have been broken into on account of sickness for the last three winters more than they have previously been. Two years ago this winter, and a year ago this winter, they were largely affected with the canker-rash and scarlet fever, so much so that the Committee had talked considerably about closing the schools at one time. We did not close them, however. That was last year, and the year before.

Q. — Have there been any cases of diphtheria? *A.* — I would not say that, as a general thing, there had been any diphtheria. No doubt there have been some cases.

Q. — Do you know of any cases reported as diphtheria? *A.* — Well, I cannot call their names. I very often go into a school, and ask why such a boy is not there; and they have told me he was sick with diphtheria: but I have no particular recollection of the names.

Q. — Is this trouble more noticeable in winter than in summer? *A.* — I think the disease of canker-rash was more general in the fall and winter than in the summer-time.

Q. (By Dr. HODGKINS.) — Do you know whether there have been any cases of typhoid fever in those schools? *A.* — No, sir: I can't say whether there have been or not.

Q. — Do you know of any cases of diphtheria? I don't understand you to say that you have known of any cases, but that you went into a school, and asked if a boy was there, and they said he was sick with diphtheria. *A.* — No, sir: I can't say there was any particular boy or girl who had diphtheria.

Q. — Do you know of a single case of diphtheria? *A.* — Do you mean, if I have seen it myself, — been to the house?

Q. — I mean, when you have been about the schools inquiring in this way, and found that certain children were absent, do you remember any single instance where the disease was diphtheria? *A.* — I cannot remember.

Q. — Or how many they have told you were sick with diphtheria? *A.* — No, sir. I see so many of them, and ask so many questions in regard to why they are not there, that I don't remember. I know that sometimes they have said that some were sick with diphtheria, and some with canker-rash; but to say how many were sick with diphtheria, I cannot.

Q. (By Mr. GOULDING.) — How many schools are there in town? *A.* — There are sixteen.

Q. — In various parts of the town? *A.* — Yes, sir.

Q. — How many are within two or three, or three or four, hundred feet of Blackstone River? *A.* — Ten.

Q. — Ten buildings? *A.* — Ten schools.

Q. — How many schoolhouses are there in town? *A.* — There are four schoolhouses.

Q. — In the whole township? *A.* — Oh! in the whole township, there are seven that I think we occupy.

Q. — Seven schoolhouses occupied? *A.* — Yes, sir.

Q. — And of those, four are in the village and on the Blackstone River? *A.* — They are on the Blackstone River, sir.

Q. — These epidemics that you speak of were general throughout the town, were they not? *A.* — I don't think they were, sir: I am not positive about it. My schools were right down on the river. I did not go up to West Millbury, nor to the Old Common, until this year. I would say, however, that I visited the Old Common yesterday, and looked up the average; but it was examination-day there, and their average this winter was better than last. That is perhaps a mile and a half from the river.

Q. — Last winter, how did you say the average there compared with the average down by the river? *A.* — Their average has always been better since I have been on the board.

Q. — Where are the other schoolhouses, besides the one in West Millbury, that are not on the river? *A.* — Well, there is the Old Common and West Millbury. I think the rest of them are on the river, that we occupy.

Q. — You did not close your schools either winter? *A.* — No, sir.

Q. — Was not that a matter of talk in the school committee? *A.* — Yes, sir.

Q. — The proposition was to close what schools? *A.* — Well, to close the schools at the Union Building and at the Providence-street school.

Q. — Where is Providence Street? *A.* — That is the one down in the lower village that we have been talking about.

Q. — That proposition was talked about? *A.* — Yes, sir.

Q. — And finally decided against? *A.* — We decided not to do it.

Q. — Any physician consulted by the board? *A.* — Yes, sir: one of the members of the board is a physician.

Q. — His advice was taken about it? *A.* — Yes, sir.

Q. (By Dr. WILSON.) — Do you know whether the mortality has increased among the pupils in those schools during the last four years? *A.* — I don't know whether it has or not. I do not keep the record: therefore I cannot tell. The scholars are coming and going. It is a manufacturing place: but the average number of our scholars in the schools is about the same; and my attention was never called in that direction.

Q. — Was that scarlet fever that you refer to epidemic? *A.* — I think it was. I don't remember how many cases there were, sir.

Q. (By Dr. HODGKINS.) — In going about to ascertain the number of scholars absent from the schools, have you noticed any difference in the number of absentees from those causes, diphtheria, and so on, in those schools near the river and the other schools? *A.* — Yes, sir.

Q. — You have noticed a larger number of scholars absent in the schools near the river? *A.* — Yes, sir.

Q. — And you think from those causes? *A.* — I am not prepared to say, for I am not a medical man; but I know there are less scholars absent from the schools in other parts of the town than from those near the river.

Q. — There is a larger number absent from those schools from sickness? *A.* — Yes, sir, from sickness.

Q. (By Mr. GOULDING.) — I would like to ask whether those schools along the river are not attended by the children of operatives in the factory, and whether it is not a general truth, as far as you have observed, that that class of scholars do not attend so regularly as in the farming districts in the back parts of the town? Is it not true that those districts show a better average attendance than the schools in the factory villages? *A.* — Well, sir, I should be of the opinion that it was.

Q. (By Mr. FLAGG.) — Those four schools that you speak of take up all the school population of that part of the town? *A.* — Yes, sir.

Q. — Not only the mill population, but all others? *A.* — Yes, sir, all others.

Q. — And the other schools of the town, in Bramanville, for instance, how do they compare? *A.* — Bramanville is about the same as Providence Street and the Union Building.

Q. (By a member of the Committee.) — What proportion of the scholars in town attend the schools along the river? *A.* — Well, suffice it to say, a very large proportion: I have not got the average.

Q. — Now, in saying that there are more scholars absent from those schools sick, do you take into account the proportion, as compared with the number of scholars? *A.* — Yes, sir. I take into account the average attendance. For instance, take the school at the Old Common: the whole number may be thirty, — the average attendance should be twenty-five. Take the schools near the river, I should say the average would not be more than perhaps twenty, with the same number of scholars.

Q. — Have you been familiar with the attendance at those schools for ten years? *A.* — I have been on the board, I think, eight years.

Q. — Has this proportion increased during the last eight years? You say the proportion of absentees from sickness is greater at those schools near the river than it is away from the river. Now, is that proportion to-day about as it was eight years ago, or has the proportion changed during the last eight years? Take it five years ago, was the proportion of absentees the same then as it is to-day? *A.* — I speak particularly of last winter, two years ago, and the present winter, owing to the diseases that have been prevalent among the scholars for the last three years. I could not swear positively with regard to previous winters.

TESTIMONY OF REV. PHILIP Y. SMITH.

Q. (By Mr. FLAGG.) — Where do you live, Mr. Smith? *A.* — In Grafton.

Q. — In what part of Grafton? *A.* — North-east from Blackstone River. It is called Wilkinsonville, which is in Sutton; but the house that I occupy is in Grafton proper.

Q. — You are a minister of the gospel, and have a parish? *A.* — Yes, sir.

Q. — How long have you been there? *A.* — Nearly seventeen years, lacking five months.

Q. — During that time have you noticed a change in the river at that point? *A.* — Yes, sir. Especially in the last five years, the change has been very marked.

Q. — Did you formerly bathe in the river? *A.* — I did, sir.

Q. — Did others in the village use it as a bathing-place? *A.* — Yes, sir.

Q. — Do you now bathe in the river? *A.* — No, sir.

Q. — Do others bathe in the river? *A.* — No, sir.

Q. — What have you noticed during the last year or two as to the condition of the river? *A.* — In crossing the Blackstone River by the bridge, in going to and from the depot, the odor is very noticeable; and especially twice a year is it very marked, namely, in the fall season, when the water is very low, and in the spring, when the water is very high.

Q. — How far do you live from the river? *A.* — About three hundred feet.

Q. — Do you notice these offensive odors there without going out of the house? *A.* — Not as heavily as nearer the river.

Q. — But you do notice them? *A.* — Yes, sir.

Q. — Do your parishioners complain of this matter? *A.* — They do, sir, especially those who live near the margin of the river.

Q. — Have some of them left town, assigning as a reason that they did not like the condition of the river? *A.* — Two or three families have complained to me, and said that was one cogent reason why they should leave town, and did finally leave.

Q. — There are some among your parishioners who are not mill-owners? *A.* — I have not the honor, sir, of having a mill-owner in my congregation.

Q. — Are you familiar with the schools? *A.* — Yes, sir: I am a member of the board.

Q. — In your opinion, does the pollution of the river affect the salubrity of the air about those schools? *A.* — In the Saundersville school —

Q. — How far is Saundersville from Wilkinsonville? *A.* — A little less than half a mile, in a straight line, from the Sutton depot.

Q. — Now, will you tell about the effect of the river upon the air at those schools? *A.* — In the months of April and May last, during the latter part of April, and two weeks in May, the schools in Saundersville were very much depleted, so that in one school, for at least ten days, there were only six scholars out of an average attendance of upwards of fifty-four; and in the upper school I think they were reduced to nine, out of an average of forty-five. The prevailing troubles there were measles and scarlatina, with diphtheria. There were two cases of diphtheria near my house. The children who were sick attended that school. Their names were Annie and Susie Redpath. They were attended by Dr. Thomas T. Griggs of Grafton.

Q. — What was the state of health among the children in the other schools in Grafton; that is, away from the river? *A.* — In the Centre, the number of scholars was not as small from similar causes as in the schools nearer the river.

Q. — That is, I understand, the sickness was not as great in the other schools as in those by the river? *A.* — That is my understanding, sir.

Q. — At this time of sickness in the schools, did you notice the condition of the river? Was the odor more offensive than usual, or as usual? *A.* — Well, sir, I cannot speak positively. I don't remember, in relation to that matter, whether the odor was more or less offensive than in common times, as we often smell it.

Cross-Examination.

Q. (By Mr. GOULDING.) — How long have you been on the school board? *A.* — Nearly five years, sir.

Q. — And your duties take you into all parts of Grafton? *A.* — Yes, sir. I am the chairman of the board, and I visit all the schools.

Q. — Your parish includes what part of Grafton? Your parishioners reside where, mostly? *A.* — My parish includes Grafton, with the villages of New England Village and Saundersville, and Farnumsville, in Grafton and Sutton. I have members in Worcester, in Auburn, and in Uxbridge.

Q. — Where is your church, sir? *A.* — In Sutton.

Q. — Well, the villages of Wilkinsonville and Saundersville, as appears by the map, are in the south part of Grafton? *A.* — My church is not in Grafton.

Q. — No, sir: I mean the villages, as appears by the map? *A.* — Yes, sir.

Q. — Now, when this depletion of those schools occurred, last April and May, I understand that you did not notice about the odor from the river, in that connection? *A.* — No, sir, I did not.

Q. — The two things were not connected in your mind at all, so as to lead you to make observations? *A.* — I was impressed, sir, that one of the chief causes of this trouble could arise, and did arise, from the effluvia from the river; but I did not realize any extra odor, other than I find every day in those two seasons, — in the spring and fall seasons.

Q. — Was there formerly a pond near Sutton station which has been drawn down by the dam being swept away? *A.* — There was, sir.

Q. — When was that dam swept away? *A.* — My recollection of it is, on the 11th of December, three years ago.

Q. — That pond has never been filled since? *A.* — Never.

Q. — And the area that was covered by the pond has lain exposed? *A.* — Yes, sir.

Q. — There was also a depletion, to some extent, of the schools in other parts of Grafton, was there not, at this same time? *A.* — In the Centre, yes, sir.

Q. — Grafton Centre is on a very high hill, is it not, away from all rivers and floods, unless it is a great flood? *A.* — Comparatively so.

Q. — It is the highest part of the town, is it not? *A.* — Yes, sir: it is the loftiest part of the town.

Q. — And you say there was a depletion of the schools there. Now, how did that depletion compare with the depletion down in your region? *A.* — There was no school in the Centre that ever reached the low attendance of nine or six.

Q. — Now, then, we understand so much. Now, how low an ebb did they reach? *A.* — I think, sir, in the primary department, we had as low as fifteen scholars in the Centre.

Q. — And what was the number belonging to the school? *A.* — The number, I think, then, was about fifty-two on the register.

Q. — The reduction was from fifty-two to fifteen? *A.* — Yes, sir.

Q. — But, in the other schools, it was reduced in one of them to nine, and in the other to six? *A.* — In the Saundersville schools.

Q. — How long did that continue? *A.* — I think that continued at least ten days.

Q. — And the diseases were scarlatina and diphtheria? *A.* — And measles.

Q. — Do you know in what proportion the three diseases prevailed? *A.* — I think that scarlatina was in the ascendant.

Q. (By Dr. WILSON.) — What was the average attendance of this school, the whole number of which was fifty-two, at the time it was reduced to fifteen scholars? *A.* — I would not wish to answer positively; but I would say probably forty.

Q. — You spoke of the average attendance of the schools in Wilkinsonville, near the river? *A.* — I did, sir.

Q. — Why can you not state the average attendance of this school? *A.* — Because I am better acquainted with the other schools. Those are nearer my home, and I am there more frequently.

Q. — What proportion of diphtheria did you get up in Grafton Centre? *A.* — I don't recollect a case of diphtheria, except the two cases in the Redpath family, of children attending the Saundersville school.

Q. — Those were the only two cases in town? *A.* — Yes, sir.

Q. — During what time? *A.* — During the whole year.

Q. (By Mr. GOULDING.) — Has there been any other time, except last April and May, when your schools have been in that condition? *A.* — The year previous, in the fall of the year, we had scarlatina; and last summer my own girl was six weeks detained at home because of malarial typhoid fever.

Q. — That was your own girl? *A.* — Yes, sir.

Q. — What does your family consist of? *A.* — Now I have three daughters at home, my wife, and myself.

Q. — You have lived there seventeen years. Have your daughters lived at home all the time? *A.* — Yes, sir, all the time.

Q. — Saundersville, where there were two cases of diphtheria, I understand is on the banks of the river? *A.* — Yes, sir; nearly so.

Q. (By Mr. GOULDING.) — How many mills are there in Saundersville? *A.* — There is one, sir.

Q. — And one in Sutton, besides? *A.* — Yes, sir.

Q. (By Dr. WILSON.) — Did you ever know of any cases of diphtheria there besides those two in the Redpath family since you have resided there? *A.* — Yes, sir, I have.

Q. — How many years ago? *A.* — In Sutton, just over the line, there were two fatal cases about three years ago, in a straight line from the Redpath house across the river: I should say about a hundred feet from the river.

Q. — Did you ever know of any case before those two three years ago? *A.* — Yes, sir, I have, and in the same locality.

Q. — How long ago? *A.* — Probably a year, making four years.

Q. — Any before that? *A.* — Not to my knowledge.

Q. — You have been entirely free from diphtheria for three years? *A.* — Until last year.

Q. — I say, between the two times you have been entirely free from diphtheria for about three years all through that neighborhood. *A.* — Yes, sir. Typhoid fever last year was very prevalent along the river.

Q. (By the CHAIRMAN.) — Are the sanitary appliances of those houses equal to those of the other houses in the neighborhood? *A.* — The house that the Redpath family lived in was near my own; and Mr. Piper, the landlord, is very careful in relation to those matters, whitewashing the house once a year.

Q. (By Dr. HODGKINS.) — You spoke about typhoid fever: can you tell how much you have had of that? *A.* — About three hundred feet from the river, two cases in the Gould family, one proving fatal last November: the other, after a few months' sickness, recovered. About six hundred feet from the same house, in the house of Mr. Weir, his son James was sick for six weeks with typhoid fever, attended by Dr. Wilmot.

Q. — When was this last case? *A.* — Last October. It began in September, and reached nearly through October, as the doctor will testify by his notes. Two cases below Mr. Chase's house, — one in the house of a Mr. Norcross: the person recovered. About the same time, a few feet from the same house, in a straight line near the river, in the house of Mr. Prentiss, his little boy died after two weeks' sickness. Down towards the village, in the French house, there was a boy sick for two months: he recovered. In a house near the Sutton depot, and very near the canal, there were two cases of typhoid fever last fall, lasting over six weeks: they recovered.

Q. — The Sutton depot is near Wilkinsonville? *A.* — Yes, sir: near the banks of the canal, facing the Sutton manufacturing establishment.

Q — How far back can you remember cases of typhoid fever? *A.* — The cases that I have now mentioned are the cases that came under my cognizance last year.

Q. — When, next previous to that, do you remember any cases of typhoid fever? *A.* — I knew, in the September before that, a Mr. Johnson had a daughter who was sick for about three months, attended by Dr. Wilmot. And also, last year, there was a case which I omitted to state, of William Boyce, a hired man of Mr. N. Chase, who is present, and who lives probably two hundred feet from the river. He was sick for at least six weeks, with typhoid fever or malaria, and was attended by Dr. Wilmot, who is present. That was last September.

Q. — Were there any cases between September of last year and October or November of the previous year? *A.* — I think, in a place they call Woodbury Village, two children attending the village school in Wilkinsonville were sick about six weeks. They belonged to my congregation. I saw them.

Q. — What time of year was that? *A.* — In the fall, — in September.

Q. — Do you remember any cases between November of 1880 and August of 1881? *A.* — Not near the river, sir. I did in my town, but not near the river.

Q. — Then you do not remember about any case of typhoid fever back of that? *A.* — Not the year before; but in other years I remember some cases.

Q. (By Mr. GOULDING.) — How many other cases of typhoid fever in the town do you remember, not on the river? *A.* — I know of the case of the wife of Mr. Andrew Corey, one of my elders, in Grafton Centre; but not near the river.

Q. — It is nothing unusual, I suppose, to have typhoid fever in all parts of the town; that is, occasional cases? *A.* — Occasional cases; but those that I have enumerated were comparatively near the margin of the river.

Q. — I understand that; but I ask you whether or not it is an unusual thing to have cases of typhoid fever all over the town in September? *A.* — Well, it is. We have more or less every year of typhoid over the whole country; but last year it was more marked, especially in those cases.

Q. — That is, they were marked as you have stated? *A.* — Yes, sir. I would state, furthermore, in relation to the river, that I remember, some five years ago, the cows of two gentlemen near my place were driven past my door during the dry season to drink at the river. Now they never go by there: they go to Champney's Brook, about a quarter of a mile distant, and drink the stream that comes from the hill. And Mr. Piper's dog (a very strange circumstance, but nevertheless true) refuses to bathe in the river. He goes along with the cows, and washes himself in the stream. The horses of Mr. Blair, Mr. Piper, and Mr. Young (these men keep stables) never drink of the water of the river. They have wells in their barns. In former times, some ten or twelve years ago, horses were known to drink the water of the river, but not for the last five or six years.

Q. (By Senator TIRRELL.) — Is there evidence of the presence of sewage water in the portion of the river that you are speaking of? *A.* — I think, from its color and weight and the odor, — those three facts establish in my mind your question's answer.

Q. — What is its color? and what else have you noticed about the river? *A.* — I have noticed, especially in the spring season, when the water is high, it has a blackish appearance.

Q. (By Dr. HARRIS.) — How would you describe the odor? *A.* — Well, probably it is better felt than described.

Q. — It is a smell that can be felt? *A.* — Yes, sir, and almost cut sometimes.

Q. — Are you familiar with the odor of sulphuretted hydrogen? *A.* — I would not be willing to state that it was quite as noxious to one's olfactories as that, especially when it is placed at them; but the smell of the river is certainly, I would almost say, tangible.

Q. — Have you ever got the odor of sulphuretted hydrogen from the river? *A.* — Well, I never assigned it under that name. I would not like to distinguish it as having a likeness to that. But sometimes, in crossing the river by the bridge, you will be compelled to put your handkerchief over your nose, especially twice in the year, — in the spring and in the autumn.

Q. (By Mr. FLAGG.) — We won't use the term "sulphuretted hydrogen;" but does, or does not, this odor that you speak of resemble that of a cesspool or privy? *A.* — Yes, sir: it is the odor of the excreta of our common privies.

TESTIMONY OF DR. THOMAS WILMOT.

Q. (By Mr. FLAGG.) — You live where? *A.* — In Farnumsville.

Q. — In what town is Farnumsville situated? *A.* — Grafton.

Q. — How far from the city of Worcester is Farnumsville? *A.* — About ten miles. I cannot state absolutely.

Q. — You are a practising physician? *A.* — Yes, sir.

Q. — How long have you been so? *A.* — Something over thirty years.

Q. — How long have you lived in Farnumsville? *A.* — Since May, 1877.

Q. — From your experience in Farnumsville, what do you say as to the effect upon the general health of the people of the present pollution of the river? *A.* — I should say it was decidedly injurious.

Q. — Your practice is not confined to Farnumsville, but extends, does it not, to Saundersville, Wilkinsonville, and other villages? *A.* — Saundersville, Wilkinsonville, Sutton, and down as far as Whitins' and North Uxbridge.

Q. — Now, will you state to the Committee any particular facts that you have noticed in regard to the effect of the river upon health? *A.* — I have noticed, that at low water, when the shores were exposed to the rays of the sun, the emanations were still more disagreeable and cogent, and also that the river was of a disgusting appearance, black and nasty, and at all seasons of the year had a certain amount of smell.

Q. — What sicknesses have you noticed during your practice there? *A.* — There is a prevailing sickness, which is scarcely worthy the full name of typhoid fever. It is more like an intermittent fever. There is no distinct medical name for it. It assumes all the appearance of a mild typhoid, without going into the extreme stage of it, *purpuræ;* without having the purple spots, which are symptomatic of the true typhoid fever, but producing lassitude and debility for some five or six weeks. It goes under the common name in the country of " slow fever."

Q. — Do you ascribe the cause of this disease to the river, wholly or in part? *A.* — To a great extent, I think it is, sir, particularly at low water. There are two cases in particular that I can state to you. I refer to two sisters in the village of Rockdale.

Q. — In what town is Rockdale? *A.* — I cannot say.

Q. — Is it not in Northbridge? *A.* — I think it is. In this village the pond was drained very low. It was drained down lower than the average of the ponds along the river, while they were making some repairs or alterations on the dam. That was none of my business, and I did not inquire what they were. The smell from the pond there was frightful. There is no modification of the word required, — it was perfectly frightful. It was worse than the wards of a hospital.

Q. — What did it smell like? *A.* — It smelt exactly like a water-closet, — " sulphuretted hydrogen " is the scientific term, — and continued for some length of time. The repairs were extensive that they were making.

Q. — When was that? *A.* — It was the latter part of the summer, or beginning of the fall, of last year. And, to prove to the Committee that these two cases were particularly caused by the emanations from the river, they both were taken with the ordinary low fever, typhoid fever, so called, and continued for some little time, until one of them, the younger one, — one was twenty-four, and the other twenty-six, — developed distinct symptoms of malarial fever; first shivering, then great heat, and then going off into perspiration, exactly like fever and ague. But it was only developed in one case. They were both in the same house, and both in the same room. And to prove that it was actually malarial fever, I will say that one was treated with quinine, and the disease yielded from day to day; but it had to be kept up a great while before it was finally conquered. The other one never had any symptoms of pure malaria.

Q. — Have you any doubt that the condition of the river had to do with these two cases? *A.* — I have no doubt of it whatever.

Q. — Are there any other cases that you can name? *A.* — Not so distinct as that. There are eight or nine more in relation to which I would make the casual remark, — when a person asked me, "What is the cause of this fever?" — "I think it is living by that nasty, stinking river."

Q. — You have been there since 1877? *A.* — Yes, sir.

Q. — Has the condition of the river, as to its offensiveness, grown worse during that time? *A.* — As far as my observation has gone, it has gradually increased from year to year. I think last year was the worst I have ever encountered at all. It was more of a nuisance than it ever was before. There was another case which I think strongly confirms my theory on the subject. My wife went with me when I visited these very patients. She did not go into the house, or near the patients. I hitched my horse a hundred feet from the river, and she remained there. She complained bitterly of the smell of this pond; and, when she went home, she was taken with a sudden attack of pure Asiatic cholera. I thought she would die before morning; and she and I attributed it entirely to the smell from this pond. I have two daughters at home now, and I constantly take them with me when I visit my patients; and one of them positively refuses to go down that road in the summer-time. She says she won't go, it smells so nasty.

Q. (By Dr. WILSON.) — What do you mean by pure "Asiatic cholera"? *A.* — It was not subject to the collapse that the pure Asiatic cholera has.

Q. — Was it Asiatic cholera? *A.* — Well, no, sir, I would not use the word "Asiatic." It was a very severe case of cholera-morbus. But that is a most indefinite term, because "morbus" simply means *sick*.

Q. — Then, what you mean is, a very severe case of cholera-morbus? *A.* — Yes, sir: I think the most severe case I ever saw. The extremities were cold, and there was the contraction of the features that you see in cases of poisoning.

Q. — Do you think that the fact that one of those cases was cured by quinine is sufficient proof of its malarial character? *A.* — No, sir: but, accompanied by the symptoms which appeared in that case, I do; having every day, as regularly as the hours came round, a shivering fit come on, followed by heat; and then having it go off and come on again at the same hour the next day.

Q. (By the CHAIRMAN.) — When you speak of that severe case of cholera having been caused by the river, do you suppose the effluvia from the river was sufficient to cause that, or was the system of the patient in such a condition as to make her peculiarly susceptible to such influences? *A.* — No, sir, I don't think it can be attributed to that; because she is a woman about my own age, and has been remarkably healthy. I have never known her to be attacked during the thirty-six years we have been married.

Q. — Not peculiarly susceptible to attacks of that kind? A. — No, sir, not at all, but the reverse. That is the only case I can trace so directly to the effect of the river: but diarrhœa is very common all the way through those manufacturing villages, along the line of the river; and those houses are all situated almost on the margin, as near as they can be built with any degree of safety.

Q. — You have a good many cases of cholera-morbus? A. — Yes, sir, mild cholera-morbus. "Mild diarrhœa" would be the more correct term. They are more frequent than they were two or three years ago, I think; but that may be owing to my longer residence and more extensive practice. I do not want to attribute it altogether to the river.

Q. (By Dr. WILSON.) — Were the abdominal symptoms very marked in those two cases of typhoid? A. — Yes, sir, they were very marked; but they were not so severe as to produce *purpuræ*, — the purple marks.

Q. — Is the character of the cholera-morbus more severe now than three or four years ago? A. — No, sir: I don't think there is any very marked difference in the degree of severity.

Q. — About the same last year as three years ago? A. — As near as I can judge or remember, I should suppose it was. I have not observed any thing to cause any alarm; nothing but simple diarrhœa, that would yield to ordinary treatment.

Q. — Have you ever seen any other case of intermittent fever? A. — Not since I have been in Massachusetts.

Q. — Have you ever known of any other case? A. — Not of my own knowledge. I have not even heard any one speak of it. In fact, I do not think that intermittent fever ever prevailed here.

Cross-Examination.

Q. (By Mr. GOULDING.) — Those two cases to which you referred were at Rockdale? A. — Yes, sir.

Q. — I don't quite understand how you prove that the disease was caused by the river, when one had one set of symptoms, and the other had another set. A. — I will tell you how I would account for that, I think: that one was more susceptible to an attack of malarial fever than the other, and the malaria which produced the attack arose from the fact that a very large surface of mud was uncovered. I believe it is generally understood and known by the medical profession, that malarial poison arises from the deposition of vegetable matter, and that typhoid arises from the deposition of animal matter.

That is making a broad assertion, not saying that it is absolutely so in every case. I think in one case the patient was susceptible to the inception of the fever; in the other, she was not.

Q. — The water was lowered a good deal? A. — The pond was absolutely empty. You could cross it. The pond was drawn down so as to enable them to put in a new bulkhead.

Q. — How large was the pond? A. — I cannot form any idea of that. It stretches over the country, I should think, nearly three-quarters of a mile, with three or four little islands in it, covered with the long grasses that grow in ponds.

Q. — How long did it remain drawn down? A. — I cannot say positively. I should say three months. It was a long time, I know.

Q. — There was no other case of intermittent fever except this? A. — Not any.

Q. — Any other people live near the pond? A. — Yes, sir.

Q. — Does your practice extend into other parts of the towns except along the river? A. — Not very widely. I go down to Grafton Centre occasionally.

Q. — Your practice, I suppose, has been gradually increasing since you went there? A. — Yes, sir, it has gradually increased.

Q. (By Dr. Wilson.) — Have you had any cases of diphtheria about the river there? A. — Three cases that I call distinct cases of diphtheria.

Q. — Where were those, and when? A. — They were very widely scattered. One, I think, was in a place called Ferry Street, in Farnumsville.

Q. — Near the river? A. — Yes, sir. It is situated in what is called Fisherville. The river, in a straight line, is about three hundred yards distant.

Q. — When was that? A. — That was two or three years ago.

Q. — Now, the next case? A. — The next case was down in Riverdale, in a frame-house. I don't think the river had any thing to do with it. I attributed that entirely to a filthy cesspool that I detected near the house. There were two or three cases in the house. One died, and the others recovered. There was another case between Rockdale and Riverdale.

Q. — Near the river? A. — Yes, sir: the house is quite close to the river.

Q. — When was that? A. — Last summer.

Q. — Do you think either of these three cases is attributable to the river? A. — I can't say that I do. I can't say that those three cases were directly attributable to the river.

Q. — I understand those were three locations: there were more than three cases? A. — Yes, sir: there were five cases altogether.

Q. — Do you remember any other case, during the last five years, of diphtheria? *A.* — No, sir. I have rather peculiar notions about that thing. I don't think that one case in fifty of what is called diphtheria is diphtheria at all. There have been a great many sore throats which people are very apt to call diphtheretic sore throats, and many other little pet names they give to it, which are no more appropriate than that.

Q. — Where did you practise before you went to Grafton? *A.* — I practised twenty years in Nova Scotia, and ten years in London.

Q. — From your experience, do you think diphtheria and typhoid prevail in this locality more than the average? *A.* — Yes, sir: more than in any locality I have been in for a number of years. But then, again, that ought not to weigh very much; because, where I lived, I could sit in my office and throw the stump of a cigar into the sea.

Q. — I understand there were only five cases in five years that you remember? *A.* — That I remember.

Q. — How many of those do you attribute to the river, and how many to other causes? *A.* — I have told you distinctly that I do not attribute any of them to the river, directly, although I believe that the polluted state of the atmosphere considerably retarded their recovery.

Q. (By Mr. GOULDING.) — I want to know whether, when you say "not directly," you mean to say that you attribute those cases of diphtheria to the river at all? *A.* — No, sir: you do not quite understand me. What I mean is this: that the cases of diphtheria occurred, — cause unknown; but I think it probable that recovery was retarded by the polluted state of the atmosphere that they were daily breathing. Just exactly as when a sick person is taken out of a little, close room, where he is half-stifled to death, and put into a good, airy room, it will do him more good than half a dozen doctors.

Q. — What is the population of the towns in which you practise? *A.* — I think the town of Grafton contains about twenty-five hundred; but I cannot be sure about that. The town of Northbridge contains a larger number than that; but around this part I am speaking of, they are very thinly scattered.

Q. — Are these the only cases that have occurred in those neighborhoods? *A.* — Those are the only cases of diphtheria and typhoid fever; but common sore throat, putrid sore throat, and all those things, are very prevalent. I have had forty or fifty or a hundred cases a year.

Q. (By Mr. FLAGG.) — You speak of only five cases. I understand you to say that you hesitate to call every case to which you are summoned, diphtheria, although it may be popularly so termed? *A.* — No, sir: I call a case diphtheria very reluctantly indeed.

Q. — When you speak of five cases, what do you mean? *A.* — I mean five distinct cases of what, upon my oath, I would say were distinct cases of diphtheritis. I was very much surprised to find those, I can assure you. It was more than I had seen for a great many years.

Q. (By Dr. HARRIS.) — Do you think you discover that breathing the air from the river has a tendency to increase diseases of the throat? *A.* — Yes, sir, I think I do.

Q. — Do you believe there is sulphuretted hydrogen in the atmosphere? *A.* — To a great extent, it is, near the river; but, when you get further away from it, it is so attenuated that you do not smell it.

Q. — What is the effect of that sulphuretted hydrogen upon the mucous membrane of the throat? *A.* — It causes irritation.

TESTIMONY OF NEHEMIAH CHASE.

Q. (By Mr. FLAGG.) — Where do you live? *A.* — In Wilkinsonville, in the town of Sutton, by the side of Blackstone River.

Q. — How long have you lived there? *A.* — Nearly twenty years.

Q. — Your business is that of a farmer? *A.* — Yes, sir.

Q. — How near to the river is your house? *A.* — My house and barn are about three hundred feet from the river.

Q. — How long did you say you had been familiar with the river? *A.* — I have lived by the side of the river nearly sixty years.

Q. — Tell what the river was fifteen years ago, — what it was used for. Did cattle drink out of it? *A.* — Cattle used to drink the water fifteen years ago, and we used to get fish out of the river fifteen years ago.

Q. — Did you use to bathe in it? *A.* — We did.

Q. — Do you now? *A.* — We do not. I do not see any one bathing in it.

Q. — Why do you not use it? *A.* — Because it is not considered a suitable place. The water is not fit: it is too dirty.

Q. — The place has nothing to do with it? *A.* — No, sir, the place has nothing to do with it: it is the water.

Q. — Do you notice the water, as to its odor and color? *A.* — I notice, that, especially after a big rain, it has a sort of roily, yellow color; and, in crossing the bridge, I notice the odor more particularly than I do back on the land, — the bridge where we cross the Blackstone.

Q. — What is the color of the water? *A.* — Sort of darkish color, a little inclined to yellow at times.

Q. — About the cattle, do they now drink the water of the river? Tell your experience with your cattle. *A.* — No, sir: it is very seldom that cattle or horses drink any water out of the Blackstone River now.

Q. (By a member of the Committee.) — Is it because cattle refuse to drink it, or because they are not taken there to drink it? *A.* — Two years ago I had no place to water my oxen except the Blackstone River; and they would not drink only once in two days at the river. Then I dug a well at my barn, where I found water; and there they would drink twice a day, when from the Blackstone they would only drink once in two days.

Q. (By Mr. FLAGG.) — When they could not get any other water, they would drink once in two days from the Blackstone? *A.* — Only once in two days they would drink water out of the river.

Q. — Then you dug a well, and they would drink twice a day? *A.* — Yes, sir: they would drink, as cattle usually do, twice a day.

Q. — About fish: tell us what fish were formerly there, and what kinds you see now. *A.* — We used to catch fish years ago in the spring of the year; but now there are no fish to speak of, but there are plenty of water-snakes.

Q. — Have you noticed any cases of typhoid fever or malaria in the neighborhood? *A.* — I had a case in my house last October.

Q. — Who was it? *A.* — It was my hired man.

Q. — Did he recover? *A.* — He did.

Q. — Was the condition of the river assigned as the cause of his sickness, or thought to be so? *A.* — It was thought to be so.

Q. — Do you know of any other cases? *A.* — There were.

Q. — Will you mention the others, or one other? *A.* — There was a boy, twelve years old, in the second house across the river, that died. Others were sick there.

Q. — When was that? *A.* — That was last fall.

Q. (By Mr. GOULDING.) — Are these the same cases that Mr. Smith was telling about? *A.* — I think he spoke of one of the cases.

Q. — He spoke of your hired man? *A.* — Yes, sir: that is one of the cases spoken of.

Q. — Do you know of any cases that Mr. Smith did not speak of, and that Dr. Wilmot did not speak of? *A.* — I don't recollect as Dr. Wilmot spoke of this case at my house: I don't know but he might.

Q. (By Mr. MORSE.) — I would like to ask Mr. Chase whether this bad condition of things in regard to the smell from the river and the appearance of the water has improved or has grown worse during the last three or four years? *A.* — I should not say that it had improved; I should say that it had increased.

Q. (By Mr. SMITH.) — You spoke of watering your cattle at the Blackstone River. In former years, was it the practice of farmers living upon the borders of that stream to water their cattle in the river? Did they use it for that purpose to any extent? *A.* — They did.

Q. — Now you say they cannot use it for that purpose? *A.* — No, sir: they do not even pretend to water horses in the Blackstone now. It was formerly a general watering-place. There was many a farmer on the stream that had no other water than the Blackstone River.

Q. — About when did they stop watering cattle there? *A.* — It has been used for that purpose but very little for the last five years, I should say.

Cross-Examination.

Q. (By Mr. GOULDING.) — What is your business? *A.* — Farming.

Q. — You have always lived at this same place? *A.* — Yes, sir: almost always.

Q. — What does your family consist of? *A.* — My family at home is a housekeeper, hired man, and mother.

Q. — How long has your mother lived there? *A.* — She has lived there since the first of last October.

Q. — Where did she live before? *A.* — In the next house above.

Q. — Is that near by? *A.* — That is near by: yes, sir.

Q. — How long did she live there? *A.* — Sixty-seven years.

Q. — How old is she now? *A.* — In her ninety-first year.

Q. — Pretty good health for an old lady? *A.* — Not very good: no, sir.

Q. — Not very robust? *A.* — No, sir.

Q. — So as to be about the house? *A.* — Some days.

Q. — Did your father formerly live there? *A.* — Yes, sir.

Q. — He is dead now? *A.* — Yes, sir: he has been dead twenty-two years.

Q. — What age did he die at? *A.* — Sixty-eight.

Q. — Have you ever had any children? *A.* — I have one son in Wilkinsonville.

Q. — How long has your wife been dead? *A.* — My last wife died the thirteenth of last May.

Q. — You have only one son? *A.* — That is all.

Q. — What is his age? *A.* — Thirty-six.

Q. — Where did this hired man who was sick last year come from? *A.* — Came from Ireland.

Q. — When did he come? *A.* — He came to my place the 28th of June.

Q. — Had he come directly from Ireland, or very recently? *A.* — He came to New York, and stopped there a few days.

Q. — Within a few weeks he had come from Ireland? *A.* — Yes, sir.

Q. — He came from Ireland to this country, and came to work on your farm, and was taken with this fever? *A.* — Yes, sir.

Q. — And was sick how long? *A.* — I think he was confined to his bed about four weeks.

Q. — Has he fully recovered now? *A.* — I think so : yes, sir.

Q. — Does he still work for you? *A.* — Yes, sir.

Q. (By Dr. WILSON.) — When was he taken sick? You say he came to your place in June? *A.* — He came to my place in June.

Q. — When was he taken sick? *A.* — About the last of September or first of October : I can't tell you the date exactly.

TESTIMONY OF REV. JOHN L. EWELL.

Q. (By Mr. FLAGG.) — You live in Millbury? *A.* — Yes, sir.

Q. — You are a minister of the gospel? *A.* — Yes, sir.

Q. — How long have you lived there? *A.* — It will be four years next month.

Q. — During that time you have seen the Blackstone River? *A.* — Yes, sir.

Q. — Smelt it? *A.* — Yes, sir.

Q. — This is a matter of common complaint in your congregation? *A.* — Yes, sir : I hear of it constantly. Before I came there, when I was about deciding to come, I heard of the trouble with the river, and made special inquiries about it; and, the more I inquired about it, the more that statement was confirmed.

Q. — What statement? *A.* — The statement that the river smelt bad was one that was made to me; and the inference was, that the village was not altogether healthy. I thought it would be pleasant to live by the river, as I had formerly lived by a river, and had depended upon it for recreation in fishing, bathing, and rowing ; but I found myself cut off from all those things.

Q. — From your familiarity with it, you would not feel like bathing in or boating on that river? *A.* — No, sir : I should not think it healthy to row on the river.

Q. — Will you describe in your own words what you have noticed about the river? *A.* — The dark color of the water, the thickness of it, and the odor. I have frequent occasion to pass the bridge over the river, near the Providence Station, and also somewhat frequently the bridge which is above Burling Mills ; and at each of them I have noticed the strong odor of the water and the dark color, uniformly, I think I may say, in crossing those bridges.

Q. (By the CHAIRMAN.) — Did you intend the Committee to understand that you hesitated somewhat about going there in consequence of what you had heard? *A.* — Yes, sir.

Q. (By Mr. FLAGG.) — In going to and from the cemetery, is the odor from the river noticeable? *A.* — Very noticeable, sir.

Q. — Will you describe how the cemetery is situated as to the river, — whether or no it is almost surrounded by the river, a sort of peninsula? *A.* — A part of it is a peninsula.

Q. — Among your parishioners this has been a matter of talk, you say. Have any thought of going away on that account, who have talked with you? *A.* — I understand that one family has moved away, and that that was a consideration with them. Another family that I have in mind, who are excellent people, connected with our church, and we were very much afraid that we should lose them, were troubled about the river,

Q. — Who was this? *A.* — This was Mr. Whitworth. He mentioned this incident to me, that, when he first came to Millbury, it was convenient for him to row to and from the sash and blind shop: but the odor of the water that was stirred up by the oars became very offensive; and, as nearly as he could judge, he had strong symptoms of typhoid fever. He was obliged to leave his work, I think; and he gave up rowing upon the river. And I know (and I presume it may have been brought out here, because it is a matter known to every one in Millbury), that, when the sash and blind shop was burned, all the people of Millbury, or a large number of them, did what they could to induce the proprietor to rebuild; and the strongest objection that he urged was the polluted condition of the water.

Q. — Did he urge that to you? *A.* — Yes, sir.

Q. — But you talked him over, and he is going to stay? *A.* — Yes, sir.

Q. — When was the fire? *A.* — It was, I should say, in the month of October, 1881: it may have been later than that.

Q. — In your opinion, from your familiarity with your parishioners, and the other people of Millbury, do you think that considerations of health, or considerations of loss of water-power, are the matters which cause this agitation? Is it either one alone? *A.* — I suppose both, sir; but, in the minds of the people of the town, it seems to me that the great consideration is the one of health.

Q. (By the CHAIRMAN.) — Do you mean that is what ought to be, or do you think it is? *A.* — I think it ought to be, and I think it is..

Q. — That is, from your conversation with them? *A.* — Yes, sir. At the same time, one can see how the river is filling up, and the injury to the water-power.

Q. (By Mr. MORSE.) — And that injury to the water-power means, not simply the loss of power, I take it, but the loss of water that is suitable for manufacturing purposes: the water is used for other purposes than for power? *A.* — Yes, sir.

Q. — And the health of the operatives in the factory also is a matter of consequence to the manufacturers as well as to themselves? *A.* — Yes, sir.

Q. (By Dr. WILSON). — Were your fears realized at all in regard to coming to Millbury? Did you suffer in body? *A.* — Our children have suffered more than before, decidedly, from throat-diseases; and we have had a good many family councils about the matter. We try to take all the pains we can as to food, air, and health generally; and still they have throat-diseases more than they did before.

Cross-Examination.

Q. (By Mr. GOULDING.) — Where is your church situated, Mr. Ewell? *A.* — It is in the village nearest Worcester, sir. The Baptist church and ours stand directly opposite each other.

Q. — That is in Armory Village, in Millbury? *A.* —Armory Village, I think, is the distinguishing name.

Q. — How near to the river? *A.* — I am not a good judge of distances: possibly it is an eighth of a mile from the river to the churches.

Q. — How near to the river do you live? *A.* — About the same distance.

Q. — An eighth of a mile? *A.* — Yes, sir.

Q. — In what direction? *A.* — The general direction would be east from the river.

Q. — What does your family consist of? *A.* — Wife and four little children.

Q. — How old are your children? *A.* — The eldest is eight and a half, and the youngest is a year old.

Q. — Had a physician considerably in your family? *A.* — No, sir; but very little.

Q. — Very little occasion to call a doctor? *A.* — No, sir: I think that my wife is a pretty good nurse, and perhaps we are not quite as ready to call in a physician as some would be.

Q. — What was this family's name which moved away? *A.* — The family's name was Johnson.

Q. — Where did they live? *A.* — They lived in what is called Blackstone Street.

Q. — How near to the river? *A.* — Well, sir, perhaps one-sixteenth of a mile: it was quite near the river.

Q. — What was Mr. Johnson's business? *A.* — He worked in the sash and blind shop.

Q. — Where did he go to? *A.* — To Fitchburg.

Q. — When did he move? *A.* — I think, sir, some time last fall.

Q. — What was Mr. Johnson's first name? *A.* — I don't recall it.

Q. — But he formerly worked for Mr. Morse in the sash and blind factory? *A.* — Yes, sir.

Q. — Did he move away about the time the factory was burned? *A.* — I think he did, sir.

Q. — How long before? *A.* — My impression is, that it was at the beginning of the fall. I cannot give a definite statement as to that.

Q. — The fire was in October, was it not? *A.* — In October or November.

Q. — And one of the reasons that he assigned to you was the river? *A.* — He did not assign that to me, as I recall; that is, I cannot make a positive statement to that effect: but I have understood that that was one of the reasons.

Q. — Then you never heard him assign any such reason, but you understood so? *A.* — I should not dare to say that he stated it to me.

Q. — Did he give any reason to you for going? *A.* — I don't recall any. I think most likely that he mentioned that to me; and yet I do not recall just what he said to me: but I have understood that that was one of the reasons.

TESTIMONY OF ESEK SAUNDERS.

Q. (By Mr. FLAGG.) — You live in Saundersville? *A.* — Yes, sir.

Q. — That is in the town of Grafton? *A.* — Yes, sir.

Q. — How long have you lived there? *A.* — Since 1835.

Q. — And during that time you have been interested in business there? *A.* — Yes, sir: manufacturing and building up the village there. I was engaged in the cotton manufacture up to last May. I sold out my business there then.

Q. — You have seen that village, then, grow up from a very small place to its present size? *A.* — Yes, sir.

Q. — How many people are employed in the mill? *A.* — About two hundred.

Q. — Will you go back fifteen years, and tell the Committee what was the state of the river then, as to the quality of the water? *A.* — Fifteen years ago we used the water for any purpose that we wanted. We could drink it, we could use it in our boilers, and for any thing that we required water about the manufacture; but it has been polluted, and growing worse ever since the sewage of Worcester was first turned into the Blackstone River.

Q. — It is not now used for bathing? *A.* — No.

Q. — Nor for domestic purposes? *A.* — No: nothing but motive-power. Cattle won't drink it; we cannot use it in our boilers; we cannot bathe in it; and we cannot use it for any thing but motive-power.

Q. — Coming now to boilers, do you consider the water in the river, as it is at present, fit to use for the purpose of making steam? *A.* — Oh, no!

Q. — For the purpose of making sizing? *A.* — No: we can't use it at all for that.

Q. — What is the condition of the water that is in the tanks in the mill? *A.* — The water in the tanks in our mill we take from springs separate from the river. Don't use any of it.

Q. — Why? *A.* — Because the river-water is polluted so. We used to use it for that purpose, but we have not for some time. We had some bath-houses there for people to bathe in: we took them out ten years ago.

Q. — You took them out for the same reason? *A.* — Yes, sir.

Q. — Going into your flumes, what do you notice about them? *A.* — Well, there is a thick sediment that adheres to the sides and bottom of the flumes, and all the irons that go across to support them.

Q. — Is that sediment offensive? *A.* — It is very offensive to go into a wheel-pit now. It used to be a part of the machinist's duty to go into the wheel-pits often; but now it is a separate job. You have to furnish him things for it, and it is very offensive.

Q. — Your house is how far from the river? *A.* — My house is probably eight hundred feet from the river, on a rise about sixty feet above the river.

Q. — Do you notice the odors from the river? *A.* — When the river is low, and the wind is in the direction to bring the odor from the river, it is very strong in the yard back of the house.

Q. — Is there any doubt about what that odor is? *A.* — Not at all: you can trace it all the way to Worcester.

Q. — When the water in the river is low, how great a part of that which flows in the river goes through your flumes? *A.* — Well, it all goes through. When the river is low, we use it all.

Q. — The flumes are in the basement of the mill? *A.* — The flumes are in the basement of the mill. There was an addition built on to the mill of ninety feet, and the water-wheels are in that addition. It is closed up; and when we go in there in the morning, and there has been no ventilation, it is very offensive: and we cannot keep it open in the winter season, because our steam-pipes run through there, and the cold would freeze them. It is very offensive where the water comes in and is confined, as it is in the wheel-pit, over night.

Q. — How great a portion of the year is the water in the river so low that the greater part of it goes through your flumes? *A.* — When the mill is running, from nine to ten months a year.

Q. — In your opinion, what is the effect of the pollution of the river upon the health of your operatives? *A.* — Well, I don't think their health is as good as it used to be. I think they lose more time, and I think it injures their health. Our mill faces to the north, towards the pond. In the afternoon, particularly in the weaving-room where they use more or less steam in the summer season, if they could raise the windows, and let a gentle breeze come in, it would be very refreshing; but now when it comes in, with the state of the river, it is very offensive, and they have to close the windows, on the windward side, at any rate.

Q. — From your familiarity with the people in Saundersville and in the other villages along the Blackstone River, do you think that considerations of health serve to keep this agitation going? *A.* — I don't think the health of the people is as good as it formerly was, particularly those that live in houses very near the river. They have not had any prevailing sickness there more than throat-diseases, and the common diseases that come; but there are a good many that are puny, and that are running down with consumption, who were from healthy parents.

Q. — Are not the people of the valley generally fearful of the effects of this pollution upon their health? *A.* — Yes: that is talked over with us all.

Q. — Do you think it has had any effect upon your health? *A.* — Yes: for two years I was quite unwell. I attributed a good deal of it to the work I used to do about the mill in the morning, before breakfast, raking out the rack, at the time the dead fish and such stuff was floated down in there. For the last year I have not had any thing to do with it. It is very offensive when dead fish and all this stuff is drawn under the gates and on to the rack.

Q. (By the CHAIRMAN.) — What is your age, Mr. Saunders? *A.* — I am in my eighty-second year; born in 1800.

Q. (By Mr. SMITH.) — Is your wheel-house situated underneath your factory? *A.* — Yes, sir.

Q. — Does the odor from the wheel-house penetrate the rooms above, so as to be perceptible? *A.* — No, it don't penetrate above; it is closed up: it would if there was any place open. It is on a level with the floor of the main mill and of the addition that has been put on. The main building is one hundred and seventy-five feet by fifty-one feet; the addition is ninety-four by thirty-eight.

Q. (By Mr. FLAGG.) — Have you noticed any effect upon the wells in the village? *A.* — Yes, sir. I have a well down near the river which was dug thirty years ago, from which I used to get a good supply of good water; but the water from the river has got into it, and destroyed the well. I took out the first pump, and put in a drive-pump, and drove it down further. That answered for a little while; but it has come in again now, so that it is entirely useless.

Q. — Come in through the soil, do I understand you? *A.* — Come in through the soil.

Q. — What is the nature of the soil? *A.* — Gravel. When the well was first dug, there appeared to be a spring from the bank across the road: it did not come from the river, and I did not suppose it would be affected by the river. Some of the other wells in the village are affected by the river-water, I suppose.

Q. — You have taken care of your village, so as to have as good health there as possible, I am told? *A.* — Yes, sir: I have built it all up from four houses to the size it is now, and looked after all the arrangements for drainage, water-closets, and every thing.

Q. — How large a village is Saundersville in population now? *A.* — There are about six hundred people there now.

Mr. FLAGG. Mr. Chairman, Mr. Saunders has been one of the prominent men in the valley for a number of years, and is a man from whom the Committee could get valuable information, if there is any point they would like to inquire about.

Q. (By Dr. WILSON.) — Do you know any thing about any cases of diphtheria or typhoid fever among your help or among your neighbors lately? *A.* — I cannot distinguish any. There has been but very little unusual sickness there: I don't think there has been any prevailing sickness.

Q. (By the CHAIRMAN.) — As I understand, you have not any thing very definite to say about sickness, except, in a general way, that in your opinion the health of the people is not as good as it used to be, but you do not desire to state that you are in any danger of an epidemic? *A.* — No, sir. I notice by our pay-roll, and notice by the appearance of the people, that there is more loss of time now than there used to be, and people are not as active as they were.

Q. — What makes you attribute the loss of time to sickness? *A.* — Well, I attribute the loss of time to poor health: I don't think you could trace it to dissipation, and I don't think the work is harder than it used to be. We work shorter hours now. When I first came to the business, we worked twelve hours and over; and now the work is done in ten, and there is every arrangement made for the health of the operatives. You look after us pretty close here. We have to have tenements; we have to have every thing as convenient for our help as we possibly can. We have to have escapes for them to get out of the mill in case of fire; we have to box all our machinery; we can't work a child only so long; we have to do all these things; and then they look into our tenements to see if they are all right. But here is a point that we cannot alter, and we come to you for that. Every thing else they come to us, and say we must do so and so; and these young men are sent up from here, and their orders are absolute.

Q. (By Mr. FLAGG.) — Is it your opinion that the legislation in favor of the operatives has been more than offset by the Legislature permitting Worcester by the Act of 1867 to pollute the river? *A.* — If they went there with the authority they usually come with, they would command Worcester to stop that sewage right off. They come to us, and say, "Take that child out." The tenements where we used to have two families, we cannot have but one in now. We are under very strict rules from the Legislature. It all emanates from this hill, and we have to obey it.

Q. (By Mr. SMITH.) — Don't you think these regulations tend to benefit the health of your village? *A.* — What regulations?

Q. — I mean the regulations from this hill. Don't they tend to protect and prolong the lives of the people of your village, on the whole? *A.* — I don't know as they really do; because we have been a little family concern, and have taken care of ourselves. We have had no constables; we have had no lawyers. We usually have a man come and preach to us on Sundays, and have established a little church there.

Q. (By the CHAIRMAN.) — You obey the rules the State lays down to prevent people from being injured by their work? *A.* — Yes, sir: we have been a law-abiding people all through, and my associates up and down the river there — I am familiar with them — I think co-operate with us. I think there is no law that has been imposed about labor, or the hours of attendance on school, or any thing, that has been intentionally violated on the Blackstone River. There may have been cases where there has been some mistake about it, and somebody has been fined; but it is not a general thing.

Q. — But you think that men do not live any longer in consequence of it? *A.* — No, sir.

Q. (By Mr. MORSE.) — I would like to ask whether you take special pains in regard to the drainage and cleanliness of your village? *A.* — Yes, sir: we have always done so. All the backhouses and every thing are fitted so that they can be replenished with loam or ashes, or some cleansing thing, after they are cleared out in the spring, or two or three times a year if they want to, until the fall. And there is every arrangement made in the mill for hot and cold water for every purpose; and the water-closets are looked to, and all the sink-drains, and every thing is kept clean. It has always been under my supervision.

Q. — Is the drainage of your factory poured into the river? *A.* — Not at all; not a particle of any thing of that kind; never, sir.

Q. (By Mr. CHAMBERLAIN.) — I wish to ask you if, within your knowledge, the people of your village feel that they are living in air that should be made purer for them? *A.* — Yes, sir, decidedly so. It is the universal feeling. The house was very desirable for operatives to live in, because it was near the mill. Now it is hard work to get the same class of people to live in it: they don't like the scent from the river.

Q. (By Dr. HODGKINS.) — Have you noticed that your people complain largely of sore throats? *A.* — That has been a general complaint: a good many of them go with mufflers.

Cross-Examination.

Q. (By Mr. GOULDING.) — You have had personal supervision of your business up to the present time? *A.* — Up to last May. I sold out my interest in the manufactory then. I have a good deal of real estate there, and carry on a farm. I attended personally to the business of the mill all the time from 1835.

Q. — Have you been in pretty good health yourself? *A.* — Yes, sir, I have enjoyed very good health, until about five years ago I had a sickness.

Q. — What was that? *A.* — I had a violent attack of a sort of fever, — sore throat.

Q. — What time of the year was that? *A.* — In the fall.

Q. — How long were you sick? *A.* — Two months, five years ago; and I was attacked two years ago this last fall. I took cold, and was hauled up nearly the whole winter. Last winter I had a little attack of it.

Q. — You had a cold this last time, and it was in the winter? *A.* — Yes, sir.

Q. — With the exceptions you have mentioned, you have been well always? *A.* — Always.

Q. — You have a family? *A.* — Yes.

Q. — Mrs. Saunders living? *A.* — My first family is all gone. When I went to Grafton, I had a wife and three daughters; but Mrs. Saunders died, and the three daughters died.

Q. — How long ago did Mrs. Saunders die? *A.* — She died in 1864.

Q. — Did your daughters die at home? *A.* — One died at home two years ago. They were all married, and lived at Worcester. I was married in 1866 to my present wife. She is living, and well.

Q. — She is a younger person than yourself? *A.* — Yes, sir.

Q. — Had this daughter that died at home lived at home? *A.* —

She had been at home since the death of her husband: her husband died some four years before she did.

Q. — What did she die of? *A.* — She died of pneumonia; took a violent cold.

Q. — How is the sink-drainage through the village disposed of? *A.* — That is carried off into cisterns. There are casks sunk into the ground, without the lower head. The water runs into the ground, and they are cleaned out two or three times a year.

Q. — That is the custom all through the village? *A.* — All through the village.

Q. — And the privies? *A.* — They are all outside of the houses, with vaults put in, and covered over on the back side, and the contents are carried off in the spring or fall: and, when they are cleared out in the spring, there is loam or coal-ashes from the mill tipped up there; and at different times scavengers go through the village, and put in these materials.

Q. — You never have had any sewers built there? *A.* — No, sir.

Q. — Your surface-water in the streets, I suppose, runs into the river eventually? *A.* — That runs into the river. We have got a great many under-drains that carry off the water. It is a flat place, along by the business places; and there is a great deal of under-drainage that carries the water down to the river from there.

Q. (By Mr. FLAGG.) — You are familiar with the towns just above Worcester, and the brooks forming the Blackstone, — with Millbury, Sutton, Grafton, Northbridge, Uxbridge, and Blackstone, being all the towns in this State on the Blackstone? *A.* — Yes, sir.

Q. — Have any of them any town system of sewerage? *A.* — No, sir.

Q. (By Mr. GOULDING.) — They all have a right to lay sewers, when they have a mind to, under the general statute of 1869? *A.* — Yes, sir.

Q. (By Mr. FLAGG.) — Is there any emptying of sewage into the Blackstone by a system of sewerage, by any town or city on the Blackstone, in this State, outside of the city of Worcester? *A.* — I don't know of any.

Q. (By Dr. INGALLS.) — What causes this pollution in your river? *A.* — The sewage of Worcester.

Q. — Entirely, do you think? *A.* — I don't think it is entirely: I think there are other causes, but I think that is the greatest cause.

Q. — What proportion should you judge came from the sewerage of the city of Worcester, in comparison with the surface drainage of these other towns, and the pollutions that come from the manufactories? *A.* — That would be guesswork with me. I have not gone into any mathematical calculation about it; but judging from the condition of the river after the sewage was put into the river, and what it was before, I should think that seven-eighths of it came from the sewage of Worcester. I talked with Mr. Blake about the time he was putting it in, and he admitted that it would be very offensive in the river, but said it would never come to Millbury: it would be utilized for some other purpose.

Q. (By the CHAIRMAN.) — Something was said about the formation of islands in the river, in a way to indicate that it was caused by sewage. Do you think it is caused altogether by the sewage of Worcester? *A.* — I think a great deal of it is caused by the sewage of Worcester. It creates weeds, — what we call "pickerel weed," — which grows there luxuriantly: it will grow up so that it will cover over a space of three or four feet sometimes; and, as the water rises and falls, it rises up, washes down, and then, when it goes on to the banks, it creates a wild weed that grows up. The ponds are filling up very fast. We have had to clear out our pond. We have a small pond, and we have had to clear it out, and it takes a good deal of time to get that clear; and the ponds are filling up now. Mr. Morse's pond is filling up very fast, and the ponds below. What we call Pleasant Falls Pond, which is between Millbury and the Sutton Manufacturing Company, has filled up two or three rods from the shore. In a few years I do not see why it is not going to fill the entire river up, unless it is cleared out.

Q. (By Mr. SMITH.) — You speak about having cleared out your own pond: what was the character of the material which you took from your pond? was it sewage matter? *A.* — It was mixed with sewage matter, and with weeds, mud, etc. It was very offensive, and I carried it away; and, after letting it dry over winter, I took it out, three years running, to the farm, and worked it into compost.

Q. — How large is your pond? *A.* — About twelve acres: it is a small pond.

Q. (By Mr. FLAGG.) — Something pollutes the water a great deal more to-day than ten years ago, or five years ago? *A.* — Oh! it has increased all the time gradually.

Q. — Has the amount of pollution from surface drainage or the manufactories increased during the last five years? *A.* — I don't see why it should, from the surface.

Q. — Has the amount put into it by the city of Worcester increased within the last five years? *A.* — Oh, yes! that has increased every day: that they don't deny. They tell me I got on the wrong end: if I had got up above Worcester, I would be well enough off; as I went below Worcester, I must take any thing that comes to me.

Q. (By the CHAIRMAN.) — Is that what is said to you personally by Worcester people? *A.* — Yes, sir.

Q. — People in high position there? *A.* — Yes, sir.

Q. — That is the common kind of talk you meet with? *A.* — That is the common kind of talk we meet with. They don't admit that they have done any thing that they had not a perfect right to do, and say that we must submit to what they have done.

Q (By Mr. SMITH.) — When gentlemen speak to you in that way, do you consider it serious talk, or a little bantering? *A.* — Well, it is talk. It is not the authorities, the mayor of the city, and those people: but it is people that bluff off these things, and say we have no claim; that if we have settled there, the natural stream ran there before we went there, and we must take the consequences; and, if we block up the stream and use the privileges, we must take what they send down to us, which is the natural drainage which the river was calculated to carry off.

Q. (By the CHAIRMAN.) — Is it what you would characterize as good-natured talk, or as serious talk? *A.* — I take it as good-natured talk; but there is something back of it. They don't intend to put their hands in their pockets to meet any thing in the way of taxes to help us out. When they are short of water, they go out and put a steam-pump in, and say they want the water to run through their sewer; and they tell us, if they use it in a steam-engine, we get it in vapor in the first shower that we get down the river.

Q. (By Mr. CAMPBELL.) — Have any citizens of Worcester acknowledged that they pollute this water to any great extent? *A.* — Oh, yes, sir! a good many of the first officers of the city have.

Q. — Did they claim that they had a right to do it? *A.* — Yes. I have talked with a number of the officials, Mr. Chapin in his day, Mr. Earle and Mr. Ball and Mr. Blake, who first put it in. Mr. Chapin always admitted that it was wrong: Mr. Earle did not admit that it went into the river to any extent, to go down as far as our place; but, in going down to his monthly meeting, I called his attention to it. He looked at it, and admitted that there was a great deal of dirt in the river; and he said, "Thee has a very dirty place about here."

Q. — What did you say to the last part of his remark? *A.* — Well, I took it as a joke: I know Mr. Earle very well.

Q. (By Mr. FLAGG.) — Do you remember when the dams on the river below Worcester were built? how early any of them were built? *A.* — I don't think I could give the date.

Q. — Not the exact date; but was it ten, fifty, or a hundred years ago? *A.* — I recollect when the Burling dam was built. It was since the railroad was built: it was since the discontinuance of the Blackstone canal.

Q. — Have any of them been built since 1867? *A.* — I should think not. They built a dam at Burling Mills last year, or a year before, across the river, where they took water formerly out of the canal.

Q. — But there was a dam there before that? *A.* — Yes.

Q. (By Mr. GOULDING.) — You have a familiar acquaintance with a good many of the leading men of Worcester, I suppose, — the bankers, mill-owners, and business men of all sorts, and have had for a good many years? *A.* — Oh, yes, sir !

Q. — Many of them are your personal friends, I suppose? *A.* — Yes, sir.

Q. — When you get together, and banter and talk, and argue on one side and the other of this question, you maintain your side of it to the best of your ability? *A.* — Yes, sir.

Q. — That is all you mean to say, isn't it? You don't mean to say that there is any organized opinion in Worcester, one way or the other, about this thing, that you know of ? *A.* — I don't think there is. I think there has been a deficiency in looking after all this thing, until it has got beyond their control. I don't think that they supposed, when it was first put in by Mr. Blake, that it would ever amount to what it has. They have had success in building up the city, they require a great deal of water to come into it, and it has got to go out in their sewer ; and it has got now beyond what they or anybody else, fifteen years ago, expected it would. I was always fearful that we should have trouble there, and always talked about it.

Q. — I want to ask you another question, and that is, whether the rate of increase in this impurity was greater between 1870 and 1875, than between 1875 and the present time? *A.* — It has been greater the last five or six years : it has increased with the city and the wants of the people.

Q. — You know the State Board of Health of Massachusetts say there has been a marked decrease in the ratio of increase, so to speak ; that it increased more rapidly between 1870 and 1875, than between 1875 and 1880, whether you agree with that opinion, or not?

Mr. FLAGG. I don't know that.

Mr. GOULDING. Let us understand each other as we go along. On p. 124 of the report of the State Board of Health, Lunacy, and Charity, for the year 1882, they say, "Comparing the results from the several examinations in 1881 with those of the State Board of Health in 1872, it is clear that the pollution of the stream has increased since that time. As compared with the chemical examinations made by the Board in 1875, there is also an increase, although much less marked." That is the expression that I refer·to, if it bears out the remark : if it doesn't, it don't, that's all. In other words, there has been an increase from 1872 up to the present time. " As compared with the chemical examinations made by the Board in 1875, there has also been an increase, although much less marked." I understand it to mean that the increase is much less marked.

Mr. MORSE. I desire the Committee to notice, in connection with that, a remark of the Board on p. lxv. In the middle of the page they will find these words: "A comparison of the chemical analyses of the waters of the Blackstone River, made in 1881, with a large number made by the State Board of Health in 1875, reveals a very serious increase in the percentages of polluting matter."

Q. (By Senator TIRRELL.) — Suppose all those dams through Millbury and those other villages were removed, and this river had its natural flow, would there be any cause of complaint then, in your opinion? A. — Yes, sir. It would not affect it but a very short time: it would all fill up. Joseph Mason of Worcester called my attention to that, and asked my opinion. In my opinion, it would grow up just as ditches do in our low ground; and it would all fill up. I do not see why this river would not fill up with those weeds and all this stuff, just the same as a ditch in a low meadow.

Q. (By Mr. GOULDING.) — What is the fall from Worcester to Saundersville? A. — I think it is about forty-five or forty-six feet.

Q. — And between Saundersville and the Rhode Island line, about what is it? A. — I have not the minutes. It is all laid down.

Q. — Between Providence and Worcester, it is four hundred and twenty-eight feet, isn't it? A. — Yes, sir.

Q. — You think that the result of taking those dams out would be, that the river would fill up? Where would the water go to, down that four hundred and fifty-eight feet of slope? A. — Flow over the land. It would go at random, wherever it found the lowest place.

Q: (By Mr. FLAGG.) — What would be the effect upon the industries of Millbury, Grafton, Sutton, Northbridge, Uxbridge, and Blackstone, with their twenty-five thousand inhabitants and thirty-two hundred operatives, of taking down the dams? A. — Well, it would depopulate that country. That is the business that they have been brought up to, the business that they are calculated to carry on: I do not see any other business that they could adapt themselves to.

Q. (By Mr. MORSE.) — Were those dams in existence long before the city of Worcester turned its sewage into the Blackstone River? A. — Oh, yes!

Q. — Do you know when the city of Worcester turned its sewage into the Blackstone River? A. — I have not the date. It was in Mr. Blake's administration. I believe Mayor Stoddard, at your right, can tell you exactly.

Mr. GOULDING. I do not understand that to be the history of the sewage of the city of Worcester.

Mr. MORSE. Do you claim that the city of Worcester had any system of sewage that was turned into the Blackstone River before the Act of 1867 authorized it to be done?

Mr. GOULDING. Long before that, the court had held that it had a right to do it.

Mr. MORSE. By a system of sewerage?

Mr. GOULDING. Yes, sir, by a system of sewerage. I do not understand that a sewer that embraces several streets is any thing less than a system of sewerage. I do not mean that the statute of 1867 was passed before 1867 : that was just the year it was passed.

Mr. FLAGG. Didn't the court hold that they had no right to put their sewage into the river to the insignificant extent that they were then doing it? and didn't they come down here and say that they must have the statute of 1867 in order to empty it into the river?

Mr. GOULDING. We shall discuss the law, hereafter, very fully; and we shall discuss the cases. Very likely our friends may not agree with us as to what the law is. As a matter of fact, the sewage of Worcester has gone into the Blackstone for a hundred and fifty or a hundred and seventy-five years.

Mr. MORSE. It has been a very different kind of sewage until within the last few years.

Mr. GOULDING. For more than thirty years before 1867 we had systems of sewers.

TESTIMONY OF JOEL SMITH.

Q. (By Mr. FLAGG.) — You live where? *A.* — Wilkinsonville.

Q. — That is in the town of Sutton? *A.* — In the town of Sutton.

Q. — Tell us how far Wilkinsonville is below Millbury? *A.* — A little over two miles.

Q. — Making five miles from the mouth of the sewer? *A.* — Yes, sir.

Q. — What is your business there? *A.* — I am superintendent of the Sutton Manufacturing Company.

Q. — How long have you been superintendent? *A.* — Three years.

Q. — How many people do you employ? *A.* — Two hundred and seventy.

Q. — How many people live in the village of Wilkinsonville? *A.* — I could hardly tell.

Q. — About how many? *A.* — About six or seven hundred.

Q. — Mostly connected, in one way or another, with the mill? *A.* — No, sir : about two-thirds of them connected with the mill.

Q. — Who owns this mill? *A.* — H. N. Slater of Webster.

Q — You say you have been there three years. In your dealing with the water there, what have you noticed, as to its purity, since you have been there? *A.* — I have noticed it is very impure.

Q. — Is it fit to use in boilers? *A.* — I am using it in my boilers.

Q. — Do you consider it good? *A.* — No, sir: I would not use it if I could get any other conveniently.

Q. — What means do you take to get along with it in your boilers? *A.* — We blow off our boilers very frequently.

Q. — More frequently than formerly? *A.* — Oh, yes, sir! that is, more frequently than I ever did at any other place.

Q. — Tell us as to the appearance of the water about the flumes. *A.* — It is very dark, and frequently has a yellowish scum on the top of it; so dark that we cannot see the bottom of the trench, — some three or four feet deep. There has been only one occasion, since I have been connected with the establishment, when I could see the bottom; and that was, I think, the very day that the State Board of Health were in Millbury. The night before, I noticed it, and called the attention of several other persons to the fact that I could see the bottom. I had occasion to go into the flume that night; and I could see the nails through the water in the bottom of the flume, where the water was two feet deep, which we were never able to do before, and have not since. I mention that as a fact: of course, I can't account for it. Perhaps some of our Worcester people can, but I cannot.

Q. — What proportion of the volume of water in the river passes through your flumes? *A.* — From June until December, the last three years, we have used nearly all of it.

Q. — Where are the flumes situated? *A.* — We have a flume on the north side of the river, and pipes to take the water across.

Q. — As regards the mill, are the flumes in the basement of the mill? *A.* — Yes, sir.

Q. — So that for several months in the year the greater part of the water in the river passes through the basement of your mill? *A.* — Yes, sir: nearly all of it.

Q. — What, in your opinion, is the effect upon the health of the operatives? *A.* — Well, during the last year we have had more sickness in our mill than any time previous since I have been there. Our pay-roll shows that there were more out, and more reported sick; and the troubles seemed to be throat and bowel complaints: nearly all the cases that I have inquired after were either one or the other. Some of our tenements are located near the trench that supplies the water for the mill, and I have frequently had persons refuse to take those tenements on account of being so near the canal. One tenant last year I was obliged to move to another tenement, on account of that: the smell was so offensive, they could not stand it.

Q. — Have you had any trouble at your rack with dead fish? *A.* — Yes, sir.

Q. — On more than one occasion? *A.* — On three different occasions, — two last year, and one the year before.

Q. — Describe the trouble to the Committee. *A.* — Dead fish collected there, and they had to be taken out.

Q. — In quantities, or only one or two? *A.* — At one time I took out three bushels; and they are coming down to the rack, half a dozen at a time.

Q. — Do you think of any thing else that you wish to state? *A.* — I don't know that I do.

Cross-Examination.

Q. (By Mr. GOULDING.) — Can you state how many more were out by reason of sickness last year than the year before? *A.* — I couldn't state positively; but it was many days noticeable in running the machinery.

Q. — Do you have the means of telling exactly by your books? *A.* — Not positively, because our books do not mark those that are sick when they are out.

Q. — Have you examined your books to see how many more were out last year than the year before? *A.* — No, sir: only from observation.

Q. — It would have been easy for you to tell by looking at your books what the fact was about that, I suppose? *A.* — Well, not easy; because frequently hands stay out when they are not sick.

Q. — I mean, it would be easy for you to tell, by examining your books, whether more, and how many more, staid out last year than the year before, for all causes? *A.* — Yes, sir: we could do that, although we have been troubled some for water, and it would be pretty hard to tell.

Q. — This, then, is an impression, or belief, or opinion, that you have from observation, not from any examination of the books? *A.* — Yes, sir.

Q. — Have you any knowledge of medicine yourself? *A.* — No, sir.

Q. — Have you any knowledge of what was the cause of those diseases? *A.* — No, sir: I only know from inquiry what the matter was, why they were out.

Q. — Have you more than one tenement that is so situated that it is troublesome to let it? *A.* — There are eleven.

Q. — And all situated equally near to the stream? *A.* — The same.

Q. — Are they all occupied now? *A.* — No, sir.

Q. — How many are unoccupied? *A.* — I think four unoccupied.

Q. — How long have they been unoccupied? *A.* — Four or five months.

Q. — Are these unoccupied in consequence of being near the stream? *A.* — Not altogether.

Q. — How many of them, if any, are unoccupied by reason of being near the stream? *A.* — I might say all, because the people take other tenements in preference to those, on account of their being there; and, as long as we have any others that they can get, they take those in preference.

Q. — They take the preferable tenements? *A.* — Yes.

Q. — Has there been any time when you have been obliged to get tenements outside for your help? *A.* — No, sir: we have plenty now.

Q. — They take the preferable tenements that are placed away from the stream? *A.* — Yes, sir.

Q. — How many times did you say you had had trouble with dead fish? *A.* — Three times; that is, three noticeable times: but we take fish out there nearly every day in the summer.

Q. — All these three times were in the summer, when the water was low? *A.* — Yes, sir: no water running over the dam.

Q. — How many did you say were taken out those three times? *A.* — Three bushels at one time.

Q. — Was that the first time? *A.* — No, sir: that was the second time; that was early last season.

Q. — Could you tell the month? *A.* — I should think about June, the last of June.

Q. — Last June? *A.* — Yes, sir.

Q. — That was the second time, and three bushels were taken out then? *A.* — Yes, sir.

Q. — And the first time was when? *A.* — That must have been the fall previous.

Q. — How many were taken out then? *A.* — I don't remember: there were a good many.

Q. — The third time was when? *A.* — That was last September.

Q. — The water was low each time? *A.* — Yes, sir, low all those times.

Q. — What kind of fish were they? *A.* — Nearly all suckers.

Q. — Those not suckers, what were they? *A.* — I don't remember that there were any but suckers.

Q. — Any perch among them? *A.* — No, sir.

Q. — Any flat-fish? *A.* — I think not. I believe they were all suckers. There might have been one or two bull-heads among them.

Q. — More or less dead fish are still coming down? *A.* — Not in the winter.

Q. — Not in the winter, but in the summer? *A.* — In the summer.

Re-direct Examination.

Q. (By Mr. FLAGG.) — I don't remember whether I asked you about the odor. You spoke of it, I believe. What sort of odor did you notice? *A.* — Well, it is as near to cesspool odor as any thing I can think of.

Q. — Where do you notice it? *A.* — All along the river.

Q. — Through the mill? *A.* — Yes, sir.

Q. — How far back from the river do you notice it? *A.* — Well, a quiet morning I have noticed it, coming down to the mill, probably three hundred yards from the river.

Q. (By Dr. WILSON.) — What diseases did you say the operatives had last summer? *A.* — They were throat and bowel diseases mostly.

Q. — What name did they call the throat-disease? *A.* — They called it sore throat, as near as I could find out. They did not call it diphtheria, because it was not severe enough to be termed that.

Q. — Beyond that, you don't know? *A.* — No, sir.

Q. — What were the bowel-diseases called? do you remember any thing about that? *A.* — No, sir.

Q. — Was there more of that last year than the year before? Did you notice any difference? *A.* — Yes, sir: my impression is, that there was a great deal more of it last summer.

Q. — Than any year previous? *A.* — Than any year previous. I have only been there three years, you understand.

Q. — When was the water lowest, last year or the year before? *A.* — I don't think there was very much difference. The low water came earlier the year before than last year.

Q. — And lasted longer? *A.* — Lasted longer.

TESTIMONY OF GEORGE W. FISHER.

Q. (By Mr. FLAGG.) — You live where? *A.* — Fisherville, Grafton.

Q. — What is your business? *A.* — I am agent there for the Fisher Manufacturing Company.

Q. — What do they manufacture? *A.* — Cotton goods.

Q. — Is that the next privilege below Saundersville? *A.* — The next below Saundersville.

Q. — You have heard the testimony of the other witnesses: what do you say as to the evil? *A.* — Well, very similar indeed.

Q. — About using the water for the boilers, is it good? *A.* — It is not good. I have had the same experience, although I have used it. I have wanted to have some other supply for the boilers, but have not as yet obtained it. We were burned out Jan. 27, 1881, and have not run, of course, since. We have been rebuilding.

Q. — How many operatives do you employ, or did you? *A.* — We did employ about a hundred and twenty-five.

Q. — How many people live in the village? *A.* — I don't know: perhaps some three or four hundred.

Q. — Tell us what you have noticed about the odor of the river in your flumes. *A.* — It has been very strong, very noticeable.

Q. — How would you describe the odor? *A.* — It is, as Mr. Smith says, as near a cesspool as any thing.

Q. — There is no mistake about it? *A.* — There is no question about that: it is an unquestionable fact.

Q. — What is your opinion as to its effect upon the health of your operatives? *A.* — I don't know as I have noticed much about that.

Q. — What is your opinion?

Mr. GOULDING. If he has not noticed, he cannot well have any opinion.

A. — Well, it is common talk and common report, that it has an injurious effect; but still I cannot point out any particular case, and say that that was caused by the water.

Cross-Examination.

Q. (By Mr. GOULDING.) — How long have you been there? *A.* — I have been there since 1843, with the exception of eight years.

Q. — How near do you live to the stream? *A.* — Perhaps three hundred yards.

Q. — What does your family consist of ? *A.* — Wife and necessary help.

Q. — How old are you, Mr. Fisher? *A.* — Thirty-seven.

Q. — Is this company named for you or for your father? *A.* — Well, no particular name, as I know of, — Fisher Manufacturing Company.

Q. — I did not know but it was named for your father. *A.* — No: it has been organized since his death. It was organized this last spring.

Q. — Are you the son of Mr. Waterman Fisher? *A.* — Erastus Fisher.

Q. — Did your father live there before you? *A.* — He did: he moved away from there in 1861, and I went back there in 1868.

Q. — When did your father die? *A.* — He died a year ago last April.

Q. — He did not live there at the time? *A.* — Oh, no! he has not lived there since 1861. He used to be back and forth while he lived. There is one point that has not been called out; and that is, the use of the water for bleaching purposes. We wanted to use it for bleaching purposes, and for starching our towels, or sizing them; and it got to be so foul that we couldn't use it, — that is, on white work, — but we could on the brown goods. If we had not been burned out, we were going to bring, this last summer, a supply from off the hill for that purpose.

TESTIMONY OF DR. WILLIAM H. LINCOLN.

Q. (By Mr. FLAGG.) — Where do you reside? *A.* — In Millbury.

Q. — Your profession? *A.* — Physician.

Q. — How long have you resided in Millbury? *A.* — Sixteen years last May.

Q. — Have you noticed in that time a change in the Blackstone River? *A.* — Yes, sir.

Q. — Have you noticed any thing which would enable you to say that there was a change in the general health of Millbury during that time? *A.* — Yes, sir.

Q. — Will you tell what you have noticed? *A.* — If the Committee will allow me, and the counsel do not object, I will make a simple statement, which perhaps will make it clearer than answering questions. I came to Millbury sixteen years ago last May. The population of Millbury in 1870 was 4,397, I think. What the census was the ten years previous, I have forgotten; but, if my memory serves me, it was 3,900 and something: but I won't be positive as to that. When I came there, there were two physicians in town; and they thought there was no more than they could attend to well, that there was no place for a new man, — that they had nothing more than they cared to do. There are now six physicians there, five of them in active practice; and perhaps it is safe to say that any one of the five is doing as much business as either of the two that were there before. I think that answers the question of the gentleman whether there is more sickness there now than formerly.

Q. — In other words, your answer is that there is? *A.* — There is.

Q. — What have been some of the kinds of sickness which you would think might be attributed, either in whole or in part, to the foulness of the river? *A.* — Well, I should say that the common sicknesses had been mostly of the zymotic type, — what we call the filth diseases; perhaps scarlet fever, diphtheria, diarrhœal troubles, dysentery, and diseases of that character. The increase would be largely of that kind.

Q. — Have you in mind any particular cases which you can call to the attention of the Committee? *A.* — No: I don't know of a case that I should be warranted in saying was the result of the sewage, or any thing of that kind. The general health-rate isn't as good among our people.

Q. — And you attribute that to the influence of the river, as I understand you? *A.* — I know of no other reason: I know of nothing else to attribute it to.

Q. — You have been in Millbury now sixteen years: will you tell us what you have noticed about the river as it was when you came there, and how it compares now? *A.* — When I came there, the boys used it for bathing. There has been a gradual increase in its foulness.

Q. — Have you heard the preceding testimony in regard to bathing? *A.* — Yes, sir.

Q. — Do you agree with that testimony? *A.* — Yes, sir.

Q. — Do cattle drink of that water? *A.* — No, sir. I have a piece that adjoins the river, and I pasture my cow there; and she won't drink at the river. I have to drive her somewhere else, twice a day, to have her get water.

Q. — In general, what do you think the effect of the foulness of the river has been on the health of the people of Millbury, and the towns along the river? good or bad? *A.* — Bad. That is the idea; but the Committee will understand me, I know of no other reason to which to attribute the amount of disease more than previously.

Q. (By Dr. INGALLS.) — Do you know what the death-rate was in Millbury when you went there? *A.* — I do not.

Q. — Do you know what it is now? *A.* — I do not.

Q. (By Senator TIRRELL.) — Do you think that the number of physicians in Millbury is larger in proportion to the population, than in other places? *A.* — It would depend altogether upon the locality of the place. To illustrate that, so that you can understand it, — if you take a place of three thousand inhabitants, twenty or thirty miles from any larger place, that would be a better place for four or five physicians than a place of three thousand inhabitants would be for three, if it was within six or eight miles of some larger place. Any physician understands the principle upon which that is based.

Q. — Is it not a fact, that doctors, like lawyers, have multiplied very rapidly within the last ten years all over the State? and do you not find, in the towns with which you are acquainted, a larger number of doctors and lawyers in proportion to the population than there was ten years ago? *A.* — I can name you three or four towns that I am well acquainted with, where the population has changed but very little in the last ten or sixteen years, where there are fewer physicians than there were at that time. I do not think that in the country the number of physicians has increased very much in the small towns.

Q. (By Mr. FLAGG.) — If it had, it would not account for the increase in the death-rate, would it? *A.* — That would depend something upon the gentleman's faith in physicians.

Q. (By Dr. WILSON.) — Did you say what the death-rate of Millbury was? *A.* — No, sir: I did not make any statement in regard to that.

Q. — What is your idea about it? *A.* — I do not know that it has increased. I said that the health-rate was not as good as it was at that time.

Q. — That means, that the increase of doctors has not done any harm? *A.* — I think so, certainly. My idea is this: that the health-rate may not be as good, the general feeling of the community may not be as healthy and as well, and yet the death-rate not increase very much.

Q. (By Mr. GOULDING.) — Is that a general proposition, doctor, that the death-rate is no criterion of the state of health? *A.* — I did not say that it was no criterion. I say I can understand that the health-rate of a community may not be nearly as good, and yet the death-rate not increase a great deal, — not in proportion to the decrease in the health-rate.

Q. — Would not such a thing be an exception to a rule? *A.* — I consider this place an exception to the rule.

Q. — I am not asking about that now. I was trying to see if I could get some general principles that do not link themselves with absolute closeness to Millbury, if there were any such general principles. If you will be kind enough to leave out Millbury for a little while: we are in Boston now. We have at present no odor from that river; and now I would like to ask you if such a state of things as that, where the death-rate does not furnish a pretty satisfactory indication as to the health-rate, would not be an exception to a general rule? *A.* — In a long series of years, it may be; but in a series of five or ten years it might not be.

Re-direct.

Q. (By Mr. FLAGG.) — Your attention was called at one time to a tank in C. D. Morse's mill? *A.* — Yes, sir.

Q. — Will you describe what you saw? *A.* — Well, there was a tank there; and water came from the river into the tank.

Q. — Where was the tank situated? *A.* — In the upper part of the mill.

Q. — How large a tank? *A.* — Well, I won't state; for I can't say. It was a tank that they pumped up water into to use for certain purposes.

Q. — Would it hold a hogshead of water? *A.* — I should say it might, and it might hold more or less. I took a small stick and passed it around the edge of the tank; and there was quite a large amount of sediment on it, that had a very strong fœcal odor.

Q. — You saw no means of any fæcal matter getting in there, except from the river? *A.* — No, sir.

Re-cross.

Q. (By Mr. GOULDING.) — When was this? *A.* — I can't say. Some two years ago, I should say.

Q. — How large was the tank? *A.* — I don't recollect. I should say it may have held a hogshead, and may have held more.

Q. — Was it about the size of a hogshead, or larger? *A.* — It was nearly a square tank, — rectangular.

Q. — And would hold about a hogshead? *A.* — Somewhere thereabouts.

Q. — And was full of water? *A.* — No, sir: there was but very little water in it at that time. I won't say there was any water in it.

Q. — Was it dry matter or muddy? *A.* — It was muddy.

Q. — And you stirred it up with a stick? *A.* — No, sir: I told you I had a stick in my hand, and I scraped it around the edge of the tank, and then smelt of it, and found that it was a pretty stinking place.

Q. (By the CHAIRMAN.) — How did the water get into the tank? *A.* — By pumping.

Q. (By Mr. GOULDING.) — Did you ever see a tank that had held water for a length of time that would not have more or less mud in the bottom? *A.* — I don't think I ever did: this was undoubtedly fæcal mud.

Q. — Have you ever made any figures or calculations for the purpose of determining what the health-rate of Millbury has been for the past five years? *A.* — I know of no way to make that calculation.

Q. — Then you have not, of course? *A.* — No, sir.

Q. — What you say about that is your impression from your general practice? *A.* — That is all.

Q. — What school of medicine do you belong to? *A.* — Well, what is called the regular school.

Q. — Are all the physicians in Millbury regulars? *A.* — No, sir: there is one homœopath.

Q. — And you say there are now how many physicians in town? *A.* — Six, — five in active practice. One is the father of the homœopathic practitioner, who does not do much business.

Q. — Has your business grown considerably since you went there? *A.* — Well, I should say that for the past few years it had not grown; rather cut off at the back end of it, somewhat.

Q. — Is it feasible for a physician to determine with some degree of certainty whether a particular case is the result of river pollution?

A. — Well, I never have seen a case that I could say, with any degree of certainty, that that was caused by the river.

TESTIMONY OF DR. GEORGE C. WEBBER.

Q. (By Mr. FLAGG.) — You live in Millbury? *A.* — Yes, sir.

Q. — And practise as a physician there? *A.* — Yes, sir.

Q. — How long have you practised there? *A.* — Eleven years last December.

Q. — As to the condition of the river when you came there, and as to its condition now, — have you noticed any change? *A.* — I should say it was very much more foul now than then.

Q. — What do you notice now about it as to foulness? *A.* — Its color is dark, muddy: the water and odor are both extremely foul.

Q. (By the CHAIRMAN.) — Have you heard what the preceding witnesses have said? *A.* — Most of them.

Q. — Do you agree with what they say, in general, in regard to the color and odor, in regard to people bathing in the river, and all that, as far as you know? *A.* — I agree with it all, sir.

Q. (By Mr. FLAGG.) — As to your practice there, in what way have you noticed the effect of the river upon the health of people, or what can you say as to that? *A.* — I should say its effect was bad.

Q. — That is stating the matter generally. Now, have you any particular cases that you would speak of? *A.* — I would state first, if I may be allowed, generally; and I will then go into some particular cases. The foulness of the stream, and its offensive odor, are generally acknowledged. Such a stream emits such exhalations as are conceded by all sanitary authorities to be the producing causes, often, of zymotic diseases. That in a general way. I will say further, before alluding to specific cases, that I think it not right to consider entirely and exclusively the death-rate; that there are injurious influences which the figures of death-rates do not show. I do not know that I can any better state that than by reading a short paragraph from a work on "Filth Diseases and their Prevention," printed in 1876, under the direction of the State Board of Health of Massachusetts. It is the third paragraph on the sixth page. It is by Dr. John Simon of England, a sanitary authority there.

> "I do not pretend to give any exact statement of the total influence which preventable diseases exert against the efficiency and happiness of our population; for it is only so far as such diseases kill, and even thus far but very imperfectly, that the effect can be represented in numbers. Of the incalculable amount of physical suffering and disablement which they occasion, and of the sorrows and anxieties, the often permanent darkening of life, the straitened means of subsistence, the very frequent destitution and pauperism which attend or follow such suffering, death statistics, to which alone I can refer, testify only in sample or by suggestion."

As to specific cases, I will allude to a few in which the proximity of the cases to the river render it at least probable that this may have been one of the determining causes of the disease. I cannot follow them in chronological order, but that is of no consequence. A year ago last fall, there occurred in one house four cases of typhoid fever, which is one of the diseases attributable to such causes. One case was very severe, and proved fatal : the other cases recovered. There was no apparent cause about the premises. I inquired as far as I could, and could determine no cause about the premises. The house was situated somewhere, I should judge, from two hundred and fifty to three hundred feet from the river, at the bridge near the Atlanta Mill. That agency was plainly there, and ready to do whatever work it was capable of doing.

Another instance was of the occurrence of dysentery in four cases in a house situated somewhere in the neighborhood of three hundred feet, I should say, from the ordinary channel of the river; but at high water it overflowed the land to within seventy-five or eighty feet of the house. There also the premises were apparently cleanly, and no discoverable source of such infection. One of those cases proved fatal. It, however, occurred in the case of a child who had been sick for some time previous, and dysentery was developed in the course of convalescence; but the other cases that occurred were with persons previously healthy.

There have been other cases, perhaps not quite as strong as those. I mention those from among a number as perhaps being the strongest evidence in reference to this matter. One other case I will allude to. The case has been alluded to sometimes hitherto. It is that of a gentleman recently deceased, who died on the 10th of January, who had lived to a somewhat advanced life. Mr. Benjamin Flagg first came under my observation and advice last May. He was one of our best known citizens, seen on the street every day; and it had been a common remark that he was looking very badly, as he had for two or three years. When he came under my observation, I found he was affected with organic heart disease ; but of course I could not attribute that to the sewage. Whether it may have had any indirect connection with that, I am not competent to say. But this much I will say in reference to that case: in the latter part of the fall he had occasion to have extensive repairs done at his mill, in some way connected with his water-power; and, while that was progressing, he was much of the time in that part of the mill where he would be exposed to emanations from the river, which is at that point very offensive, the river being narrow, and the water pouring through a comparatively small space. While he was engaged in overseeing and giving directions about these repairs, he was attacked with diarrhœa. I think it is generally conceded that affections that are characterized by disturbances of the bowels, diarrhœal affections, are peculiarly liable to be occasioned by such causes. He was attacked by diarrhœa, which was not controllable; and in his somewhat debilitated condition it continued until the fatal end, which occurred Jan. 10th last. He was ill about six weeks, if I remember rightly.

Q. — Were you familiar with the case of Mr. Howard, who died a short time ago? *A.* — I knew of the death of Mr. Howard. I had no professional connection with the case. I know he lived in a place where he must breathe those poisonous gases, being within a hundred or a hundred and fifty feet of the canal, which in times of low water must carry the greater portion of the water running through the river.

Q. — How old was Mr. Howard? *A.* — Forty-six or forty-seven, I should say. I don't know his age.

Mr. FLAGG. The Chairman of the Committee will remember that Mr. Howard testified before the Committee last year.

WITNESS. In connection with the fact of such disease being occasioned by the foul condition of the river, it may be worthy of remark, that in 1877 I was called to see a number of cases of diphtheria, several very severe, at Quinsigamond Village, in the square brick house which is not far distant from the river; and it is well known by gentlemen from Worcester that diphtheria prevailed extensively in that village that season.

Q. — Do you know how far the mouth of the sewer is from Quinsigamond? *A.* — I do not. It was not then near the Washburn & Moen Manufacturing Company. I think it was at that time up in the neighborhood of Cambridge Street.

Q. — But these cases occurred near the Blackstone River? *A.* — Yes, sir: these four cases in one house.

Q. — What can you say about the prevalence of sore throats in Millbury? *A.* — Sore throats are common there, and they are plainly not simply local diseases. They are not simply diseases of the throat; for there is in the majority of cases a good deal of constitutional disturbance and prostration, headache, back-ache, low fever, and a general debilitated condition which takes much longer to relieve than the local trouble. The throat gets well, leaving the patient weak for a considerable time; and that is one of the things which are attributed by many authorities to such influences as these.

Q. — Have you an opinion on the subject? *A.* — I believe that they are more or less attributable to such causes.

Q. (By the CHAIRMAN.) — Do you think that matter is yet fully understood? *A.* — I think there are very many things yet to be known about it, sir.

Q. (By Mr. TIRRELL.) — In order to get all the facts, I would like to know whether the nationality of the population of Millbury has materially changed during the last ten or fifteen years? Whether there are more of what we call foreign-born population now there than ten or fifteen years ago, and, if so, what proportion? *A.* — I should not be willing to venture any opinion as to the proportion. I should say that there was perhaps a slight relative increase in the foreign population, but not large.

Q. — Whether disease is more prevalent among that class in your town than among the native born? *A.* — That class generally live in tenement-houses, and do not take the same care as to their surroundings that the native born do, and with results such as are seen in other places.

Q. — You do not think that the increase in the amount of sickness is accounted for by the increase of that class of people? *A.* — Not to any large extent.

Q. (By Dr. WILSON.) — You spoke of several cases of typhoid fever and diphtheria. I want to ask you where the people living in those two houses get their drinking-water, if you know. *A.* — Both from wells, but in both instances remote from privies or sink-drains or stables.

Q. — Not located so that they could by any possibility be affected by bad drainage? *A.* — Not that I could see.

Q. — You examined into that? *A.* — I examined into both those cases carefully.

Q. (By Dr. CAMPBELL.) — Have you had under your care any considerable number of the employés of the manufacturing establishments referred to? *A.* — I have more or less under my care all the time. I cannot say at any particular time any particular number that I had under my care.

Q. — Any particular epidemic among them of any kind? *A.* — Not recently.

Q. (By Dr. HARRIS.) — Do you consider the odor which you speak of largely charged with sulphuretted hydrogen? *A.* — I have never applied any chemical tests. I should presume it would be found to be so.

Q. — What is the effect of that gas upon the mucous membrane of the throat, and other portions of the body, so far as you have observed? *A.* — I should express an impression simply, that it was an irritant; but the throat affections which I spoke of, I expressly stated, were not simply local affections, but there was a great deal of constitutional disturbance accompanying them.

Q. (By Dr. WILSON.) — What do you mean by "constitutional disturbance"? Do you think this constitutional disturbance was the result of the throat-disease? or do you think the throat-disease was the result of the constitutional disturbance? or do you think there were two separate causes? *A.* — I think there was one cause which produced both effects.

Q. — Do you think there was any blood-poisoning? *A* — I think there was.

Q. — You think there was blood-poisoning, together with the local affection? *A.* — Yes, sir.

Q. — Should you think that was due to sulphuretted hydrogen? *A.* — I should not want to say it was due to sulphuretted hydrogen. I should say it was probably due to poisonous emanations, not necessarily chemical. I have noticed an odor which resembled the odor of sulphuretted hydrogen often.

Q. (By Mr. FLAGG.) — Do you think, from your knowledge of the river, and the work that the people of the town are engaged in mostly, that the condition of the river is such as is very likely to produce an epidemic? *A.* — I should perhaps wish to modify the question a little. Is it likely to become such? In the condition in which it now is, I should hardly venture an opinion: but my observation has been, that, with the growth of Worcester during the last ten years, there has been an immense increase in the filth carried down the river, and the odor has become increasingly disgusting, and, at times, well-nigh unendurable; and if the city continues to grow at the same rate, when it shall have reached a population of a hundred thousand, as I have no doubt its citizens believe it will, that will be immensely increased; and it will be very likely to be a cause which might produce epidemic diseases.

Re-cross.

Q. (By Mr. GOULDING.) — Do you know any thing about what the death-rate of Millbury has been for the past five years, as compared with the five previous years? *A.* — No, sir.

Q. — You have not investigated it at all, or looked to see? *A.* — I have not looked at the figures.

Q. — Do you belong to the regular school of medicine? *A.* — Yes, sir.

Q. — Have you ever had any typhoid fever in Millbury, except these cases to which you have referred? *A.* — Yes, sir.

Q. — Whereabouts? *A.* — There have been cases where similar influences were found prevailing.

Q. — I don't ask you to argue the case; I ask you to answer my question: the counsel will argue it fully when the time comes. Now, can you tell me in which localities you have had typhoid fever, without arguing any thing about it? We want to get at the facts. *A.* —

Yes, sir: I can answer the question. I have had some cases in the vicinity of what is known as Brierly's Mill.

Q. — Is that in Bramanville? *A.* — Yes, sir.

Q. — How many cases had you there? *A.* — I cannot tell you: it was some years ago. I should have to count them up on my book.

Q. — I only want a general idea. *A.* — I have had quite a number in that vicinity.

Q. — Can you tell about the number? *A.* — No, sir: I should not venture an opinion; it may have been half a dozen, and it may have been a dozen.

Q. — Within how many years? *A.* — Since my residence in Millbury, which is eleven years.

Q. — Any other localities where you have had typhoid fever, except those you have already mentioned? *A.* — There have been cases of typhoid fever occurring in various localities in the town.

Q. — Some remote from the river? *A.* — Some remote from the river, and several that I have not mentioned near the river.

Q. — How many remote from the river? *A.* — Those in town that have been remote from the river have been where there is the most population, and therefore in Bramanville, which is more remote than the lower village.

Q. — I did not ask you about Bramanville, because you had already stated about that. Are there any other localities except Bramanville, remote from the river, where you have had cases of typhoid fever? *A.* — I do not recall any now within the limits of the town. I have had them in the adjoining towns.

Q. — Have you had any other cases of dysentery except those four cases you have mentioned? *A.* — Yes, sir.

Q. — Where have you had them? *A.* — In various localities in town, some in Bramanville, and some in other places.

Q. — How old was Mr. Benjamin Flagg when he died? *A.* — I don't remember just his age, — seventy-five, seventy-six, or seventy-seven.

Q. — Was he taken sick in November or December? *A.* — His last sickness was in November.

Q. — And he died in the winter? *A.* — He died in January.

Q. — I don't understand that you attended Mr. Howard? *A.* — I did not, sir.

Q. — Do you know who was his physician? *A.* — I think Dr. Slocomb, the last part of his illness. He had several during his illness.

Q. — Had he been a man of pretty robust health? *A.* — I had never had any professional acquaintance with him, and had no occasion to talk with him about health matters.

Q. — Then, you don't know about it? *A.* — Not specially, sir.

Q. — I don't understand that you undertake to testify that there is any such condition of things there now as is likely to produce any epidemic, but you think that in the future there may be? *A.* — I think in the future there probably will be. I think there are injurious influences there now.

Q. — I understand you to say that; but the question I asked you was, whether you undertake to say that there is at the present time such a condition of things as will be likely to produce an epidemic? That is a perfectly simple question, and need not be coupled with any thing else. *A.* — I am not ready to say that there is an epidemic threatening us from the river.

Q. — Whether there is any cause for an epidemic? *A.* — I say there are causes capable of producing it.

Q. — Likely to produce it? *A.* — I won't say *likely;* I say *capable* of producing it.

TESTIMONY OF WILLIAM H. HARRINGTON.

Q. (By Mr. FLAGG.) — Where do you live? *A.* — In Worcester.

Q. — Your business is where? *A.* — In Millbury.

Q. — What is your business in Millbury? *A.* — I run the Atlanta Woollen-Mill. I was interested in the Burling Woollen-Mills; but I sold out my interest in the Burling Mills about four months ago.

Q. — How long have you been familiar with the river in Millbury? *A.* — I went to live in Millbury in 1840. I bought the Atlanta Mill, or became part owner, in 1856, and became a part owner in the Burling Mills in 1869.

Q. — Going back to that time, do you remember the river-water? *A.* — Perfectly, sir.

Q. — It was used then for bathing and domestic purposes, was it not? *A.* — At the Atlanta Mill in 1856 (it was then used as an ironworking establishment, making edge-tools), the water used to come into the shop, and was used on the trip-hammers; and in the winter the workmen never went out of the shop to get any water to drink, but drank that water. I have drank it myself thousands of times, and should in the summer, if it had not been for its being warm.

Q. — It is impossible to do that now? *A.* — It is impossible.

Q. — You have heard the testimony of the other witnesses as to the odor and color of the water, — is your experience the same as theirs? *A.* — Only more so, sir.

Q. — Will you state, in your own way, how more so? *A.* — Because I think I have had more experience, and been there more. Last summer I relined the Burling Mills flume; and there was about half an inch of sediment settled on the inside of the flume, which, when we went to repair it, we had to stop three days, I think, to let it dry, before we could go in there; and then we went in and scraped it off before we could commence repairing it at all. It was a sediment of filth collected on the inside of the flume; and it was so offensive when we were repairing the flume, that, when I would go down to examine and see what the men were doing, I would stay there as short a time as possible, and go away, and then come again. It was so bad, that, as I stood upon the floor, I couldn't stand it. It was difficult to get the men to go in to do the repairs.

Q. (By Mr. SMITH.) — Ten or fifteen years ago was there any thing of this same kind? *A.* — Not the slightest. We put in a new flume in 1857, and took out the old ones; and the flumes which we took out were just as clean as they were the day they were put in. We took out the old wheel and put in a new one in 1857, and repaired the flume; and the lining was worn thin, but free and clean.

Q. (By Mr. FLAGG.) — As to the fitness of this water for use in boilers, scouring, and so forth? *A.* — At the Burling Mills we had to abandon the use of it entirely for scouring wool or cloth. We do as yet use it for the boilers, but blow our boilers out three times a week. We are driving wells now, for the purpose of getting clean water to put into the boilers.

Q. — Have you any hesitancy in saying, from your familiarity with the river, that it is in such a state of pollution as to be a nuisance to you, both in your business and health? *A.* — There is no doubt but what it is to me, and everybody above and below me.

Adjourned.

THIRD HEARING.

TUESDAY, March 14, 1882.

THE hearing was resumed at 10.15.

The Chairman stated that the Committee desired, if possible, to close the hearing this week, and suggested to the counsel for the petitioners, that they should confine their testimony to new points, and refrain from introducing merely cumulative evidence.

Mr. FLAGG. Mr. Chairman, our only embarrassment in proceeding with the hearing is to determine where to draw the line. We understand very well that the Committee ought not to be burdened with simply cumulative evidence; and we had supposed, until my brother, the City Solicitor, gave us to understand to the contrary, that the fact of nuisance was one that would not be seriously controverted. But the Committee will remember, and it appears in the report, that my brother said, "We deny that there is any nuisance. We shall offer to show, from evidence that has already been put in, that there is no nuisance by necessary implication." That drove us to the introduction of testimony which we had supposed would not be necessary. We had thought that the evidence which we put in, put in from the Reports of the State Board of Health, more particularly that of last year, would be sufficient upon that point. But still, in the evidence that we propose to offer, we shall endeavor, not to make it simply cumulative, and shall offer evidence only from parties who have some particular facts to bring before the Committee which have a bearing upon that point.

I will say a word further. Assuming that it is not necessary to pile up more evidence as to the nuisance, the remaining question will be, "What ought to be done?" We desire to offer evidence upon that point; and we shall, in the course of the hearing, offer the evidence of Mr. Waring. I understand that Dr. Walcott, one of the commission making the Report to the State Board of Health, expects, from some intimations from the Committee, to appear; and we desire to have his testimony. Dr. Folsom, another of the commissioners, desires, if he appears at all, to appear as called by the Committee, not as a witness offered by us. If asked by the Committee, I understand he is willing to appear. I presume the Committee will desire to have the testimony of experts as to the feasibility of such plans as have been adopted in other countries; and I hope, before the hearing is over, that the Committee will see fit to ask Dr. Folsom to appear.

Mr. GOULDING. On two or three different occasions the counsel on the other side have tried to lay the blame of the protraction of this hearing upon the counsel for the city of Worcester. I have on one occasion before, and I now repeat, said that we shall not controvert any of the facts stated in the last Report of the State Board of Health, Lunacy, and Charity. I said once before that the counsel in his opening remarks had made a great many vague statements, and drawn conclusions which perhaps, as we lawyers say, are not strictly traversable: we could not plead to them, but we most emphatically deny their conclusions. We most emphatically deny their inferences and their vague generalities. If they can find in the Report of the State Board of Health, Lunacy, and Charity, any statement that a nuisance, which is injurious to the public health, exists in connection with this river, they will have the benefit of it. We deny the fact, and say they cannot find it in that report. If they can, they can point it out in their closing argument. We do not controvert any facts, that I am aware of, that are stated in that report. We disclaim all responsibility for any protraction of this hearing by the inhabitants of the town of Millbury.

Mr. FLAGG. I do not understand whether my brother calls the statement on p. lxv of the Third Annual Report of the State Board of Health, Lunacy, and Charity, a statement of fact, or a conclusion which he will deny. Upon that page you will find, referring to the report of this commission, the following *statement of fact*, as we call it: —

"As the consideration of this report, in connection with one to be subsequently publicly noticed, will bring more directly to public attention than ever before the rapidly increasing pollution of streams not used as sources of water-supply for domestic uses, but which, as in the case of the Blackstone at Millbury, are becoming too foul, even for manufacturing purposes, and as objectionable to residents on their banks as open sewers would be," etc.

Mr. GOULDING. I suppose that counsel understand the difference between a collection of matter that emits offensive odors that may be disagreeable, and a public nuisance or a private nuisance. When I use the term "public nuisance" or "private nuisance," I use it in its accepted sense. I find, on looking over this Report of the State Board of Health, Lunacy, and Charity, that they were men who knew just exactly what they were talking about. They were not making a report for one side or the other. They were guarded in their language. They have made a perfectly fair, honest, and square report. In the closing argument on this case I shall have occasion to call attention to that report. My friends on the other side seem to confound the distinction between the conclusions the men who work in Mr. Morse's factory, and who "curse the river," come to, and the judgment of a commission that has investigated the subject, and undertakes to state facts.

Mr. FLAGG. It seems, then, we shall agree only on one point (I am glad to agree on that), — that the members of the commission knew what they were talking about.

Cross-Examination of WILLIAM H. HARRINGTON.

Q. (By Mr. GOULDING.) — How long is it since you ceased to use the water of the Blackstone River as a beverage down at your mill? *A.* — A good many years.

Q. — About how many? *A.* — I don't remember of using it for fifteen years.

Q. — You were here and testified last year, I believe? *A.* — I think so: yes, sir. I have been here before, and I think it was last year.

Q. — Did you not testify in substance last year, with regard to the washing of your wool at the Burling Mills, that it seemed to you that it was iron in the water that caused the difficulty? I do not undertake to put into my question your exact words, but I simply recall the impression that was on my mind. *A.* — To explain whatever I might have said, it is, in my opinion, that, with many other things.

Q. — Do you remember whether that was what you said before? or don't you remember? *A.* — I don't.

Q. — Whatever your opinion was, what is your opinion now in regard to the proportion of iron that produces this effect? *A.* — I have no opinion as to the proportion. I think there are iron and vitriol in the water.

Q. — Do you know the amount of chemicals that is used in the Burling Mills, when they are in full operation, for the purpose of scouring? *A.* — We use, for scouring, salt and soda-ash.

Q. — I speak about the quantity, whether you know the quantity? *A.* — I do not.

Q. — Do you know the quantity of dyestuffs that is used when the mill is running full? *A.* — Not in pounds or tons.

Q. — Have you at any time, in the Burling Mills, scoured wool brought there for that purpose to be carried away again, — the wool of other parties brought there to be scoured? *A.* — I have not, sir.

Q. — Do you know of that being done there before you owned those mills? *A.* — No, sir: I don't know that they ever scoured any.

Q. (By Mr. FLAGG.) — So far from scouring wool for other parties, have you not at times felt obliged to buy scoured wool? *A.* — We have, for the last number of years, bought scoured wool in the room of scouring it, whenever we could find what we wanted. We have abandoned the use of the water now for the purpose of scouring either wool or goods, and are making preparations for abandoning it for the purpose of making steam: it is so impure that it cannot be used for that purpose.

Q. — You are familiar with what is poured into the river at the woollen-mills from dyeing and scouring. How does that compare in offensiveness, and as to its effects upon health, in your opinion, with sewage matter? *A.* — It is not offensive to the smell: it is offensive to the sight, because it is dark-colored; but there is no odor that arises from it, and nothing that I can see that would affect the health at all.

Q. (By Mr. GOULDING.) — You are not a physician, I suppose, so as to know what the effect upon health would be? *A.* — Not a practising physician.

Q. — But it does produce an effect upon the color of the water, you th'nk? *A.* — Yes, sir.

Q. — Like that, for instance [showing bottle marked "No. 3"]? *A.* — Well, you can get some like that. I can get some like that any time.

Q. — So far as the color of the water is concerned, you admit that the dyestuffs might produce the effect? *A.* — Of course they would. I can get it of a color like that. I can get you sediment like that in our race any day.

Q. — I was not asking you any thing about your race or the pond. That is not responsive to my question; but I don't want to restrict your answers at all. I understand that, so far as color is concerned, the dyestuffs make a difference? *A.* — We have no woollen-mills above us at all.

Q. — Kettle Brook is full of woollen-mills, is it not? *A.* — What I mean by "above us," is between us and the mouth of the sewer. The water is so impure there, that that mill cannot use the water for any purposes other than for power. We have driven seven wells now.

Q. — How much does it cost to drive one of those wells, pipe and all? *A.* — There are different sizes. I think one of them cost us $250.

Q. (By Mr. FLAGG.) — Have you any thing more that you wish to inform the Committee about? *A.* — As regards the unhealthiness of the village, I cannot say that it is directly traceable to the sewer; yet it seems to be the general impression that it is unhealthy.

Q. — At any rate, you live in Worcester rather than Millbury? *A.* — I do, sir: I should not want to live on the stream all the time.

Q. — On account of its impurities? *A.* — On account of its impurities.

Q. (By Mr. GOULDING.) — Not if you had a handsome residence on the main street in Worcester? *A.* — No, sir: I think the air is better, Mr. Goulding, where we live than down there.

Q. — Have you any statistics, or have you made any observation of the facts, so as to know whether the health-rate or the death-rate of Millbury compares favorably or unfavorably with other towns similarly situated? *A.* — No, sir, I have not.

Q. — You speak from your general impression; but you know nothing about the facts, do you? *A.* — We hear these complaints among the operatives.

Q. — Have you ever smelt that tripe-factory over on the west side of Blackstone River, between Worcester and Millbury? *A.* — I have smelt something a great many times. In fact, in the summer-time you can hardly go down the road unless you smell something. Whether you can trace it directly to the tripe-factory, I can't say.

Q. — You never noticed any thing particular along there opposite that tripe-factory? *A.* — I notice it particularly all the way from Burling Mills to Quinsigamond in the summer-time. I don't think the tripe-factory would add any pleasant odors.

Q. — It is not really a tripe-factory, is it? *A.* — No, sir: I believe not.

Q. — Don't you know that all the dead horses in Worcester are carried down to that factory? *A.* — No, sir: I know they don't carry any horses there.

Q. — Don't you know that all Kendrick's horses are carried to that factory? *A.* — I know there isn't one of Kendrick's horses carried to that factory. They go below there, and are buried on what is called the Ewing farm, which is a mile below the tripe-factory.

Q. — Don't you know that there is a factory (I don't know whether it is properly called a factory, or not) on the west side of the Blackstone River, down below Quinsigamond Village, which you approach by a turnout in the field from the old Millbury road, where all, or a large part, of the horses that die in Worcester are carried; and that all Kendrick's horses are carried there, and cooked up, and turned into glue, and whatever else they can make out of them? Don't you know that fact, Mr. Harrington? *A.* — I don't know it.

Q. — Do you know it is not so? *A.* — I am positive it is not so.

Q. — How do you know that it is not so? *A.* — Why, because this Ewing, who lives on the place below what we call the tripe-factory, is the man who disposes of all the horses that have died.

Q. — What does he do with them? *A.* — They are buried there.

Q. — Buried? *A.* — I think that he cuts them up; but this smell from what we call the tripe-factory don't come from that.

Q. — Then, you think that the dead horses are not carried there?

A. — They are not carried to this tripe-factory: they are carried down about a mile below, and buried there.

Q. — Wherever they are carried, are they not disposed of by being cooked, and turned into glue, or whatever is done with them, at some factory establishment? *A.* — I think this Ewing has got a set kettle, and he cuts them up and boils them.

Q. — You never were more mistaken in your life: there is plenty of evidence about that. *A.* — If that tripe-factory is cutting up horses, it is beyond me. I never saw any thing of the kind there, and I never heard of it before. Those horses are carted to the Ewing farm, which is about a mile below the tripe-factory.

Q. (By Dr. HARRIS.) — This Ewing has a rendering establishment, as they term it, hasn't he, where he takes the dead horses that come down, and takes off their hides for leather, their hair for other purposes, and utilizes every part of them, as far as possible? *A.* — He has no factory: he has a set kettle that is set out of doors, but there is no mill and no building.

Q. (By the CHAIRMAN.) — There are two different establishments about there of some sort, I take it: that is all there is to it. *A.* — Yes, sir: the other one makes glue, or some substance.

Q. (By Mr. GOULDING.) — What do they make the glue out of? *A.* — Well, I have seen them carrying down there the refuse from the meat-markets of Worcester.

Q. — Have you ever been over that factory yourself where they make glue? *A.* — I have.

Q. — Did you ever see that pile of horses' skulls as large as a small mountain? *A.* — I think you are mistaken: those were the heads of cattle. You are mistaken in the kind of heads.

Q. — Possibly; but cattle up our way have horns. *A.* — They were taken off before you saw them. The horns come off with the hides, you know.

Q. — I suppose it to be a fact that there is a rendering establishment out there; and I know from information from men who have knowledge of it, that all Kendrick's horses are taken there. I was not there when it was done. *A.* — They are taken to this Ewing.

Q. — Taken down to this rendering factory? *A.* — That is Ewing's, not the other man's place: I have forgotten his name.

TESTIMONY OF PETER SIMPSON.

Q. (By Mr. FLAGG.) — You live in Millbury? *A.* — Yes, sir, I do.

Q. — You are engaged in manufacturing in Millbury? *A.* — Yes, sir.

Q. — How long have you lived in Millbury? *A.* — Twenty years.

Q. — Where did you live before coming to Millbury? *A.* — Woonsocket, R.I.

Q. — In what business were you engaged there? *A.* — Woollen manufacturing.

Q. — Are you also engaged in business in Farnumsville? *A.* — Yes, sir: four miles below Millbury.

Q. — The location of your mill in Millbury is next below the Cordis Mill? *A.* — Yes, sir: the last one in Millbury.

Q. — And you manufacture what? *A.* — Woollen goods.

Q. — In manufacturing woollen goods, you find it necessary to use large quantities of water for certain purposes, do you not? *A.* — We have to, sir. We have to scour our wool, and scour our goods, and color them, which requires a large quantity of water.

Q. — What has been your experience in using the Blackstone-river water at your mill in Millbury? *A.* — We have had a great deal of trouble with it, more particularly for the last three years. By the way, I have a minute which I took of a quantity of goods which we made up in 1880. There were something like seventy thousand yards, which we made in one batch, one hundred cases, on which we made an allowance of five cents a yard, amounting to twenty-three hundred dollars, for the reason that we were not able to get them clean. I presume when the word "cadet" is used, it is understood by you, gentlemen: it was a lot of mixed — black and white. The water being so impure, it was impossible for us to get them clean and get them bright. At first, perhaps, they might appear partially clean; but the stain would work through, and turn the white yellow or drabbish, perhaps.

Q. (By Mr. WILSON.) — When did you say this happened? *A.* — Those goods were made in the year 1880. We had had some trouble before that; but that year we had the most trouble of any year. The goods were sold by Pomeroy & Palmer, in New York, a house well known. After the goods had been sold some time, somebody made a claim; and it was settled by allowing five cents per yard on about seventy thousand yards of goods, amounting to a little over twenty-three hundred dollars. Since that, we have not made that class of goods, for fear we should have the same trouble. It is a class of goods on which, when we make them, and make them properly, we think there is a margin.

Q. (By Mr. FLAGG.) — The water is not so bad for scouring some kinds of goods as others, I take it? *A.* — It is just as bad, only it will not look as bad. There is the same trouble in dark goods that there is in light goods, and it will show itself afterwards. You cannot get a good black, nor a good brown, even, unless you have pure water: that is well decided. Then, again, the water will rust — corrode — our wheels; so much so, that we are obliged to scrape our wheels at least twice a year in order to get the power out of them. Now, in regard to corroding a wheel, it not only spoils the wheel, but reduces the power. You cannot get so much power from a wheel if it is corroded as you could if it was clean and bright. Not only does it destroy your wheel in time, but it destroys the power of the wheel. It will cause the gates of a wheel to corrode, and work hard; so much so, that we had to give up iron gears and get metal gears cast on account of the gates working so hard. A common cast-iron gear would break, would not be sufficiently strong.

Q. — Is the Blackstone-river water at your mill in Millbury suitable for manufacturing purposes? *A.* — Not woollen.

Q. — As to the effect upon health at your mill in Millbury? *A.* — We have attributed a good deal of sickness to the water being so foul, so impure.

Q. — Within a few days, there has been a case of typhoid fever there, I am told? *A.* — Yes, sir: the man died.

Q. — Will you state the facts to the Committee? *A.* — A Mr. Wilmarth, who had charge of our mill at Farnum, four miles below Millbury, was taken sick a week ago yesterday morning, and died yesterday morning. He had been with me about four months: he came from Oxford. Mr. Wilmarth, in his own mind, attributed the cause of his sickness to the water; and I have no doubt that Dr. Gage of Worcester would also say that the cause was the water.

Q. — That was at your mill in Farnumsville. *A.* — Yes, sir, that was at Farnumsville.

Q. — At that mill is the water fit for boiler purposes or making sizing? *A.* — It is not. We put in a slasher — I presume you know what a slasher is: it is something that takes the place of a dresser in a cotton-mill, that we dress our yarn with. It was a slasher built by the Lowell Company. I am perfectly satisfied in my own mind, if we had had pure water to make our steam which we heat with, the pipes would have lasted perhaps twenty years. It has been running about four years, and we are now replacing the pipes in that slasher. The pipes were all eaten up with rust. We never had any trouble with our pipes rusting from steam made from pure water.

Q. — You carry on a farm, Mr. Simpson? *A.* — Yes, sir: I farm it some.

Q. — Do you produce milk to sell? *A.* — Yes, sir.

Q. — Do your cows drink the water of Blackstone River? *A.* — They have: yes, sir.

Q. — State your experience as to the effect upon the milk. *A.* — In 1879 our customers found fault with our milk. We mean to average about a dozen cows to milk the year round, to supply our customers; and in the fall of 1879 our customers found fault with our milk. The Blackstone River runs through my land. I own each side of the river: and I couldn't think what could be the matter with them; but I took them out from that place, and in ten or twelve days the trouble was removed.

Q. (By Mr. CHAMBERLAIN.) —Were they fed just the same? A. — Exactly. I never feed my cows any thing but corn-meal and hay: I have fed them with nothing else. I think that is the cheapest food there is to feed a cow. In 1880 I thought I would not pasture the cows where we mow, next the river; so I kept them on the upper land, and we had no trouble. In 1881 our land produced pretty good grass. We mow pretty much all we have got twice a year, and then we get pretty good feed afterwards. I thought I would try again, and see if I couldn't have my cows eat that grass off, for more than one reason. It is good feed for our cows; and the next year, if we do not have it eaten off, the dead grass makes bad cutting if it is not removed: and we pastured our cows there in 1881, and the same trouble came up again with our milk. We have regular customers who take our milk; and they found fault, and gave up taking the milk. I went to work and fenced the river away from the cows each side, up and down, so that the cows could not get to the river. In two or three weeks we had no more trouble with our milk, and our milk is good to-day.

Q. (By Dr. WILSON.) — What fault did your customers find with the milk? A. — Bad taste, and the milk would really smell: small babies would not drink it.

Q. — And you had to fence the river away from them? A. — I fenced the river away from the cows, and the river stands to-day fenced on each side.

Q. (By Mr. SMITH.) — It seems your cows would drink the water of the river? A. — They had no other water to drink, sir: they were obliged to drink it.

Q. (By the CHAIRMAN.) — Would they drink it right along? A. — I presume they did: they had nowhere else to go. I presume they did drink it. We had nowhere else to water them, when we pastured them there.

Q. — Some of the witnesses have said that their cows, when starved into it, would not drink the water more than once in a day or two. Do you know how often your cows drank it? A. — I can't say how often they did drink it. It was all the water they had when we pastured them in that place.

Q. (By Mr. FLAGG.) — You have heard the testimony of others as to the ponds filling up, as to the smell from the river, and as to the grass which grows in the river. Has your experience been the same as theirs? *A.* — Yes, sir, it has.

Q. — Have you noticed a great change since you first came to Millbury in that respect? *A.* — Very much.

Q. — What is your opinion as to the effect upon the woollen business in the Blackstone Valley if this sewage continues to contaminate the river, and increase in quantity? *A.* — Oh! we must give it up; there is no doubt, we must give it up. If you will allow me, I will say that these water-powers, from Worcester away down to I should say Uxbridge, are not very large powers. You might say, "Why not go to work and make cotton goods on those water-powers if we cannot make woollen? we have got to make something for a living." The great trouble is, that those water-powers are too small for cotton-mills; they will do for small woollen-mills, but the time has come when a small cotton-mill cannot live with the large ones. Then, you might say, "Why don't you run by steam?" Fall River can buy coal about two dollars a ton cheaper than we can: that is the reason that we cannot compete with them, or with New Bedford, or any of those places. We are too far up in the country to run with steam, and compete with our neighbors.

Q. — What is the distance from the mouth of the sewer at Quinsigamond to the State-line at Blackstone? *A.* — Not far from twenty-four miles. I am not positive, but I should say twenty-four or twenty-five miles.

Q. — I have a map here on which are laid down all the dams, as I understand. First, Burling Mills dam; next, Morse's dam; next, Atlanta dam; next, Millbury dam; next, Cordis dam; next, Simpson's dam, — these are in Millbury. Then, in Sutton, Wilkinsonville dam; in Grafton, Saundersville dam, Fisherville dam, Farnumsville dam; in Northbridge, Rockdale dam and Riverdale dam; in Uxbridge, the North Uxbridge dam, the Centerville dam, a place called Shankbone dam, Millville dam, and Blackstone dam? *A.* — Yes, sir.

Q. — Seventeen in all? *A.* — Yes, sir: that includes all the dams there are on the Blackstone in Massachusetts, I believe.

Q. — Those were all there were when you came to Millbury? *A.* — Yes, sir: I don't think of any new dam built on the Blackstone River for the last forty years. Some of the dams have been repaired, and some of the old ones have been replaced by new. No new water-power has been taken, to my knowledge, on the Blackstone River for the last forty-four years.

Q. (By Dr. WILSON.) — I suppose your wheel is iron, as you speak of its having corroded? *A.* — Yes, sir: a turbine wheel, so called.

Q. — Have you any idea what caused it to corrode? *A.* — When we used the water twenty years ago, or fifteen years ago, it did not corrode. We were using iron wheels then, as we do now. I take it for granted it is the impurity of the water: I don't know of any thing else.

Q. — Would sewage corrode it? or would it be more likely to be the acids, or any thing of that kind, used in manufacturing? *A.* — All I can say is, that the wheels do corrode. I don't pretend to say what causes it.

Q. (By Mr. CHAMBERLAIN.) — Do you mean by your testimony that you deducted five cents a yard for the damage to the cloth? *A.* — We allowed five cents a yard on seventy thousand yards of goods.

Q. — Then, you allowed more than twenty-three dollars, didn't you? *A.* — Yes, sir: some over twenty-three hundred dollars.

Q. — That is not five cents a yard. I did not know but you meant that you compromised for twenty-three hundred dollars, as seventy thousand yards at five cents per yard would come to more than twenty-three hundred dollars. *A.* — The claim was for over twenty-three hundred dollars on that lot of goods.

Q. — I understand you that that was settled for twenty-three hundred dollars; and you also said that there were seventy thousand yards, and you allowed five cents per yard, — what I want to know is, whether the claim was compromised for twenty-three hundred dollars? *A.* — I would say, if you will allow me, that there were a hundred cases in that lot, some seventy thousand yards; and my book-keeper tells me we had to allow twenty-three hundred dollars. The goods were sold by Pomeroy & Palmer, so that it comes from good authority.

Q. (By Mr. FLAGG.) — The term "low water" has been used a good many times. Will you explain what those who are familiar with the river understand by "low water"? *A.* — It is at the time of year when we use all the water there is in the river. All the water there is in the Blackstone River will run through our wheels; and then, as a matter of course, the water will not run through the bed of the river. Take the Burling Mills: their dam is some ways up from the mill; and the water will go down through their canal, and the bed of the river will be dry. Then, the people who live on the banks of that river will say it is low-water mark; when, if the same quantity of water ran through the bed of the river, it would show quite a stream, comparatively; but it is all used over the wheels, and the bed of the river would appear to be dry in that case. Several of our mills have dams placed in that position, — not so much right in Millbury. That is what they call "low water," when they use all the water over the wheels.

Q. — That is usually in the summer-time? *A.* — Yes, sir.

Q. — How many months? A. — Four or five months. Two years ago, it went into January.

Q. — You were present the day the Committee were there? A. — Yes, sir.

Q. — What do you say as to the height of the water then? A. — I should think there was at least thirty times as much water as we can use with our present arrangement of wheels.

Q. — How long since the river has been as high as it was that day? A. — I think it must have been five years ago. I think we took notice of it at our dam five years ago, and it was about as high as it was that day. Since that, I don't think we have had so large a quantity at one time.

Cross-Examination.

Q. (By Mr. GOULDING.) — How large is your mill? A. — We run four sets of woollen machinery at Millbury.

Q. — What water do you use in your boilers? A. — From the Blackstone River.

Q. — What water do you use for scouring? A. — From the Blackstone River.

Q. — For all purposes, you use that water? A. — Yes, sir, we do.

Q. — You have given the Committee a pretty full account of the troubles you have had with it? A. — Some of the troubles; yes, sir.

Q. — Now, what quantity of chemicals does it require to do the scouring for your mill? A. — We scour our wool with sal-soda and soda-ash, — that is what we use to scour with.

Q. — About what quantity? A. — Do you mean to be understood per a hundred pounds of wool?

Q. — Put it in any way you have a mind to, so that we may know how much chemicals you use per day, per month, or per year? A. — I hardly know how to answer that. You have got me. Mr. Goulding never scoured much wool, I think. You take California wool, where they don't get twenty pounds from a hundred pounds, and it will take more soda-ash than other wools will take. We can scour a hundred pounds of Ohio wool with five pounds of soda-ash. I don't know how I can get at it.

Mr. GOULDING. I never scoured any wool.

WITNESS. I did not suppose you had: if you had, you would not have asked me the question.

Q. — You can answer the question, or you can state that it is not possible to answer it. A. — Now, you just put it, and I will see if I can answer it.

Q. — What quantity of chemicals do you use per day, or per year;

or per month, or any other period that you have a mind to select, in scouring your wool at your mill, as a matter of fact? *A.*—That is a pretty nice point. As I told you, if I scour greasy California wool, it will take twenty pounds of soda-ash to get forty pounds of wool.

Q.—Do you scour California wool, or don't you? *A.*—I have scoured California wool: I told you I had.

Q.—Then, it would seem that you could tell how much chemicals you used? *A.*—When I scour California wool, I can tell you how much I use. If I scour Ohio wool, it won't take twenty-five per cent of alkali to scour it.

Q.—Can't you tell what proportion of the different kinds of wool you use, so that we can tell the quantity of chemicals that you do use, as a matter of fact, or what quantity you have used for the past five years? *A.*—I could if I had my books here. If you will wait until to-morrow, I will bring them down and tell you. I want you to know exactly: I don't want to dodge it. I can tell you, we scour our wool: we don't work any grease.

Q.—What do you do with the water in which you scour your wool, after it is scoured? *A.*—It passes down the Blackstone River.

Q.—What kind of dyestuffs do you use in your mill in dyeing your wool? *A.*—We don't color any wool at all: we work it all white.

Q.—Have you ever colored wool? *A.*—Never. I have been there twenty years, and never colored a pound of wool.

Q.—You don't use any dyestuff, then? *A.*—Yes, sir: we use logwood.

Q.—What do you use that for? *A.*—To burr dye. We stain our goods with it.

Q.—How much chemicals do you use for the purpose of staining your wool? *A.*—I stain my goods, not the wool.

Q.—How much do you use for that purpose? *A.*—Well, I should think we used a ton of extract of logwood per month; and we use half as much soda-ash as we do logwood.

Q.—What is done with this dyestuff after it is used? *A.*—It passes down the Blackstone River, sir.

Q.—Do you use about the same quantity of chemicals for scouring that the other mills in Millbury do, in proportion to the size of your mill? *A.*—I presume so.

Q.—All the woollen-mills in Millbury do scouring, of course? *A.*—You cannot make woollen goods without scouring the wool: it has all got to be washed.

Q.—The water, after it is used for scouring, with the chemicals, passes down the river in all cases? *A.*—Yes, sir: as far as I know.

Q.—You know about this matter as well as anybody, and I will ask you this question: How many mills are there in Millbury on the

Blackstone River, and on Singletary Brook? *A.* — Seven, I believe, in all; although I won't be positive.

Q. — What is the size of those woollen-mills? *A.* — I think the largest one has eight sets. I won't be positive.

Q. — Yours has four? *A.* — Five.

Q. — How many of them are there as large as eight sets? *A.* — I think there is not but one: that is the Burling Mills. The Atlanta, I think, has four sets.

Q. — How many has Mr. Lapham? *A.* — I think he has six.

Q. — That is at Bramanville? *A.* — Yes, sir.

Q. — How many has Mr. Walling? *A.* — I won't be positive, but it strikes me it is six.

Q. — How near to the Blackstone River do you reside? *A.* — I should say a quarter of a mile.

Q. — Is it above Morse's Mill, or below it? *A.* — I should say below it.

Q. — You have a family? *A.* — Yes, sir.

Q. — Have you any aged grandmother? *A.* — No, sir: my grandmother is dead.

Q. — You have pretty good health yourself? *A.* — Very good: never saw a sick day in my life.

Q. — You had a bullet shot into you a few years ago? *A.* — Yes, sir. You don't call that sickness, for a man to be shot at, do you?

Q. — You wrestled with it, and recovered? *A.* — Yes, sir.

Q. — This Mr. Wilmarth, who died, how old a man was he? *A.* — Sixty.

Q. — How long had he lived in that region? *A.* — He had been with me about four months. He came with me about the fore part of last December.

Q. — You understood his sickness was typhoid fever? *A.* — Yes, sir.

Q. — You spoke of Dr. Gage. Did Dr. Gage attend him? *A.* — We called him the last day, which was last Sunday.

Q. — You first noticed this effect upon your milk in 1879? *A.* — In 1879, sir.

Q. — You pastured your cows near the Blackstone, and they drank this water? *A.* — Yes, sir: they must have drank it, because they had nothing else.

Q. — Was any analysis ever made of that milk? *A.* — No, there was not.

Q. — You distribute your milk in Millbury to families around? *A.* — Yes, sir.

Q. (By Mr. FLAGG.) — The refuse from the use of those chemicals, as compared with the sewage, which is the worst for the water? *A.*

— You cannot get any smell from this. We intend to get all the life there is in the logwood, and all the other dyestuff, into our goods, if we can. We mean to throw away as little as possible. We do not put any more logwood into our dyestuffs than what our goods will take up. It will stain the water.

Q. (By Mr. GOULDING.) — Don't you think that the water from those dyestuffs and scouring-machines rather improves the river, than otherwise? *A.* — Oh, no! you cannot believe that, Mr. Goulding. What I wish to be understood is, that we mean to get all the color out of the dyestuffs that we use into our goods, not into the river: that is what I mean to be understood.

Q. (By Dr. WILSON.) — You say that the bad smell from the river caused that case of typhoid fever: why do you say that? *A.* — Because the doctor said so: that is all.

Q. — Both doctors? *A.* — Both doctors. We had two doctors. I expect one of them will be here to-morrow.

Q. (By Mr. GOULDING.) — What doctor said so? *A.* — I wish I could remember his name.

Q. — Dr. Maxwell? *A.* — Yes, sir; and Dr. Gage agreed with him.

Q. — Did you hear Dr. Gage say so? *A.* — I did not: they so told me.

TESTIMONY OF THOMAS HEAP.

Q. (By Mr. FLAGG.) — You live in Millbury? *A.* — Yes, sir.

Q. — At Burling Mills? *A.* — Yes, sir.

Q. — What is your business? *A.* — Superintendent of Burling Mills.

Q. — How long have you been superintendent, at this time, of the Burling Mills? *A.* — Ever since the first of last September.

Q. — Were you there previously? *A.* — Yes, sir.

Q. — When? *A.* — Between seven and eight years ago.

Q. — Will you state your experience with the water, when you were there previously, as to dyeing? *A.* — We did not have much trouble with it at that time, although there were remarks made by the dyer that it was troubling him somewhat, but not a great deal.

Q. — You succeeded, then, pretty well, in using it for dyeing and scouring? *A.* — Yes, sir.

Q. — What problem did you have presented to you when you came back as superintendent in '81? Was the mill in trouble about its goods at that time? *A.* — The mill was in trouble about its goods. They had somewhere about two hundred pieces in the finishing-room all stained. They wanted me to find out what the trouble was. I went in there and examined the goods, smelt of them, and they smelt bad: the white was turned yellow. In order to find out what was the cause of that, I went back to first principles. I took some wool and scoured it in clean spring water, and it came out all right and white. Then I took the same grade of wool, and scoured it in the Blackstone-river water, and I found that it turned out all yellow. All the stuff that I could put to it, I couldn't get it back: it seemed to be a fixed color of itself. Then I went into the dye-house; and I found that the dye-kettles had been standing there some few days with water in them from the river, and on the sides of the kettles was a yellow slime about half an inch thick, that smelt very bad. Then I went into the office and reported to Mr. Harrington and Mr. Barker, that, if they intended to run fancy cassimeres in that mill, they had got to get some water from somewhere: they couldn't use the Blackstone River. I tried a few pieces, and it came out just the same as the pieces that were in the finishing-room. So we went to work and sunk artesian wells, and got over the trouble in the finishing-room by using that water.

Q. — What would apply to the Burling Mills would apply to any other woollen-mill on the stream? A. — Yes, sir.

Q. — You say the water would not be fit, as it is at present, to advantageously, at any rate, scour or dye woollen goods? A. — No, sir.

Mr. GOULDING. "Fancy cassimeres," he said. A. — No, sir: that mill is a fancy cassimere mill. If we had taken the water of the Blackstone River, it would have been washed out of existence as a cassimere mill.

Q. — Have the operatives in the mill complained to you? A. — In the finishing-room they have complained to me about the smell. I told them I would take measures to see if I couldn't obviate it. I cut off the water-pipes that led into the finishing-room, and turned them into the flume. Then I boarded over the flume; and that, in a measure, took away the smell, but not wholly, but so they could get along.

Cross-Examination.

Q. (By Mr. GOULDING.) — The Burling Mills are stopped now, are they not? A. — Yes, sir.

Q. — That is owned substantially by Turnbull & Co. of New York, is it not? A. — I don't know.

Q. — A very large amount of the goods that were manufactured are in their hands? A. — Yes, sir.

Q. — It is not shut up for want of water to scour wool, I take it? A. — No, sir.

Q. — How many artesian wells have you bored there? A. — Five.

Q. — What does it cost to bore one? A. — I cannot say.

Q. — How deep are they? *A.* — About twenty-nine feet.

Q. — How do you do it? *A.* — Drive them.

Q. — You get excellent water there, don't you? Very good water.

Q. — Very nice water indeed? *A.* — Yes, sir.

Q. — And plenty of it? *A.* — Well, that will be a matter of time to prove.

Q. — So far? *A.* — So far, we have.

Q. — When was the first one sunk? *A.* — In September last.

Q. — Was there not a well there before? *A.* — There was a well there, yes.

Q. — Was that an artesian well? *A.* — Yes, sir.

Q. — When was that sunk? *A.* — I cannot tell you any thing about that. That was sunk before I went there the last time.

Q. — Was it there when you were there the first time? *A.* — No, sir.

Q. — It has been sunk since '74, then? Yes, sir.

Q. — And there is a supply of water from that? *A.* — Well, at that time they did not seem to think that that well was worth any thing: so I had them apply a force-pump, to see if we could exhaust the well; and I found we couldn't. So I used that for finishing purposes.

Q. — That is good water? *A.* — That is good water.

Q. — And you could not exhaust the supply? *A.* — We couldn't at that time, nor we haven't since we have been running.

Q. — You have tried more than once to exhaust it? *A.* — We have been running it ever since we tried to exhaust it.

Q. — So far as appears, it is inexhaustible? *A.* — It seems so now.

Q. — And that is true of the other wells that have been sunk since, so far as they appear now? *A.* — Yes, sir.

Re-direct Examination.

Q. (By Mr. FLAGG.) — You say that you do not know that the mill is closed on account of the impurity of the water? *A.* — I don't know any thing about it at all.

Q. — The impurity of the water has been a great damage to its business, has it not? *A.* — It has been a great damage to the business.

Q. — You don't know why it is closed, I suppose? *A.* — No.

Q. — Are you familiar with the Atlanta Mills? *A.* — Yes, sir.

Q. — Can artesian wells be driven there? *A.* — No, sir.

Q. — Why not? *A.* — It is built on a ledge.

Q. — Is there the same trouble with the water there? *A.* — There is the same trouble in summer-time, more than they have now. Last summer we had a few pieces spoiled at the Atlanta Mills, on account of the water making the goods yellow.

Re-cross.

Q. (By Mr. GOULDING.) — Have you any knowledge of chemistry, so as to understand what it is that produces this effect? what the chemical ingredients in the water are that produce this effect on the wool? A. — Well, when I was there before, we didn't have this effect; but since I have come back this time we have this trouble, and I should say it arose from the sewage.

Q. — I know you would say so, because you come from Millbury; but my question was, whether you have any chemical knowledge, so as to know the chemical ingredients that cause the effect? A. — I don't pretend to be an analytical chemist. All I can judge from is the smell.

Q. (By Mr. FLAGG.) — Have you such a nose, Mr. Heap, that you can recognize the smell in the goods? A. — Yes, sir.

Q. (By Mr. GOULDING.) — Do you dye your wools at the mill? A. — Yes, sir.

Q. — Do you know what quantity of dyestuffs you use per week? A. — Well, I can't say any thing about that. It depends upon the color we are making. Some colors require more than others.

TESTIMONY OF JOHN GEGENHEIMER.

Q. (By Mr. FLAGG.) — You live in Millbury, Mr. Gegenheimer? A. — Yes, sir.

Q. — How long have you lived there? A. — Seven years the thirty-first day of this month.

Q. — What is your business? A. — I am superintendent of the Cordis Mills at the present time.

Q. — State where the Cordis Mills are situated. Are they the next mills below the Millbury Cotton-Mills? A. — Yes, sir.

Q. — In the lower part Millbury Village? A. — Yes, sir.

Q. — How long have you been at the Cordis Mills, and familiar with them? A. — Ever since I have been in the town.

Q. — Who was the agent who preceded you? A. — Mr. B. B. Howard.

Q. — Will you state your experience there with the Blackstone river water? A. — At the time I went there, seven years ago, I went there to take the time in the finishing or cloth room. At that time all the water we had to use throughout the mill, for any purpose whatever, was taken from our pond, back of the mill, which is supplied by the Blackstone River. Very soon after that, there began to be complaints of our goods. We were on ticking, and they complained of our whites being a very dirty yellow; so much so that Mr. Howard had to take some means to remedy it. To do this, he dug a well in the yard, and pumped out of that into a tank in the attic, and used that for sizing purposes; and the next year, or the year after, they complained of that also; and Mr. Howard was at a loss what to do. Finally, he had some of the water analyzed by Professor Thompson of Worcester, with the result which I think you have, or had here, at the last hearing; and he had to give up the use of that well, and had some driven wells put in, that we are now using; so that the water that we use in our sizing comes from those driven wells. Two years ago he reset his boilers, and put in two new ones; and we had no facilities for feeding those, only from the river-water; and it had become so filthy, they had to blow them off two or three times a week partially, and every two weeks wholly; and he thought that was rather expensive. And a year ago last June, I think, he put in a new engine, and put in a heater in combination with that; and the water was so dirty, that the past season he was induced to put in a Crocker water-filter, to filter water for the boilers; and we are now running that filter. The water in our boilers that we use in our sizing comes from these driven wells. You have samples of the sediment and stuff that we got out of the filter: you can see what we take out.

Q. — I see here a sample labelled "Feb. 14, 1882. Washings of filter after running one hour, with twenty-five strokes per minute at the pump." Will you take that, and explain it to the Committee? *A.* — We clean out our filter, when the water is very bad, every hour. It has a reversible cage on the inside, filled with animal charcoal; and, by reversing and starting the pump, we force the water right through; and it forces out this sediment and stuff through the waste-pipe, and that was collected in a barrel; and then the top-water that was used in washing it out was filtered off, and that is the sediment.

Q. (By Dr. WILSON.) — That is the sediment of the barrel? *A.* — The sediment of the water that went into the barrel, in washing out the filter. The barrel was about half-full.

Q. (By Mr. FLAGG.) — Is that a fair sample of what was collected in the filter? *A.* — Yes, sir.

Q. — Explain those other samples, if you please. *A.* — There is a sample that was taken in the same way, with the pump running only ten minutes. Of course it doesn't show as large a proportion of sediment, but the character of the sediment is just the same. There is one in a smaller bottle that was taken the day after, on the fifteenth day of February. It shows a large proportion of sediment. The washings were drained into a larger vessel; and then the clearest water was siphoned off, so as to preserve the sediment. I won't ask you to smell it, because I don't think you would like it. It is the concentrated essence. It is a fair specimen of what we get in the summer season. There is a specimen that was taken yesterday afternoon from our filter: I had our engineer catch that. Here is one that was taken Feb. 1.

Q. — State generally, Mr. Gegenheimer, whether these are fair samples of what the filter takes out of the water. *A.* — They are, sir.

Q. — What is your opinion and your experience as to the effect of this water upon the general health of the operatives? *A.* — Well, I don't know. We think it is necessary, to get a fair average in each department, in every room, to keep several spare hands. We have been driven up with our work, and have tried to keep all of our machinery running every day. In order to do this, we have to keep several spare hands in each department, to take the places of those who are out from day to day, from any cause whatever. Very often our overseers come to me in the morning, and say that they are short: such a hand is out. "What is the matter?" — "I don't know: they are sick."

Q. — You have heard the testimony of the other witnesses as to the smell of the water, its appearance, and the filling up of the ponds. Do you agree with what they have said? *A.* — Yes, sir.

Q. (By the CHAIRMAN.) — Mr. Chamberlain, who is a prohibitionist, wants me to ask if there are more out on Monday than other days. *A.* — I don't think there are. They are mostly women and children, who are not that kind of help.

Cross-Examination.

Q. (By Mr. GOULDING.) — Do you know whether there are any other filters in Millbury, at any of the other mills? *A.* — I don't know, sir.

Q. — Up at Bramanville or elsewhere? *A.* — I have heard that Mr. Rhodes was using a filter; but what it is, or how it works, I don't know.

Q. — Where is his mill? *A.* — That is the first mill on the Singletary Stream, as you go up. I have heard that they had a sort of filter rigged up in a barrel. I can't give you any description of it, or tell you any thing definite about it.

Q. — Did you understand what they filtered the water for, — whether it was for the boiler, or what? *A.* — No, sir: I don't know any thing at all as regards it. I never saw it, or made any inquiries.

Q. — I understand that what is in these bottles was procured by reversing your filters? *A.* — Reversing the cage on the inside, that contains the charcoal.

Q. — How long does it take to clean the filter when you do that? *A.* — I should say from three to five minutes, perhaps.

Q. — And that produces how much water? *A.* — I should say it takes, perhaps, from twenty to forty gallons to wash out the sediment that collects there.

Q. — And clean the filter? *A.* — And clean the filter; yes, sir.

Q. — From twenty to forty gallons? *A.* — I should say so; thereabouts.

Q. — These specimens were procured by letting it run a certain length of time, and then cleaning it, and some of these smaller bottles are the settlings of that result? *A.* — That one that was run an hour was procured by saving the washings of the filter in a barrel, catching them as they came from the waste-pipe in a barrel, and letting them set, and then drawing off the top of the water so as to save the sediment. The others were caught simply by holding a pail under the waste-pipe as the water ran out, and catching a pailful of the water, sediment and every thing, and then let it settle. The small bottle contains the sediment and every thing, just as it was caught running from the waste-pipe.

Q. — As I understand, what is in this bottle was taken from the waste-pipe without allowing it to settle at all? *A.* — Yes, sir.

Q. — How was it with the other bottle? *A.* — The washings of the filter were collected in a barrel, and then allowed to settle, so as to leave only that quantity of water and sediment.

Q. — How much water was in the barrel? *A.* — At the time this February 14th specimen was taken, we measured the water as we took it off, and there were twenty-one gallons.

Q. — Leaving this as the residuum? *A.* — No, sir: counting that in with the twenty-one gallons.

Q. (By Mr. FLAGG.) — How much does that bottle hold? *A.* — About a pailful. We call them ten-quart pails.

Q. — How was that bottle obtained? [No. 3.] *A.* — That is what was caught in the pail and allowed to settle; then the water was drawn off, so as to get what went into that bottle.

Q. — You mean that that is the sediment from a pailful of water? *A.* — A pailful as it runs from our filter. When the filter is turned over, the water begins to come clear; then there comes a dark sediment, and then it gradually comes clearer, and our engineer lets it run until it becomes gradually clear.

Q. — You caught a pailful of water from the filter? *A.* — Yes, sir.

Q. — Then what did you do? *A.* — We let it settle until there was what is in that bottle in the pail, and the rest we threw away.

Q. (By Mr. GOULDING.) — How many hands do you employ in that mill? *A.* — About a hundred and sixty.

Q. — How many spare hands do you employ? *A.* — Well, we have from two to three in each department. There are three departments. In addition to that, we have in our weaving department, for our looms, — we have quite a number of hands who live in our tenements. Parts of the families are at work in the mill, and there are others who come in occasionally for a day, or two or three days, if they are needed to help us out. All we have to do is to send out and tell them that we want them, and they come in.

Q. — Do your employés live near the mill? *A.* — Yes, sir.

Q. — That is the custom with all mills everywhere, is it not, to have some spare hands? *A.* — I think it is, sir.

Q. — Do you know whether there are any larger number at your mill than is usual at other mills? *A.* — I think we have, perhaps, full as large a proportion as ordinary, on account of running all our machinery. We intend to keep all our machinery running every day.

Q. — Do you think you have any larger proportion of spare hands than other mills similarly situated, which intend to run all their machinery? *A.* — I can't tell you any thing definite.

Q. — You don't mean to say that you do? *A.* — I don't know whether we do or not. I can't say. I have not had that experience, or had a chance for observation to know.

Q. (By Mr. SMITH.) — I would like to ask the gentleman if his goods are in demand? *A.* — Yes, sir.

Q. — Then the quality of the goods has not been unfavorably affected by the water of the river? *A.* — It has, at two or three different times, very unfavorably, so much so that we have been obliged to go to the expense of driving wells, and getting pumps, and digging wells, and putting in tanks, and every thing of that kind, in order to keep up and sustain the reputation of our goods.

Q. — And yet you have kept it up, so that your goods are in demand, and meets the wants of the market as well as other goods of the same character made by other mills? *A.* — I can't say as to how other mills are, but only with reference to our own.

Q. (By Mr. HAMLIN.) — You are very busy all the time? *A.* — Yes, sir: our mill is small.

Q. (By Mr. FLAGG.) — I have here the result of an analysis of that well-water, which shows that it contains one-hundredth of one per cent of free ammonia, seven-hundredths of one per cent of albuminoids; other impurities in small quantities. Is that analysis as you remember it? *A.* — I cannot speak with any degree of certainty in regard to the analysis, for at that time I was not in a position to know much about it. I saw the analysis at the time it was sent in to Mr. Howard, and have not seen it since. It came down here, and at that time I had not thought of ever coming here to testify to any thing of that kind.

Mr. FLAGG. I understand the counsel on the other side to admit that that was the analysis.

Q.—How far from the river was the water that was analyzed taken? *A.*—I should say about seventy-five or eighty feet. I don't know: I never saw it measured, or knew of its being measured.

Q. (By Dr. WILSON.)—Since you sunk those wells and got this new supply of water, are you suffering from any trouble, as far as washing the goods is concerned? *A.*—You understand that our goods are cotton goods, and are not washed. The water where we had the trouble was in our dye-house, in making our dyes, and in our sizing.

Q.—You do not have any trouble now, since you have got those wells? *A.*—Comparatively we do not have so much; but the water from those wells is put into a tank in the attic of one of our mills, and we have to clean out the tank very often, and take pains to keep it clean, in order to get along there as we do.

Q. (By Mr. FLAGG.)—By whom are those mills owned now? *A.*—They are owned by a corporation. Bliss, Fabyan, & Co. are the selling agents.

Q.—Is this getting of pure water attended with much expense? *A.*—The expense of pumping, and the expense of power.

Q.—About how much? *A.*—I am not prepared to say. The pump is attached to our shafting in the mill; and about how much power it takes, I have no idea.

Q. (By the CHAIRMAN.)—Is it steam-power? *A.*—We have both steam-power and water-power. We run the steam-engine all the time in connection with our wells.

Q.—Is this pump run by steam-power or water-power? *A.*—By both. The wheels and engine are connected together.

Q.—Whichever you happen to be running, I suppose? *A.*—We run them both all the time.

Q.—Both steam and water? *A.*—Yes, sir.

Q.—Then, it takes no extra steam to run that pump when you are using steam? *A.*—Well, it must take some, because it must take power to run the pump; and, if the pump was not running, the steam would be shut off from our engine, because our engine does the regulating. When we start in the morning, we hoist our gates wide open, and let the engine make up what is lost; and any steam that is thrown off or on during the day comes off or on our engine practically.

TESTIMONY OF HERBERT A. PRATT.

Q. (By Mr. FLAGG.) — You live where, Mr. Pratt? *A.* — Worcester.

Q. — What is your business? *A.* — Civil engineer.

Q. — You are familiar with the pond near Morse's Mill in Millbury? *A.* — I made a survey of that pond.

Q. — At whose request? *A.* — Mr. Morse's.

Q. — When was this? *A.* — The survey was made in March, 1881.

Q. — State what survey you made. *A.* — I made a survey of the pond there, and of the manufacturing property as a whole, looking at the pond and the buildings.

Q. — Is this the plan you made? *A.* — Yes, sir.

Q. — Now, will you state to the Committee what the result of your survey showed? *A.* — The survey shows the reservoir and buildings. This line through here represents the general course of the Blackstone as it is at present. These spots that are colored, and marked "deposit," are the spots appearing upon the surface of the water, to be seen. These points are on a level with the surface of the reservoir, or above it. This dark line running down here, on each side of the stream, and running across here, is a line marking the average line of deposit. I made soundings all through here, going over it, since making the survey, in a boat. I found the general depth of that stream, below the surface of the water, to be eight feet, striking what appeared to be a gravel bottom. Through here, and outside of that line, I found the greatest depth to be five feet, striking what appeared to be a gravel bottom. From this point, running up here, I found it varied from five feet, running up to the level of the surface of the pond. I found that to be the case through this line; and following along Ring Island, as it is called, I found that to be somewhat less than five feet; and through here, following along that island, I got the general depth of five feet. As I went out towards the bank, it grew less. That whole spot marked there as "deposit" is covered with a rank growth. I do not know what it is: I never saw any thing like it before. This outline is an exact survey, made last March, of the reservoir, as it was found, — the high-water line.

Q. — About what is the area of the whole pond? *A.* — About thirty-two and a half acres. That does not include Ring Island.

Q. — What part of that area appears to be filling up with this deposit? *A.* — Passing that area upon this line, as shown here, and following that line out until it strikes high-water mark, there is an area in there, above those portions marked as "deposits," showing on the surface of the water of eleven acres.

Q. — That is, there are eleven acres of deposit that shows above the surface of the water? *A.* — No, sir: it reaches from this point where I got my five feet, running up to a level with the surface of the water.

Q. (By the CHAIRMAN.) — What are we to understand about that eleven acres? I am not quite clear about that. *A.* — Those two points are not included in this area of eleven acres.

Q. — Eleven acres, then, of that area shows deposits not reaching the surface, from various depths to near the surface? *A.* — Yes, sir.

Q. (By Mr. MORSE.) — Which are the eleven acres? *A.* — That portion along west of that line there, following the general course of the stream until it reaches the bank: this line the same, lying west of that, up to this point here, the limit of the survey. That does not include the natural islands, apparently, found there, and those three portions colored in.

Q. (By Mr. FLAGG.) — You say that the eleven acres are either wholly or partially filled up? *A.* — Yes, sir.

Q. — And you assume that the natural pond will be gradually filled until it is entirely covered? *A.* — With this Ring Island, which is hard land, lying there, I do not see how it can be otherwise. The current tends that way. That little colored spot there is covered with a growth of alders.

Q. — There is a deposit in there, above the water? *A.* — Yes, sir.

Q. — What portion of the other part of the pond appears to be filled up with deposits? *A.* — At high water, until it strikes high water on the bank there, not including that large deposit, there is an area in there of 2.15 acres, or about that. In following this line out, not including those marked "deposits," and that natural island, there is an area of about four and a half acres, more or less filled.

Q. — Tell the Committee how you know this is a deposit. *A.* — Well, from the soundings that I made.

Q. — Through the deposit? *A.* — Through the deposit, until I reached what appeared to be a hard gravel.

Q. — What was the nature of the deposit? *A.* — I don't know that I can answer that question; but it appeared to be a filling. In making the soundings, it gave forth a very unpleasant odor.

Q. (By Mr. MORSE.) — What sort of odor? *A.* — I don't know that I can answer that exactly.

Q. (By Mr. FLAGG.) — An offensive odor reminding you of cesspool odor? *A.* — Yes, sir.

Q. — Have you any doubt in your mind that that is the result of Worcester sewage? *A.* — To a great extent, it unquestionably is.

Q. — Is the deposit also made up of a quick growth of weeds, which accumulate there? *A.*—Yes, sir: in most cases I found a quick growth, wherever it reaches the surface; and over a large portion of this, at the time the surveys were made, and later,—over a large portion of this there is a rank growth of weeds, which now shows upon the surface.

Q. (By Mr. HAMLIN.)—You cannot tell what weeds? *A.*—No, sir.

Q. (By Mr. FLAGG.)—Do you recognize these photographs? *A.*—Yes, sir.

Q.—Do they show the growth of which you speak and the places where the deposit appears above the surface? *A.*—They do, very clearly.

Cross-Examination.

Q. (By Mr. GOULDING.)—Have you undertaken to give the depth of this deposit from the natural surface of the ground? *A.*—Yes, sir.

Q.—How deep is the deposit on the average? *A.*—It would vary in different parts of the reservoir.

Q.—Between what extremes would it vary? *A.*—There might possibly be an average, covering the entire reservoir, of eighteen inches or two feet.

Q.—In the deepest place, how deep is this deposit? *A.*—There are places where it is four or five feet deep.

Q.—Did you explore the quality of this deposit, clear to its bottom, in these deep places? *A.*—I did, at a number of them.

Q.—Was the deposit substantially of the same material at the various places? *A.*—I considered it such.

Q.—Has it been subjected to any chemical analysis, to your knowledge? *A.*—Not so far as I know.

Q.—Did you get any specimens of it? *A.*—I did not.

Q.—Have you any information as to the length of time required to make a deposit of that sort, four feet deep, in such a pond as that? *A.*—I can't say that I have. I was not asked to prepare myself on any such question.

Q.—Now, with regard to this vegetable growth: did that grow in the water, or on that part of the deposit which was above the surface? *A.*—In both.

Q.—Where it grew in the water, how deep was the water? *A.*—About eighteen inches below the surface, or less than that; perhaps not more than twelve.

Q.—Then, it would grow in places where the water was not deeper than eighteen inches? *A.*—It showed on the surface only at that depth.

Q.—Did it grow where the water was deeper? *A.*—I cannot answer that question positively.

Q. — Was there a deposit in the vicinity of the mill? *A.* — It was less as you approached the mill, — less in quantity and depth.

Q. — How long have you known that pond, — that is, how long have you observed particularly these characteristics? *A.* — About twelve years.

Q. — You observed these places where the deposit, as you call it, was above the surface, as long ago as that? *A.* — No, sir.

Q. — Did you notice whether they were there or not at that time? *A.* — I can't say.

Q. — What has this Ring Island to do with the deposits? I observed that you connected that in your answer to the question as to the tendency to fill up. *A.* — It would form a barrier in the direction of the general current; and, as it extends nearly across a portion of the reservoir, it would aid in the filling up.

Q. — It is not an unusual thing, is it, to find deposits in a pond raised by damming a stream? *A.* — There is always some deposit.

Q. — On both sides of the channel? *A.* — Yes, sir.

Q. — Have you ever had occasion to examine any other ponds or reservoirs, the deposits to be found in the different parts, and the amount and depth of such deposits? *A.* — Within the last thirteen years, I have made surveys of a large portion of the reservoirs on ponds extending from Millbury as far as what is called, or was at that time, the Leicester Water-Power Company. I have never found any thing filled to that extent, or in which there was any such odor or smell.

Q. — That was not the question I asked you; but it was whether you had made any other surveys with reference to determining the extent and character of the deposits in other ponds? *A.* — I have made soundings.

Q. — For that purpose? *A.* — No, sir: not for that purpose. I have made soundings in other ponds to ascertain the condition of the bottom.

Re-direct Examination.

Q. (By Mr. FLAGG.) — If those deposits that you found in other ponds had been of the same nature as those you found in this pond, you would have noticed it, even if you had not been on the lookout for it? *A.* — If there had been such an offensive smell as there was in taking the soundings in this reservoir, I should, most certainly.

Q. — You say that it is true that rivers that are dammed always collect some deposits. You do not mean to say deposits of this nature, unless there is something of the nature of sewage there, do you? *A.* — No, sir.

Q. — A clear stream dammed would not furnish such a deposit as this? *A.* — Not of that nature.

Q. (By Dr. HARRIS.) — You say you noticed an odor in those deposits. Can you describe the character of the odor in any way, — what it was like? *A.* — It was very offensive, and such as would be noticed in connection with a cesspool.

Q. — What you would term a cesspool odor? *A.* — Yes, sir, I should. I don't know as I can give a better answer than that.

TESTIMONY OF CHARLES WHITWORTH.

Q. (By Mr. FLAGG.) — You live in Millbury, Mr. Whitworth? *A.* — I do.

Q. — How long have you lived in Millbury? *A.* — Nearly nine years.

Q. — Where is your house in reference to the river? *A.* — About ten or fourteen rods from the river.

Q. — In what part of Millbury? *A.* — In the north part, on the Worcester road.

Q. — In reference to the Burling Mills, where is it? *A.* — About a mile below the Burling Mills.

Q. — Where do you work? *A.* — I work for C. D. Morse & Co.

Q. — How far from your house is that establishment? *A.* — By the river, a quarter of a mile probably.

Q. — Had you been accustomed to go from your house to your work by boat? *A.* — Yes, sir: two years next May, a man who lives in the same house with me, and I, got a boat; and, as we worked down at the blind and sash shop, we thought it would be a saving of time in going to the shop if we got a boat, and went down by the river, as it was a nearer way than by the road. We got this boat, and sailed to and from our work for about six weeks; and then we were obliged to discontinue it on account of the stench of the river, and since that time we have never used it. That was the reason why we discontinued it.

Q. — Now, what can you say as to the effect upon the health of either yourself or those who went with you on the river? *A.* — Well, I thought at that time I suffered from its effects.

Q. — In what way did you feel the effects? *A.* — I was troubled a good deal with headache, and a general feeling of weakness and lassitude; and I attributed it at that time to the effects of the odor from the river. In addition to that, the same summer I took my little boy and sailed up the river about a quarter of a mile, towards Burling Mills; and in using my oars I stirred up the water some, and such was the odor arising from it that I had to return. I could not proceed any farther, so disagreeable was the odor from the water.

Q. — When was that? A. — The same year, 1880.

Q. — How long have you been familiar with this pond? A. — Ever since I came to town: about nine years.

Q. — You have heard the testimony as to its filling up? A. — I have.

Q. — You see the map opposite? A. — I have not seen the map.

Q. — You now see it; and what do you say as to its filling up? Have you noticed its doing so? A. — Yes, sir: I know it is filling up. I know, without looking at the map at all, from my own observation, that it is filling up; because the water is more shallow than it was when I went there, nine years ago.

Q. — Have you noticed the nature of the deposit with which it is filling? A. — Not particularly: only I know it is mud, and smells badly.

Q. (By Dr. WILSON.) — What month was this in the year 1880 that you tried to row on the river, and couldn't? A. — That was in the month of July.

Cross-Examination.

Q. (By Mr. GOULDING.) — What is your age. A. — My age is forty-seven.

Q. — You live how far from the river? A. — About ten or twelve rods.

Q. — What does your family consist of? A. — I have three children and a wife.

Q. — What are the ages of your children? A. — My oldest is seventeen, the next is eight, and the youngest four.

Q. — Do they all live at home? A. — They all live at home.

Q. — Is the oldest a girl or a boy? A. — Girl.

Q. — How long have you lived in that house? A. — About four years.

Q. — Where did you live before that? A. — I lived down near the sash and blind shop.

Q. — Near the pond? A. — Nearer the shop: not so near the pond as I am now.

Q. — It was in July that you started out with your little boy to row on the river, and returned: now, how long before that was it that you undertook to sail to and from your work? A. — About the beginning of May.

Q. — And continued it how long? A. — A few weeks; six or eight weeks probably.

Q. — Who was your companion? A. — A man by the name of Packard, who lives in the same house I do, — lives in the other half of the house. I may say that he continued to go alone perhaps a few days after I discontinued it. He was a stronger man than I was, and could stand it better; but I discontinued it on that account.

Q. — Did he discontinue it on that account? *A.* — I don't know whether he did or not; but I did.

Q. — How long did he continue it after you discontinued it? *A.* — For some little time; perhaps not more than a week or two probably.

Q. — Did you own the boat jointly? *A.* — It belonged to another man; but, as he didn't want to use it on the river, he gave us the privilege of using it; and we laid out some expense, and fitted it up, and used it.

Re-direct.

Q. (By Mr. FLAGG.) — Were any of those parties who were accustomed to boat on the river made sick so as to send for medical advice? *A.* — Yes, sir: I believe there was one man made sick, who used to sail upon the river as we did. He was a well man, and lived a little higher up the river than we did; and he was taken sick.

Q. — What medical advice did he take? *A.* — He consulted Dr. Gates or Dr. Gage of Worcester. I am not sure about the name: I know it is either one or the other.

Q. — Do you know what the doctor told him? *A.* — He asked him where he lived, and he told him; and he also told him that he went to his work in a boat every day and returned. And Dr. Gage told him that he was to discontinue that, and leave the river, and not to use it any more. He said it was not fit for either him or anybody else to use in that way: and he had to discontinue it; and the man is now as healthy as I am, perhaps more so.

Q. — When was this? *A.* — That was the same year, I believe, 1880.

Q. — Who was the man? *A.* — His name is Joseph Gendreau, a Frenchman.

Q. (By the CHAIRMAN.) — Did you get the story from him, or how do you know the doctor said that? *A.* — He told me himself.

Q. — The man told you? *A.* — The man. I might say, that we worked together. We are on very intimate terms, and are neighbors; and he told me that.

Re-cross.

Q. (By Mr. GOULDING.) — When was it you say he went to Dr. Gage? *A.* — If I mistake not, it was the latter end of 1880, or the fall, rather, — somewhere round about there: I don't know.

Q. — Where did he live, in what house? *A.* — He lived in a house about ten or fifteen rods away from the river; nearer the river than I do.

Q. — How long was he sick at that time? *A.* — Oh! three or four months.

Q. — Was he away from his work all that time? *A.* — He was away from his work a considerable part of it.

Q. — You mean he was away from his work nearly three or four months, or off and on during three or four months? *A.* — Off and on.

Q. — Ailing? *A.* — Ailing.

Q. — He went and had this consultation with the doctor, and repeated to you what you say? *A.* — He did.

Q. — And he followed the directions of the doctor, and got well? *A.* — He followed the directions of the doctor so far as discontinuing boating.

Q. — And he got well, you say? *A.* — He got well.

Q. — And remained well? *A.* — He is well now, as far as I know.

Q. — Does he live in the same place? *A.* — No, sir.

Q. — How long did he live where he was living at that time? *A.* — Only a few months, as near as my memory serves me.

Q. — That is, you mean he lived there only a few months after he got well? *A.* — He lived there a few months after he got well.

Q. — Where did he go then? *A.* — He went a little lower down the river, just opposite where I live.

Q. — How near the river? *A.* — About as near as before.

Q. — He has been living there ever since? *A.* — He has lived there ever since.

Q. — He is perfectly well now? *A.* — Seems to be.

Q. — Rowing did not agree with him? *A.* — Rowing did not agree with him, I suppose.

TESTIMONY OF HENRY L. BANCROFT.

Q. (By Mr. FLAGG.) — You live in Millbury? *A.* — Yes, sir.

Q. — How long have you lived there? *A.* — I was born in Millbury. It has always been my place of residence.

Q. — You have been familiar with town affairs, and matters about town generally? *A.* — Yes, sir.

Q. — You were a member of the State Senate of '76? *A.* — '73 and '74, sir, I believe.

Q. — You were moderator of the town-meeting in '74, and appointed a committee on this matter of the pollution of the Blackstone River by the Worcester system of sewerage, did you not? *A.* — I have been moderator of the town-meeting, and I presume I might have been that year.

Q. — You don't remember that you were that year? *A.* — I cannot testify as to that particular year. I presume it was so. I remember appointing such a committee in some year. I have kept no dates. Mr. Flagg has been moderator some of the time, and myself: we have been so for a great many years.

Q. — Whatever committee you appointed, you appointed looking out for the interests of the town generally, didn't you, and not of the few manufacturers? *A.* — Yes, sir.

Q. — Your business has been what? *A.* — My business in the early part of my life was mill-work, — about water-wheels and flumes, and has been for several years past.

Q. — And you have become familiar with the flumes of the different mills along the Blackstone? *A.* — Yes, sir, quite extensively: that is, quite a large number of them.

Q. — Have you any hesitation in saying, from your experience, that in Millbury the river is in such a state of pollution as to be detrimental to health and business both? *A.* — I should think there could not be any doubt on that question at all.

Q. — And there is a general fear in the community as to its effect upon health, is there not? *A.* — Yes, sir, a general fear.

Q. — Have you made yourself familiar with the stream above the mouth of the sewer, and its manufactories, and what was done on those mill privileges in 1870 and 1880? *A.* — Yes, sir. At the request of this committee, I have been through these different streams and mills.

Q. — Whether the difference in business that is carried on there to-day from what it was in 1870, if there is any difference, tends to purify the streams forming the Blackstone above Worcester? *A.* — So far as the mills for fabrics are concerned, those manufactories which have been spoken of, cotton and woollen mills, the amount of business done is less than it was in 1870.

Q. — From your familiarity with the mills, can you say whether or not the pollution is probably less? *A.* — The pollution would be considerably less.

Q. — Have you prepared a list of those mills? *A.* — Yes, sir. [Paper produced.]

Q. — This statement was prepared under your direction? *A.* — Yes, sir.

Q. — And from data which you satisfied yourself were true? *A.* — Yes, sir.

Q. — State generally what the paper shows. *A.* — It is intended to show the amount of business done at the different privileges above Millbury and in Millbury — that is, on the streams that enter the Blackstone River — in 1870, and about the amount in 1880 and since.

Q. — Those mills are all above the mouth of the sewer? *A.* — No, sir: those that are on the Singletary Stream are not above the mouth of the sewer.

Q. — This includes also Bramanville, does it? *A.* — Yes, sir: I think it does.

Q. — There is a distinct part of it relating to the mills above the mouth of the sewer? *A.* — Yes, sir: on Kettle Brook and Mill Brook.

Q. — Now, whether or not that paper shows that the number of sets and the amount of business is less, and whether you would infer, being familiar with business of that kind, that from manufactories the pollution was less in 1880 than it was in 1870? *A.* — Yes, sir.

	Name of Company.	1870.	1880.
Kettle Brook Stream.	First privilege on stream	Saw-mill.	Now discontinued.
" "	Mann's woollen-mill.	2 sets, scouring and coloring.	2 sets, no wool scouring or coloring.
" "	Kents.	Shoddy-mill.	1 set, satinet, no wool scouring or coloring.
" "	Cherry Valley Manufacturing Co.	4 sets, fancy, with scouring and coloring.	4 sets, satinet, no wool scouring or coloring.
" "	S. Pratt & Co.	2 sets, satinet, with scouring and coloring.	2 sets, satinet, no wool scouring or coloring.
" "	E. Collier, satinets.	1 set, with scouring and coloring.	1 set, no wool scouring or coloring.
" "	Olney, formerly S. L. Hodges.	5 sets, fancy cassimeres, scouring and coloring.	6 sets, flannel (part cotton), no scouring or coloring.
" "	J. A. Smith.	6 sets, fancy (day and night), scouring and coloring.	2 sets, satinet, no wool scouring or coloring.
" "	W. Bottomly, satinets.	2 sets, low grade, very dirty.	Washed out and not rebuilt.
" "	Ashworth & Jones, beavers.	4 sets, scouring and coloring.	4 sets, scouring and coloring.
" "	Darling, satinets.	2 sets, scouring and coloring.	2 sets, no wool scouring or coloring.
" "	J. A. Hunt, satinets.	2 sets, scouring and coloring.	2 sets, no wool scouring or coloring.
" "	Cunningham, satinets.	2 sets, scouring and coloring.	Burned and not rebuilt.
" "	B. James, fancy cassimeres.	5 sets, scouring and coloring.	Not running for some years.
" "	Stoneville, cotton.	96 or 98 looms.	Same, 96 or 98 looms.
Ramshorn Stream [1].	E. Hoyle.	Grist-mill and shingle-mill.	Small mill with 1 wool-washer.
" "	Griggs.	Tannery and saw-mill.	Saw-mill.
" "	B. Larned, satinets.	3 sets, scouring and coloring, a saw shoddy-mill.	4 sets, no wool scouring or coloring, and shoddy-mill.
Kettle Brook Stream.	Trowbridgeville, satinets.	3 sets, satinet.	3 sets, satinet (same).
" "	Albert Curtis, satinets.	11 sets, satinet.	11 sets, satinet (same).
Tatnuck Stream.	———, satinets.	2 sets, satinet and shoddy-mill.	2 sets, satinet, and shoddy-mill.
Kettle Brook Stream.	Hopeville Manufacturing Co.	3 sets, satinet, scouring and coloring.	3 sets, satinet, no wool scouring or coloring.
" "	South Worcester Carpet Co.	Nothing.	10 sets, with scouring and coloring.
Mill Brook.	Water Street, woollen.	Nothing.	2 sets, for filling worsteds, no colors.
"	Fox Mill, fancy.	16 sets, scouring and coloring.	8 sets, scouring and coloring.
"	Adriatic Mills.	13 sets, scouring and coloring.	13 sets, scouring and coloring.
	Totals.	88 sets.	82 sets.

Holden Reservoir filled for the first time in 1867. Capacity 450,000,000 gallons.

[1] Ramshorn Reservoir Dam was raised ten feet in 1873. The pond covers an area of one hundred and forty-five acres, water to be drawn one-sixth in July, one-third in August, one-third in September, one-sixth in October. For acreage see State Board of Health Report, 1873.

Q. — Ramshorn Brook is a brook running into the Blackstone River? *A.* — Yes, sir.

Q. — There is a reservoir on that brook? *A.* — Yes, sir.

A. — Largely increased in size lately? *A.* — Yes, sir.

Q. — What sort of a stream is that as to purity? That is a very pure stream. The dam was raised in '73, some ten feet, increasing the capacity of the pond, Mr. Curtis says, three times. Three times the quantity of water-power comes from there that formerly did.

Q. — Kettle Brook is another brook forming the Blackstone? *A.* — Yes, sir.

Q. — What sort of a stream is that? *A.* — I should think it was very pure. It comes down from the Paxton Hills. There are a good many mills on it now, but not so many as formerly.

Cross-Examination.

Q. (By Mr. GOULDING.) — You are not a chemical expert, are you? *A.* — No, sir, I am not.

Q. — Ever examined any analyses of these streams, so as to know any thing about their comparative purity. *A.* — No, sir: I am not a chemist.

Q. — What induced you to make this schedule? *A.* — I did it at the request of the committee.

Q. — When did you do it? *A.* — Last week.

Q. — Have you got all the mills there are on any of the streams? *A.* — I think I have.

Q. — How many do they number? *A.* — I don't remember: I can't tell for certain. I can go through with the list, I guess; but I don't remember the number particularly. I think that schedule shows them all.

Q. — Have you included in this list all the manufactories of all sorts, or simply the woollen and cotton mills? *A.* — I have not been to the wire-mills.

Q. — Well, machine-shops? *A.* — There are no machine-shops on the stream, except one, in your city.

Q. — What is Mr. Coe's shop? *A.* — I have not been to Mr. Coe's shop.

Q. — Have you been to any of the factories in the city? *A.* — No, sir.

Q. — All the mills that your schedule includes, then, are mills outside of the city of Worcester? *A.* — No, sir: I do not mean to be understood so. I have been to the mills that manufacture cloth or yarn, or fabrics of that sort: I have not been to the shops, including those in the city. I think the schedule shows precisely where I went.

Q. (By Dr. WILSON.) — Where is this Ramshorn Brook? Where

does it join the other stream? *A.* — It joins the other stream in the town of Auburn, just above New Worcester.

Q. — Joins Kettle Brook? *A.* — Yes, sir.

TESTIMONY OF CHARLES D. MORSE.

Q. (By Mr. FLAGG.) — You live in Millbury? *A.* — Yes, sir.

Q. — How long have you lived there? *A.* — I have lived there thirty-two years.

Q. — What has been your business there? *A.* — Sash, door, and blind work, — wood-work.

Q. — During all this time at the same place? *A.* — My business has been at the same place: yes, sir.

Q. — Will you state what you have noticed as to the condition of the river at your place of business — first, as to the pond filling up? *A.* — My pond has filled up very largely. I presume what the engineer has said will cover that very largely. Perhaps it is not necessary to say much more; perhaps not any thing. But my pond has filled up very largely for the past eight years, so as to be noticeable, the grass growing very rapidly, and filling in at the upper part more particularly than at the lower part. As the water comes in, the sewage, or sediment, is deposited on each side: the current goes directly through the pond. Last fall, when I went out in a boat with the engineer and another man to take the soundings, we found considerable trouble in rowing the boat. Although it was in November, I think the sixth day of November, we found considerable trouble in rowing the boat, on account of the odor from the river. The man who rowed us called it sculling. He stood in the hind end of the boat, and sculled it, without getting up so much odor from the river as would be the case if two oars were used. We found that the filling, at times, especially in the upper part, had a regular sewage smell. The heft of the filling is in the upper part. The first that I noticed the odor from the river particularly was some six years ago. What called it to my attention more especially was this: I had a tank perhaps eight feet long, four feet wide, and three feet deep, or something like that, which I put into the shop, for shop and fire purposes.

Q. — Is that the tank about which Dr. Lincoln testified? *A.* — Yes, sir. It ran along perhaps a year and a half after that, and I called in Dr. Lincoln and Dr. Webber to look at the tank. They were not both there at one time, but I asked them both to meet me at one time to look at the tank. Dr. Lincoln took a piece of panel, or shingle like, and scraped the excreta off the sides, and also from the bottom of the tank: and both Dr. Lincoln and Dr. Webber said that I must discontinue taking the water from the river, for fear it would create disease and make it unhealthy; and I did so. I had to wash that tank perhaps half a dozen times before I could get it so that the odor was fairly gone.

Q. — Could water get into this tank from any other source than the river? A. — It did. I put a conductor on my building, and took what rain-water I could get from the roof; and, when I was short, I took from the river.

Q. — So that there was nothing in the tank but rain-water and water from the river? A. — Rain-water and water from the river. Some two years ago we were obliged to close the windows quite often in the shop, and last year even more, on account of the odor from the river. We find more loss of time the past few years than formerly. I attribute it to the effects of the river upon the workmen. There is more lassitude, and lack of energy, — a feeling, as some of them express it, of goneness, weakness, sinking.

Q. — Have you looked into the matter to see about it? A. — Yes, sir.

Q. — What can you say as to the number of days that were lost by a certain number of men in any one year, formerly, as compared with the time lost now? A. — I took 1872. I took forty men, in their order, as they appear on my time-book. Those forty men made 110,249 hours in the year. During that year business was not over and above driving: we did not work in November or December evenings; we usually work evenings when we are busy. Had we worked in November and December the same as we worked in 1881, we should have added — calling those two months the same as in 1881 — 2,631 hours, making the total 112,180 hours. The same number of men, working from Oct. 1, 1880, to Oct. 1, 1881 — the reason I take these two dates, and not, as in the previous year, from January to January, was because I was burned out the fifth day of last October; and so I took a year back from that. In that year the same number of men worked 105,561 hours, making a loss of 7,329 hours in one year's time of forty men. At twenty cents an hour, that would amount to $1,465.80. We average about 55 men, which, at the same ratio, would increase the amount of loss to over $1,800 per year.

Q. — Loss in wages, you mean? A. — Yes, sir, over and above the time of 1872. We find that we have more or less diphtheria among our help. I have it in my own family. I don't know as I should call it *diphtheria*, — " diphtheretic sore throat," the doctor calls it. My wife was sick three years ago with diphtheretic sore throat.

Q. — How far is your house from the river? A. — About an eighth of a mile. She was sick nearly all the winter with diphtheretic sore throat, and quite sick, at times, from the smell. My oldest daughter was sick last year, and my clerk was sick with diphtheretic sore throat last season.

Q. — You speak of your mill having been burned. When was that? *A.* — Oct. 5, 1881.

Q. — You have sprinkling-pipes in your mill? *A.* — I have sprinkling-pipes: yes, sir.

Q. — Will you explain to the Committee the operation of those sprinklers? *A.* — I had a building one hundred and thirty-six feet long, three stories high, forty-two feet wide; and I put in three lines of sprinklers in each story. The basement had three lines; and in the upper story, which had a mansard roof, I put three lines, to protect the roof in case fire should go up under the ceiling in any way.

Q. — When you speak of a *line*, what do you mean? Whether or no it is a piece of steam or gas pipe? Won't you explain it to the Committee? *A.* — That is just what it is, — gas-pipe.

Q. — How large is it? *A.* — Perhaps at one end it is two inches; and, at the other end, it is reduced down to three-quarters of an inch, perforated with holes.

Q. — And this pipe is connected with the pump? *A.* — Yes, sir.

Q. — So that when you start the pump, the water is forced into this pipe? *A.* — Yes, sir: and forced out through those holes.

Q. — There were three lines of this pipe? *A.* — Three lines in each story.

Q. — Now, will you tell the Committee the experience you had in using this at the time of the fire? *A.* — On the 5th of October my building caught fire; and the first thing was to start my pump, after shutting down my wheel. It seemed at first that we had got the fire substantially under control: but, the pipes being smaller at the farther end, the sediment was forced out to that end, and it filled the holes in the pipe; and there was where the fire got its headway.

Q. — You mean that this pipe acted as a strainer? *A.* — As a strainer: yes, sir. There is where the fire took its headway: it was at the farther end, although most of my shavings were in the middle of the shop.

Q. (By Mr. CHAMBERLAIN.) — Is that pipe open at the end? *A.* — No, sir.

Q. — Then, there are holes towards the end of the pipe? *A.* — There are holes through the whole length of the pipe: small holes.

Q. — How large are those holes? *A.* — Perhaps a little larger than a pin; not so large as a pin-head, perhaps.

Q. (By Mr. FLAGG.) — How far apart are the holes? *A.* — Perhaps ten inches. It makes a perfect rainstorm, when every thing is working in good shape.

Q. — Have you a photograph showing what you saved of the mill? *A.* — Yes, sir. [Photograph shown.]

Q. — Did you have the sediment taken out of those pipes? *A.* — Yes, sir.

Q. — Won't you show that to the Committee, and state what it is? [Witness produced a bottle.] *A.* — That did not come out of the pipe in the mill, but it came out of a pipe that we had in the machine-shop to protect the end of the building in case the other shop should burn. It was an upright pipe, with a stop-cock put where you could reach it. I found the lower end of it filled with sediment. We used that only a very few minutes.

Q. — I understand there was a sprinkler on the machine-shop near your mill? *A.* — Yes, sir.

Q. — That sprinkler ran during that night? *A.* — Yes, sir: for a few minutes.

Q. — And it was from the pipe of that sprinkler that this material came? *A.* — Yes, sir.

Q. — You have photographs showing what of the mill was saved, and the looks of the mill after the fire? *A.* — Yes, sir.

Q. — By the photograph, one end of the building appears not to have been burned? *A.* — No, sir.

Q. — It was in that end that the sprinklers worked well? *A.* — That was the end where the larger pipes were: yes, sir.

Q — The large end of the pipes was the last part that would fill up, if any did? *A.* — Yes, sir: the end that burned, where it took its headway, was where the small end of the pipe was; where the sediment was thrown in, as I have reason to believe.

Q. — Have you any doubt that if those sprinklers had worked properly, much less damage would have been done that night? *A.* — I think we should have saved at least half.

Q. — Have you any doubt that the reason the sprinklers did not work was on account of the material of this sort that was forced into the pipes? *A.* — I have no doubt that that was the cause of the building burning, from the fact that there was evidence of it in the sprinkler outside.

Q. (By Mr. SMITH.) — Whether or not you have had any personal knowledge or experience with regard to the operation of this contrivance that you speak of, for putting out fires? *A.* — I tried it myself, but not on a fire.

Q. — I mean, have you any knowledge of what would have been the effect, if you had poured water upon the fire through those sprinklers, as it occurred in your shop? *A.* — I tried it in my basement without any fire, and it worked perfectly. That is, when I put it in, seven or eight years ago.

Q. — I take it that you have a great many shavings scattered round in your shop. Now, if that had worked perfectly, have you any assurance, from your own experience, that the fire would have been put out? *A.* — I have reason to believe it would, for it gained its headway where there were the least shavings.

Q. (By the CHAIRMAN.) — Do you know absolutely that the sprinklers did not work? *A.* — Well, at that end of the shop.

Q. — You assume that they did not, because the fire was not put out? *A.* — I found burnt sediment in the ends of those sprinklers.

Q. (By Mr. HAMLIN.) — Are you sure your apparatus was in perfect running order before the fire? *A.* — Yes, sir.

Adjourned to Wednesday at 10 o'clock.

FOURTH HEARING.

STATE HOUSE, BOSTON, March 15, 1882.

THE hearing was resumed at 10.30 o'clock.

TESTIMONY OF DR. ROBERT BOOTH.

Q. (By Mr. FLAGG.) — You live in Millbury? *A.* — Yes, sir.

Q. — How long have you lived there? *A.* — Almost seven years.

Q. — During that time, you have been a practising physician there? *A.* — Yes, sir.

Q. — From your acquaintance with the people and with the river, what do you say as to the general effect of the pollution of the river upon the general health of the people? *A.* — Well, I think it has a deleterious effect upon the health of the community living upon the stream.

Q. — How far down the stream does your practice extend? *A.* — As far as Blackstone.

Q. — So that you are familiar with the Blackstone River from below Worcester to the State-line? *A.* — I practised medicine in Blackstone about eight years.

Q. — Will you state in your own words to the Committee in what way you have noticed this deleterious effect upon the health of the people, — the kinds of sicknesses you have noticed, and ascribed to the river? *A.* — Typhoid fever principally, diphtheria, dyspepsia, and general debility. These are the diseases I attribute principally to the influence of emanations arising from filth or any thing of that character.

Q. — Have these diseases been noticeably common? *A.* — Yes, sir : I have attended quite a number of cases of typhoid fever, a few of intermittent fever, intermittent neuralgia, isolated cases occasionally of diphtheria. Almost all, if not all, of these cases I attribute to the emanation arising either from the Blackstone River, or, if they were not in the vicinity of the river, from like causes around the premises. Most of the cases were in the vicinity of the Blackstone River.

Q. — Under what general term are such diseases classed? *A.* — Zymotic diseases.

Q. — Can you say any thing to the Committee that will show them why it is that you ascribe the prevalence of these diseases to the pollution of the river? *A.* — I can relate to you some cases. While I lived in Blackstone, I observed that when children went much into the water — as they are accustomed to there more than they are in Millbury, because in Millbury they know what the water is: down there they are not so familiar with it — typhoid fever was very prevalent among those who went much into the water. I remember the case of one young man who was less than a year in this country, — very healthy. He went into the water to bathe; and, a short time after, he was taken down with typhoid fever. It was the most intractable and the worst case I ever met. This, at the time, was attributed to going into the water, which I had no doubt in my mind was the case. The young man died. Amongst boys it was very common while I lived in Blackstone. I forbade parents to allow their children to go into the river, telling them the consequences; as I had observed that typhoid fever frequently attacked those that went into the river to bathe. At Millbury there was one case a year ago last summer of a young lady who worked in the Mill Brook cotton-mill. She was taken with something like intermittent fever. At first the chill and fever returned every second day, finally it came every day, and then twice a day. I was called to attend her, and I used all the means in my power to arrest it; but it finally went into consumption, to which she was predisposed: and she died. I made some inquiry in relation to the river; and she said that she worked over the wheel in the mill, and that, during the summer, the stench was almost unbearable. From what I could find out from my inquiries and my investigations into the case, I came to the conclusion that it was the stench from the water of the Blackstone River that was the cause of her sickness and death. Cases of typhoid fever are very much more common in that part of the village that we call Millbury, than it is in the part designated as Bramanville; although in location Millbury is much better situated than Bramanville, as Bramanville is situated between hills, and doesn't get the free air as Millbury does. You would expect that we would have more such diseases there than in Millbury; but experience teaches the contrary.

Q. — Would your opinion as to the effect upon the health-rate by this pollution of the river be modified any by the fact that the death-rate in Millbury, for the last two years, say, has decreased? *A.* — Well, no: I should think not. The health-rate and the death-rate are very different. There may be a great deal of sickness, and very few deaths. It depends a great deal, in my experience, upon the epidemic. Sometimes we have epidemics which are very mild in their character; at other times we have those which are very severe. Whether these are caused by the poison, or some modification of that, or whether they are caused by any local troubles, I am not prepared to say.

Q. — Isn't this the theory maintained by sanitary authorities, that the number of deaths in a given district bears no constant ratio to its healthiness or unhealthiness? *A.* — I believe that is the general opinion.

Q. — There was an epidemic at one time in Maplewood Seminary, — have you looked into that matter at all? *A.* — I have the report of that case.

Q. — You have it with you? *A.* — Yes, sir.

Q. — Will you read the title-page? *A.* — "A Report upon the Epidemic occurring at Maplewood Young Ladies' Institute, Pittsfield, Mass., in July and August, 1864; including a Discussion of the Causes of Typhoid Fever. By A. B. Palmer, M.D., and C. L. Ford, M.D., and Pliny Earle, M.D."

Q. — Are there any quotations from medical authorities in that pamphlet bearing upon the relation of pollution of streams and air to zymotic diseases? *A.* — There are a number of such quotations.

Q. — Are there any that you desire to read, to substantiate your opinion? *A.* — I think some of them would substantiate what I have been speaking of. I will read some if you wish.

Q. — You may read them, doctor. *A.* — Here is one by Dr. Carpenter, an English physiologist: —

"The injurious influence of decomposing azotized matter, in either predisposing to or exciting severe disease, and particularly typhoid fever, is universally admitted among high medical authorities. The views of Dr. Carpenter on this subject are too well known to medical men to need full elaboration. His doctrine, so clearly stated, and so amply illustrated by facts, is, that decomposing materials in the system, whether generated and retained there, or taken in from without, either in water or food contaminated with foul matters, as sewerage, etc., or in the air by night-soil and sewerage emanations, either themselves produce disease, or serve as the *nidus* for the operation of specific or zymotic poisons, such poisons as produce fevers, cholera, diarrhœa, dysentery, and the like."

There is a paragraph here from Dr. Williams, author of "The Principles of Medicine," — a volume that is in the hands of almost every physician who knows of it; it is a high medical authority: —

"The soil which drains from habitations, contains, in addition to excrement, dirty water, the washings and remnants of animal and vegetable matter used as food, and other offal. All these are mixed together and stagnant in the corrupting slough that is retained in cesspools and privies, or that is carried into sewers. Every ill-drained house has a Pandora's box ready to pour forth its evils when occasion offers, and always oozing them out in degrees sufficient for the impairment of health.

"These materials continually poison both air and water; and typhoid fever, diarrhœa, cholera, dysentery, dyspepsia, inappetency, general weakness, and malnutrition are the results of their pestiferous operation acting in different degrees."

It might be stated that the reason why the same cause will not produce the same effect in two different individuals, is because of predisposition. Dr. Watson, in his work on "The Practice of Medicine," gives an illustration of it. He says, Let half a dozen men be in a boat which becomes capsized, and all the men precipitated into the water. They will remain in the water some time before being rescued, and on returning to their homes they will not all be affected alike. One of the individuals may have rheumatism, another may have pneumonia, another may probably have typhoid fever or something else, and all the rest may come off free. He attributes this to a predisposition.

Q. — Has the name "night-soil fever" been given to typhoid fever? *A.* — It is generally known by that name.

Q. — It is stated so under the authority of Dr. Murchison? *A.* — He calls it by that name, or "pythogenic."

Q. — Whether or not it is your opinion, that if the unpurified sewage of Worcester, as it increases, continues to be poured into the Blackstone River, there will be a cause there capable of producing epidemics throughout the valley? *A.* — I have no doubt of it whatever.

Cross-Examination.

Q. (By Mr. GOULDING.) — How old are you, doctor? *A.* — I am forty-three.

Q. — Where were you educated? *A.* — I graduated in medicine at Ann Arbor, Mich.

Q. — When? *A.* — In 1867.

Q. — Are you a native of this country? *A.* — No, sir, I am not.

Q. — You practised in Blackstone how long? *A.* — Eight years.

Q. — Beginning when? *A.* — The year that I graduated, — 1867.

Q. — That is the last town in the State? Then, you have been since then in Millbury? *A.* — Well, most of the time since then.

Q. — What is your school of medicine, — regular? *A.* — Yes, sir.

Q. — Cases of typhoid fever are common in all towns, I take it, in this State? *A.* — Yes, sir: very common.

Q. — Have you investigated the question of the death-rate of Millbury for the past ten or a dozen or twenty years? *A.* — I have not.

Q. — Suppose you should investigate it, and find that the death-rate had increased very rapidly since the sewage of Worcester was emptied into the river: would you, or not, think it important to put that in as evidence here? *A.* — I should rather think, if the death-rate had increased, that I would.

Q. — Why, if it wholly disconnected from the question of the health-rate? *A.* — Oh! I don't know that it is.

Q. — Didn't you so testify? *A.* — I don't understand that I testified in that way. I said that the death-rate being increased would not go entirely to prove — that was my idea — that there was more sickness.

Q. — But as a general proposition, doctor, wouldn't the death-rate for a period of years, say fifteen or twenty, be a pretty fair criterion of the health-rate? A. — That would be a difficult question to answer, because I know that the death-rate does not always correspond with the health-rate.

Q. — I understood you to say that before; but now the question that I put is, whether the death-rate for a series of years — say ten or twenty — would not, in your judgment, be a pretty fair criterion of the health-rate? A. — Well, I could not say that it would or would not. I could scarcely answer that question.

Q. — What do you understand is the object, in our State Board of Health reports and other reports of that kind, in ascertaining with so much care the death-rate? Has it any relation to the public health, or is it simply a matter of curiosity? A. — Oh, no! it is not a matter of curiosity. I presume it has some relation to the public health, and also to the prevailing diseases.

Q. — And that is another way of saying that it has relation to the health, isn't it? A. — Yes, sir, if you wished to construe it so.

Q. — I do not wish to construe any thing. I desire to get at the facts. A. — I understand the object of the reports is to get the cause and the name of the disease the person died of, and, I suppose, to know what the prevailing diseases are in different localities.

Q. (By the CHAIRMAN.) — What is the use of knowing? A. — It might be very essential: as much so, I presume, as to know the death-rate, — to know what diseases are prevailing in different parts of the country. It gives us statistics which we have in all our works. In relation to small-pox, for instance, it is necessary to know whether small-pox as a disease is on the increase since vaccination came into use or became compulsory, or whether it is on the decrease.

Q. (By Mr. GOULDING.) — Why don't they, then, ascertain, if they can, the health-rate directly, and have reports upon that? A. — That might be a difficult matter.

Q. — It is a difficulty that they don't think they can wrestle with? A. — I presume so.

Q. — I will ask you in candor if you do not understand that these death-rates are collected and published for the purpose of ascertaining the condition of the public health, and with a view to improve the public health; if you do not understand that to be their object, as an intelligent man and a physician? A. — I think that is one of their objects, certainly.

Q. — Do you think it is an object that is not likely to be obtained by any such method? *A.* — I think to a certain extent it may. I do not wish to deny at all that it would not. My statement before was, that I thought that the death-rate, say for a season, or for a number of seasons, would not indicate the state of the health of the locality; but, take it for a great many years, it might.

Q. — You think a well man would be quite as likely to die as a sick man? *A.* — Oh, no! that is not my idea.

Q. — Then, a sick community would be just as likely to live as long and have a low death-rate as a well community: that is your proposition, is it? *A.* — The proposition that I made, a good deal of it, is conveyed in that article that I read last. Those are my views, my ideas.

Q. — You did not originate those ideas? *A.* — I did not originate them, but I gave them to you as my own.

Q. — When did you first read these opinions? *A.* — I read them a great many years ago.

Q. — When did you first read this report that you have read here? *A.* — Ten or fifteen years ago.

Q. — That case of epidemic at Pittsfield occurred at that time? *A.* — Yes, sir.

Q. — Fifteen years ago? *A.* — I cannot give the date.

Q. — I understood your answer to be, with regard to the cause of the cases of typhoid fever that you have had, that they were due either to emanations from the river or from sewage? *A.* — If you will permit me, I should like to give an explanation of what I mean. I made the statement here that I thought that sewage and filth produced a certain class of diseases. This class of diseases may not necessarily be fatal; not nearly so fatal as another class of diseases which I do not attribute to sewage. That other class of diseases may come into a place and become epidemic and sweep off a whole community, whereas it has nothing to do with the question before us to-day. That is my idea, and I think you understand it.

Q. — Now, suppose it should appear that the death-rate of Millbury for the past ten or a dozen or twenty years, from what are called filth diseases, was low in comparison with towns otherwise situated like Millbury, should you then think that was a fact of any importance as determining the question of the state of the public health as affected by the local conditions of the river? *A.* — I should think, if you can prove that, you have a strong point.

Re-direct.

Q. (By Mr. FLAGG.) — In your experience in medicine, can the health-rate be put down in tables? *A.* — I don't well see how it could be.

Q. — Are there not some people who are sick who don't come to you and don't go to anybody? *A.* — A great many; and I would call your attention again to what Dr. Williams says, and you will probably understand my meaning much better than you seem to. My opinion is, that the death-rate has nothing to do with this, because we attribute only one class of diseases to this sewage. And I would like to make another remark: A year ago last winter we had the most fatal epidemic of scarlet fever that I ever met. Twenty-five per cent, I think, died of all that were affected by the disease. This epidemic I do not attribute to sewage at all; I do not attribute measles, nor any of the exanthematic diseases, to sewage; although, from what was here stated, this sewage, or these foul emanations rising from decomposing organic material, may so affect the system, that when a person is taken with any of these exanthematic diseases, or any thing which is not in this class that I have spoken of as arising from this trouble, he may not be able to bear up in his sickness. Persons may be taken with whooping-cough, measles, or small-pox; and their systems may be in such a low state, owing to the depressing influence of this poison upon the blood, that they may succumb to a disease that they would recover from if this was not the case. Those are my views.

Q. (By Mr. CHAMBERLAIN.) — You were just speaking of a case of consumption: do I understand that you think these polluted waters produce consumption? *A.* — Yes, sir: indirectly, if a person is predisposed to consumption. Any thing that will depress or lower the vitality will act as an exciting cause to bring on a disease to which a person is predisposed.

Q. — Lung diseases? *A.* — Yes, sir. This case which you refer to was a case of malarial fever, or intermittent fever, which ended in consumption.

Q. (By Mr. SMITH.) — Do you think that the number or skill of physicians in a community has any thing to do, one way or the other, with the death-rate where the health-rate is low, where sickness prevails, I mean? *A.* — I think it might.

Q. — You think physicians do some good where — *A.* — They may do some good, or they may do some harm.

Q. — I mean in the average now. This is a matter where the death-rate and the health-rate, you say, differ. Now, where the people are sickly, do the character of the physicians, their number, and their skill, have any thing to do with regard to the number of deaths? *A.* — I might answer that in one way: where there is very much sickness there is a great deal to do, and the physicians will probably be more numerous. In Millbury we have five in active practice in that small village, and all seem to have plenty to do.

Q. — They cure the people who are sick, so that they do not die?

A. — I don't suppose it speaks very well for the physicians if they do not.

Q. (By the CHAIRMAN.) — Would you say, from your experience taken as a whole, that there is a lower condition of health and vitality among the people as a whole in those towns than prevails elsewhere? *A.* — I think there is in the main. I have always practised along the Blackstone Valley; but I lived in northern villages previous to that, and I think there is a very much lower condition of health along the Blackstone Valley than in northern villages, where I have lived previous to my engaging in the practice of medicine.

Q. — What should you say, so far as you know, of the condition that prevails in Blackstone, as compared with Millbury? *A.* — Well, I think, on the whole, it might be worse.

Q. — Worse where? in Blackstone? *A.* — Worse in Blackstone than in Millbury.

Q. — Do you get as much of this sewage in Blackstone as in Millbury? *A.* — There is a great deal in the location of a place. Blackstone is a low, sunken place, and any emanations that may arise from the river remain stagnant. Millbury is higher: we get a good sweep of wind down that Blackstone Valley which blows a great deal of it away, — at least where I live, and in most parts of it. But, if Millbury was situated as Blackstone is, it would be very much more unhealthy than Blackstone. In this report which I have here, there is something, I think, bearing on that subject. In speaking of Maplewood, it says, that, at the time when this took place, there was very little wind: it was very calm and sultry. It says,—

"Absence of winds was, doubtless, the worst condition for the inmates of Maplewood. Strong winds would have tended to carry away and dissipate the vitiated atmosphere."

That has a great deal to do with it. If a village is situated on high land so that it gets the full sweep of the wind, the miasm that arises from the stagnant water will be blown away, and the people will escape; but if the village happens to be located on a low place between hills, and these emanations arise, the whole atmosphere in the vicinity becomes impregnated, and the people become much more affected.

Q. (By Dr. HARRIS.) — How long have you lived in Millbury? *A.* — I shall have been seven years in Millbury next June.

Q. — Your circuit of practice includes about how many inhabitants? *A.* — My circuit — I could not exactly say: I never thought about it.

Q. — Four or five hundred? *A.* — I should say more than that.

Q. — You spoke of having a great many cases of diphtheria: how many do you think you have had, annually or collectively, since you have been there? *A.* — That would depend on the season. Some seasons we do not have any. I think I have not had a case of diphtheria within a year. Some years I have probably had not more than two or three cases, and they were isolated: I should not attribute them to the river.

Q. — You spoke of intermittent fever: how many cases of that character of fever have you had? *A.* — I believe I have had but one case, — that is, that I would attribute to the river. Of course I have had other cases of sickness, which I would not consider as caused by the river.

Q. — You do not think that the sewage, or the emanation from the Blackstone River, had any thing to do with those cases? *A.* — No, sir. Many of the cases were not in the vicinity of the Blackstone River: some of them were.

Q. — Has typhoid fever been prevalent within a year? *A.* — No special cases. I have had a number of what we usually call slow fevers, — mild typhoid.

Q. — Was that during the last fall? *A.* — That was during the last fall: yes.

Q. — It was a general thing over the county? *A.* — Yes.

TESTIMONY OF DR. CHARLES F. FOLSOM.

[Mr. Morse suggested that the Chairman should question Dr. Folsom, inasmuch as he did not appear as a witness for either side.]

Q. (By the CHAIRMAN.) — Dr. Folsom, you have been invited to come in here by the Committee; and we shall be very glad to have you, in as brief a way as you can, give us the results of your experience in regard to the condition of things in the Blackstone-river Valley as growing out of the emptying of the sewage of the city of Worcester into that river. *A.* — I don't know exactly what points have been gone over by the Committee, as I have unfortunately been unable to hear what has been said; and I don't know exactly what points they would like to ask me about. So far as the nuisance is concerned, I should think that was unquestioned. As to the quantity of the nuisance, and extent to which the people of the vicinity are disturbed by it, I should think that the Committee could satisfy themselves fully as well from people who are living there as from any of us who have been there but a few times. The number of times I was there I was satisfied in my own mind that it was a serious economical trouble to the mill-owners, and that it was very offensive; and it is not at all impossible that the degree of smell may have had some influence on the public health. I should think it would be very likely to, although of course it would not be likely to have so much influence in a country district, where the air is pure, as it would in a city.

Q. — There is no doubt in your mind that there is a nuisance to a greater or less extent existing in consequence of the emptying of the sewage of the city of Worcester? *A.* — Oh! I should think one might state that fact beforehand without seeing the conditions. You have there the sewage of a city of over fifty thousand inhabitants emptying into a stream, the greatest flow of which in the driest weather is seven hundred and fifty thousand gallons a day. Of course, I should say beforehand that that amount of sewage, coming into a stream of that sort, would necessarily be a nuisance. Of course, the degree and extent of the nuisance would be determined by the number of people living in the vicinity of the stream, and their nearness to the stream.

Q. — Have you personally observed the smells as far down as Millbury? *A.* — One day when I was at Millbury it was quite offensive: the other days that I happened to be there, I did not happen to notice very much smell as far down as that.

Q. — Was it your opinion that that was a sewage smell? *A.* — I think that there is no question about that.

Q. — What is the amount of sewage that the city of Worcester turns in there? *A.* — About three million gallons a day, I think.

Q. — If there should be a larger flow of water, an immense flow, such as there was a week or two ago, and any one should say to you that it could not be a sewage smell that you detected because there were only three million gallons of sewage going in there, what would you say to that? *A.* — I don't think that would alter the fact of the sewage smelling. If sewage exists in a considerable quantity, and decomposes, it will make a considerably offensive smell. Of course, the more diluted it is with water, the more it is extended over a large space, and the less the nuisance would be likely to be.

Q. — Is this sewage liable to be deposited on the banks of the river? *A.* — I satisfied myself that there is a certain amount of deposit; although I think it is not so great as it would be in a climate unlike ours, where they do not have heavy spring and fall freshets. Of course, that scours it out to a certain extent; but there is a certain amount of deposit from the sewage. Of course, the first dam intercepts considerable; and the amount below that is probably less proportionately to the area of the bed of the stream than that which is above. I think that there is no doubt that there is a certain amount there.

Q. — Would a freshet scour down the sewage deposits that have been planted on the banks of the river? *A.* — To a very great extent. Of course, the fact how much they do scour it out is only to be decided by experience and observation. There is a certain amount of deposit there evidently, and unquestionably there would be very much more if it were not for the freshets; but it is a very difficult matter to ascertain, when there is a deposit in a stream, to what it is due absolutely. You cannot say positively by chemical analysis whether it is sewage, or whether it is simply vegetable mould and decaying organic matter generally. If it has been there for a considerable length of time, and if you have a deposit half a foot or a foot deep, then you can tell by chemical analysis; but, with a deposit of an inch or two, it is so much washed out by the rains any way, that I do not think there would be any way of saying positively, by chemical analysis, what that deposit is. But the fact of there being so much sewage there, and the fact that one knows that sewage always does deposit, it is very safe to infer that the deposit is largely due to the sewage.

Q. — Have you made such examination as to satisfy you that the public health of Millbury and the region around there may be impaired in consequence of this? *A.* — My opinion would be that it is to a certain extent. I could not say how far without more thorough examination. I should want to go about there pretty minutely; and, in fact, I should want to have lived there during a season to be able to judge on that point.

Q. — Suppose the death-rate of Millbury should be shown to have rather improved on the whole for a period of ten years: what should you say that indicated? *A.* — I should not think it necessarily indicated any thing. For instance, here is a case in point which was reported somewhat fully a little while ago. They were having a heavy death-rate from typhoid fever in Paris, where, as actual observation showed, that the number of cases was smaller than usual; so that the death-rate does not necessarily indicate the amount of sickness. I should not expect any fatal disease from a nuisance of this kind. I should expect something which would cause, perhaps, temporary and slight troubles, or a certain amount of impairment of the health, which would make people more subject to serious troubles when they came along. I should doubt very much indeed whether a thing of that sort could be traced in the death-rate. In London, for instance, one year when the Thames was most offensive, — so offensive that the House of Commons were actually driven out of their room, and had to give up their session, — the death-rate from typhoid fever and that class of diseases in London happened to be smaller than usual; so that I think you cannot infer absolutely from the death-rate what the condition of a town is with regard to the minor diseases, which perhaps prevent people from working, or make them uncomfortable, but do not kill them.

Q. — Have you examined much into the matter of the practicability of any thing being done by the city of Worcester to dispose of its sewage by irrigation or otherwise? *A.* — I think I have seen all the methods that are in use in different parts of the world; that is, nearly all, and a number of illustrations of most of them. With regard to Millbury, when I was a member of the State Board of Health, it was a matter which the board looked into a good deal, — the purification of streams. I visited Worcester and Millbury a number of times; and, of course, last year I went there quite a number of times.

Q. — Do you consider such a plan practicable at no very inordinate cost? *A.* — I think there is no doubt about that. I think the matter has been successfully accomplished in so many different ways, in so many parts of the world, and for so many years, that it is a matter upon which I can positively say that there is no doubt but that it can be done.

Q. — With regard to the matter of cost, I suppose that is a matter that you would be likely to say engineers could tell us better about than you? *A.* — I went over all the matters of the cost with regard to this, very minutely, with Mr. Davis. That was the part he was most familiar with, from an engineering point of view; and I happened to have seen more of the practical working of the thing in the farms. Of course, as to the details of engineering, I should not express an opinion absolutely; but I have been over so many estimates of expense with Mr. Davis, and I have seen so many times that his estimates come rather under than over the fact, that I feel very confident that the figures in that report are not overstated.

Mr. MORSE. If I may make a suggestion, Mr. Chairman, the doctor probably assumes that the Committee have had opportunity to read his report; but I judge from the Chairman's statement, at the beginning of the hearing, that has not been done. So, that if Dr. Folsom would be kind enough to state substantially what the recommendation is that the Committee make, as to the mode of purifying the sewage, it may enable the Committee to understand it.

The CHAIRMAN. It is my intention, as well as the intention of the rest of the Committee, to read that report, every word of it; but I have not had time to read it as yet.

WITNESS. In the case of the precipitation scheme, the estimated cost was $343,840, with a probable deficit of from $10,000 to $15,000 a year; that is, it would probably cost from $10,000 to $15,000 a year more to keep the thing going than would be brought in. In the case of the irrigation method, the estimated cost was $408,490; and the deficit in that case would vary very much indeed, depending upon the season. I think, under favorable circumstances, there might be no more deficit than the interest on the cost of the works; but it probably would vary from $1,000 to $6,000. I think in unfavorable years one would expect — including the cost of pumping, which would have to be done for part of the sewage — that there would be that amount of deficit. I see by Col. Waring's method the estimated cost is very much less, — $206,500.

Mr. MORSE. Now, doctor, if I may interrupt again, would you be good enough to state to the Committee what are the features of the two plans that you have given estimates of, — what are the methods proposed?

WITNESS. The first consists of a method which has been adopted in a few places in the world, where the cost of land is so enormous, and where the owners of land have such absolute rights over it, that land cannot be got. As an illustration, I might cite the case of the city of Birmingham. Sir Charles Adley, who was a member of Parliament, was one of the greatest land-owners in the vicinity of the city. The sewage was thrown into a small stream, and it was by his exertions that a bill was passed through Parliament requiring the city of Birmingham to take their sewage out of the river. He happened to own the only land that could be used for purposes of irrigating, and he refused to sell that land at any price. The city was then in the awkward position of being compelled by Parliament to do something which then was thought impossible to do. In order to get over that, they had to use some one of these precipitation schemes. Wherever this precipitating process has been used, there has always been some such difficulty as that; and the reason of its not being used is, that it is more expensive than irrigation, and that a large quantity of "sludge" is deposited every day which contains about ninety per cent of water, and is very much in the condition of street-mud, after a heavy rain, in a town where the streets are not paved. Any process of drying that is expensive, and then it has to be carted away; and, when it is carted away, it is of so little value that farmers will give almost nothing for it. At some time of the year, when there is nothing doing, they will give a shilling or two a ton for it; but, commonly speaking, it has to be given away.

The first method was to carry the sewage to a point so far from Worcester, where precipitating tanks could be constructed for precipitating this sewage, that the smell would not annoy the Worcester people. It was then to be mixed with lime; and the solid part of the lime, by mixing with the organic matter, when it settles carries the organic matter down with it, and also the sand and dirt, and every thing of that sort. There is to be a series of tanks for that purpose, so that after the sludge has settled to a depth of about a foot and a half in one, another set can be used, and the sludge from the one can be carted away. I should say, as an illustration of the condition of that sludge about that time, that from some of the precipitating tanks, instead of being shovelled out, it is pumped out, showing what a fluid condition it is in; and it then contains, as I said, about ninety per cent of water. In some parts of England, where this is done for small places, they can carry it out on land, and let it dry in the sun; but it is offensive, and of course it requires a very large area of useless land to be used for that purpose.

We considered that scheme as of very much less value than the other; because, although the first cost would be a little less, the annual deficit would much more than make up the difference between the two. And I think one may say that that has been pretty thoroughly abandoned everywhere, excepting in some such exigency as that of Birmingham, where an Act of Parliament has driven people to emptying the sewage out of the stream, and where they have but a very limited amount of land. On the continent of Europe experiments have been made in France, Germany, Austria, and Belgium; but they have universally thrown it aside as impracticable compared with the others.

The other plan which was suggested was, to carry the sewage of the city proper down to small tanks which would simply remove the heaviest part of the sewage — very small indeed — by gravity flow, and that from the high district by pumping, uniting the two, and carrying them to a point far enough down to prevent any smell being offensive, — I think there would be very little indeed, if any, — and disposing of it on about seventy-five acres of land specially prepared, so that a very large amount of sewage can be used to the acre, — some forty thousand gallons to the acre. On ordinary land, not specially prepared, only about three thousand gallons can be used. It was thought, if this proved successful from an agricultural point of view, that the farmers owning the land in the vicinity would desire to have a certain portion of the sewage carried on by conduits to their land, which they could use. The whole plan would then consist in taking this area of seventy-five acres, which would dispose of the whole of the sewage of Worcester, and allow farmers in the vicinity to take any of the sewage on their land, if they found it for their interest to do so. I think in the course of a year or two they would very decidedly prefer to have the sewage carried on their land. There is very little land of much value about there: it is mostly poor land. There is a little good land down by the stream. I think in a very short time the farmers would find it to be for their interest to have a portion of this sewage carried on to their land; and my own impression is, that that would more than supply the increase in the amount of sewage of Worcester from year to year for the next twenty-five or fifty years.

Q. (By Mr. MORSE.) — Let me ask this also, doctor, whether your plan assumes the building of a separate sewer or sewers to conduct the sewage proper distinct from the ordinary flow of the stream? *A.* — Yes, sir: I think that would be absolutely necessary. The greatest measured flow of Mill Brook is 110,000,000 gallons a day. Mr. Davis and I went over the estimate very carefully, — the area of the water-shed and the size of the sewer; and we supposed that there must be, in extreme freshets, 1,000,000;000 gallons a day go down the stream. The driest flow which has been measured is 750,000 gallons; but of course that is for a very few days: it would be an exception. If even 40,000,000 gallons a day — which is not an uncommon flow — had to be disposed of on the land, it would be simply impracticable: it could not be done with any reasonable amount of land which could be got. We went over that estimate very carefully. I see that Col. Waring differs with us on that point; but our chief reason for recommending an entirely separate system of sewers was, that in our opinion, after going through the Mill Brook sewer pretty carefully, and estimating its size, etc., it was taxed now to its full capacity in freshets, and that any thing which would retard the flow of the water at those times, or any thing else which would diminish the size of the sewer, such as a culvert running through the centre, would run too much risk of so interfering with the function of the sewer as to choke up the lateral sewers and give some trouble in the cellars. Of course, I say, there is a difference of opinion on that point.

Q. — Does your estimate which you have already given of $409,000 include the cost of the new sewers? *A.* — It includes the cost of the new sewers, which would be about $181,500. My impression is, that the city of Worcester will be compelled to do that in the course of time for their own safety, or at least something similar to that, because the present condition of things cannot go on. There is no doubt about that. They may be able to do it in a modified way at less expense, but they certainly cannot run sewage through that present Mill Brook culvert without doing something different. The bottom is very uneven; the sewage deposit is enormous every year; and they certainly cannot put that into the condition that it ought to be, to carry sewage alone, for less than $100,000. That would be my opinion, perhaps. Mr. Davis's opinion was, that a very large part of that, if not the whole of that, would have to be done by the city of Worcester at some time or other at all events, or at least some improvement upon their present method of carrying their sewage down through their main sewer. The present condition of things cannot go on.

Q. (By the CHAIRMAN.) — That is to say, if they were not looking at all at its effect on the people below them, the necessities of their own case would require this to be done? *A.* — Entirely independent of the disposal of the sewage, the simple necessities of the city of Worcester itself would require this to be done. A good deal of it is badly constructed.

Q. (By Dr. CHAMBERLAIN.) — I want to see if I understand you correctly. I understand you personally have no doubt with regard to the injury to manufactories; and I also understand you, that you have no opinion to give, or that you don't know that it is detrimental to the public health in the town of Millbury? *A.* — That is not exactly what I want to say. I could not say there is no question in my mind. The evidence to me is conclusive that the mill-owners are very seriously injured; and my own opinion is, that this condition of things is, to a certain extent, injurious to the health of the people below Worcester. The point in regard to which I should not want to express a positive opinion would be the amount of injury to health. I should not be able to decide that point without living in Millbury continuously during the summer, or spending a couple of weeks in examining pretty carefully the situation; but that there is a certain amount of injury to the public health I have no doubt.

Q. — From what little you do know, doctor, do you suppose that the amount that goes into the river is a detriment to the public health? *A.* — I have no doubt of it. I think that is a point which the experience of other countries has settled pretty conclusively. In Croydon irrigation has been used fifteen or twenty years, — I do not remember the exact number of years, — and the condition of the public health has improved there amazingly since its introduction. The sewage and water-supply were introduced about the same time. Of course, they had their influence; but it is universally known, that, whatever amount of influence such things have on the public' health, it is something. As I said before, I do not think such things usually show in the death-rate. They are of that kind of things that do not kill people.

Q. (By the CHAIRMAN.) — I understood you to say, that, although the effect was deleterious, it was in a measure offset by the general purity of the country air? *A.* — I think I did not say exactly that. I think I said, although there was a certain amount of injury to health there, it could not be so great in a country town as it was in a city, on account of the purity of the air.

Q. (By Dr. CHAMBERLAIN.) — What do you mean by the statement that Worcester cannot long do this? Do you mean because of its injury to the manufactories or to public health? *A.* — No, sir. I say they cannot long continue using their present main sewer in its present condition; for the present main sewer will need reconstruction, entirely independent of what they do with the sewage, when they get it to the outlet. Their main sewer will need a good deal of money spent on it.

Q. (By the CHAIRMAN.) — I will trouble you with only one more question. What is the general condition of that stream as affected by the sewage of Worcester, as far as your observation goes, compared with what it was ten or fifteen years ago? is it growing worse? *A.* — No doubt of that at all.

Cross-Examination.

Q. (By Mr. GOULDING.) — I suppose, doctor, your opinions upon this subject from your investigation are very fully stated in the report; that is, you did not come here to modify your statement? *A.* — No.

Q. — Or add any thing to it at all? *A.* — I came at the request of the Committee. I did not know what questions they were to ask me.

Q. — You have nothing to add to, or take from, or to modify that report? *A.* — No, sir. If I have information the Committee desire, I am very glad to give it.

Q. — I want to know whether you wrote the expert report yourself? *A.* — It was written by the commission. I wrote parts of it.

Q. — It expresses your opinion on the subject fully and completely, and you don't wish to modify it. *A.* — I don't know. Probably there is scarcely a single paragraph that was written entirely by one person. The MS. was gone over and corrected.

Q. — Carefully revised, with a view of expressing the exact opinion of the Committee? *A.* — Yes, sir. I think it does express the opinion fully. I think there is one typographical error, however. I did not happen to see the proof. In one place it says the annual cost of treating the sewage is between ten and fifteen thousand dollars, and in another between five and ten thousand. That is simply an error.

Q. — How long ago, doctor, did anybody anywhere, any city, undertake to dispose of its sewage in some other way than by emptying it into a running stream or the sea? *A.* — The first where any thing systematic was attempted was the city of Edinburgh very nearly two hundred years ago.

Q. — Where next? *A.* — My impression is, the next place where it was done on any large scale was Croydon.

Q. — When was that? *A.* — That was not far from twenty years ago, somewhere from fifteen to twenty. I should say perhaps that the attempts in the two places were made for entirely different reasons. In Edinburgh the owners of the land received some concession from the Scotch Government, which it is impossible to get rid of at the present day, simply from an economical point of view. They had very large dairies; and this sewage was used to irrigate and fertilize the land, and it has been enormously profitable. The matter in Croydon came up first as the result of the parliamentary investigations, and the pressure to purify the streams. In Milan there has been irrigation for a great many years, with a stream running through the city; but that is entirely independent of its containing sewage. The land needed the irrigation; and the amount of sewage in the stream was very small, consisting of street-washings, because it was not a water-closet town.

Q. — The Croydon method is by irrigation, I believe? *A.* — Yes, sir.

Q. — Is that as fair a specimen of the favorable results that you get from this kind of treatment as any that you know of? *A.* — Do you mean from a sanitary or economical point of view?

Q. — From the sanitary point of view. *A.* — Yes, I think it is fully, where so large an amount of sewage is used. It is the only city of that size where its sewage is disposed of in that way in England. The amount of sewage in the city of Paris is much larger; but that is the largest town in England that uses the whole of it.

Q. — How much did you give as the flow of the Blackstone River? *A.* — I think the largest gauged flow was 110,000,000 gallons a day.

Q. — In Blackstone River or Mill Brook? *A.* — No: I think that was in Mill Brook. I won't be quite sure of that.

Q. — Isn't it Piedmont district you are thinking of? *A.* — No: I think we are right about that. I remember that was Mr. ———'s statement. I think that was right.

Q. — The flow of Blackstone River is 110,000,000 gallons a day? *A.* — No: I think that is Mill Brook at the pond, sir. The average flow for the year was about 18,000,000 gallons.

Q. — Now, with regard to the death-rate, doctor, what is the object of collecting these death-rates, and reporting them year by year? Is it not with reference to ascertaining the condition of the public health? *A.* — Undoubtedly. The conditions, of course, which affect the public health are so various that it is almost impossible to pick out any one factor and say the death-rate is lower from this cause, or higher from that cause. The only thing a person can do with regard to that is to estimate as nearly as he can.

Q. — The Blackstone River, I take it, is not in any such condition as the Croydon sewage before it was treated? *A.* — The Blackstone River or Mill Brook?

Q. — Blackstone River in Millbury? *A.* — The sewage of Croydon, if I remember correctly, discharged into a stream which must be considerably smaller than Blackstone River, — it did before the sewage system, — but still it is quite a considerable stream. I remember I saw it in the summer-time, and it was then nothing but a mere brook.

Q. — It originally discharged into that stream without treatment, and twenty years ago they began the treatment? *A.* — Yes, sir: about that time. I wouldn't be sure of the number of years.

Q. (By Dr. HARRIS.) — Is it, or not, a pretty well-settled principle now that sewage should not be turned into any running stream? *A.* — Well, I think that is very largely a question of practicability. The German Government has passed very stringent laws upon that point; and some of the larger cities of Germany have held off from introducing sewerage on that account, because they cannot discharge into the rivers. It seems to me that each city must judge by itself if the evil is less from discharging it into the stream than by not having sewerage at all. If the stream is large enough, and the conditions are proper, it should be discharged into it. It is only a question of practicability.

Q. — You stated, I think, that you noticed that the breathing of this sewage air by the people of Millbury had a depressing effect upon the inhabitants, and did affect their health to some extent? *A.* — I did not state that as a matter of fact. I stated that as a matter of my opinion. I don't think there are any facts which one can show to positively prove that; but in going about among the people there, seeing the conditions, and talking with them, as we did one morning, I convinced myself that it had a certain amount of influence on the health: how much I should not care to say.

Q. — And probably indirectly upon the death-rate? *A.* — I don't know about that: possibly.

Q. — Would it not be the natural sequence, that if a person was not in a good condition to resist disease, and was exposed to this atmosphere, and various causes which induce even acute diseases, he would be less able to resist them if his system was depressed? Would it not necessarily result, in your opinion, that there might be more deaths? *A.* — I think that is quite possible.

Q. (By Dr. WILSON.) — Did you, in the investigations of your committee or board, go into the question as to what constitutes a nuisance, — whether caused by sewage, or caused by something else, — some other substance in the river? *A.* — I think there are two features which the committee observed in looking over it. I think there is a certain amount of deposit, independent of sewage, that cannot help being very impure in a river of that sort, and water-shed of that kind, of course.

Q. — Did you undertake to estimate the proportion of each? *A.* — There is no way of doing it. You cannot tell by chemical analysis, and you cannot tell by the appearance of the deposit. There is no way of doing it, as far as I know.

Q. — From what you saw, should you judge that was mainly due to the sewage, or mainly due to other material? *A.* — I think the trouble in the mills is chiefly due to the sewage; and, of course, in the sewage of Worcester is included their manufacturing refuse. And what has caused the filling up of some of the dams, diminishing the area, I think must be due to other causes besides the sewage, as well as to the sewage. In a city like Worcester, there must be an immense filling of sand, gravel, earth, etc., carried down the stream independent of there being sewers connecting with water-closets, etc. It is difficult to estimate how far it is the product of one or the other.

Q. — Perhaps I should say I mean by "sewage" the house-sewage, not washings from the street, and not the material from manufactories, but sewage proper. *A.* — Well, it is impossible to separate the factors in such a case as that, and say how far the trouble is due to one part of the sewage, how far to the manufactories, how far to street-washings, and how far to sand, loam, or earth. I don't know of any way of doing it.

Q. — You have no doubt all these different elements, of course, do contribute in producing the condition of things that exists? *A.* — Every thing excepting the smell. Of course the smell is produced by decomposing organic matter, and that comes from the sewage proper.

Q. (By Dr. HODGKINS.) — Did you make any examination of the river for the purpose of ascertaining what proportion of the pollution was caused by the manufactories? *A.* — That varies a good deal from time to time, and the same mills would discharge very variable quantities of refuse matter. At the time Singletary Brook discharges into Mill River, the pollution is so variable that there are six or seven kinds of impurity. We were not able to make any chemical examination which would show how much manufacturing refuse there was in the sewage, — a point rather difficult to show from chemistry.

Q. — One other question. What effect upon the water of the brook does the coloring matter used in these mills, in your judgment, have upon the pollution of the stream? *A.* — Well, is there a good deal of coloring matter used there?

Q. — There is some: yes. *A.* — I don't remember that there were many that used a great deal of coloring matter. There were some where there would be a certain amount of refuse. From the iron-works, — twenty-six, I believe, all together, — there would be a certain amount of acid, of course, from them.

Q. (By the CHAIRMAN.) — It has been testified here that they cannot get cloth as white as it ought to be; and some question arose whether the trouble was due to the Worcester sewage, or perhaps the iron-works. *A.* — I should think the mill-owners themselves could answer that question better than I can. They know the kind of water they need, and the precise character of the trouble.

Q. (By Dr. HARRIS.) — Do the chemicals which they use in their coloring — soda-ash and potash — necessarily pollute the water? *A.* — Used in the mills by them?

Q. — Yes, sir. *A.* — Any thing of that sort wouldn't injure the water, perhaps, for cleansing. Sulphuric acid might, and some other chemicals than soda and potash.

Q. — I mean its effect upon the health, producing this bad smell in the water: whether those chemicals would necessarily produce a bad smell in the water? *A.* — I should say from the character of the mills there, that, if any thing, those in Worcester would rather have an opposite effect, — an effect as disinfectants. I am not sure upon that point. I think in our report we don't state the ingredients used. I think we state the character of the materials the mills use.

Q. — The Washburn Mills, I understand, have used a great deal of sulphuric acid in cleansing wire. I wish to inquire about that being in the water: whether it would render it impure in relation to public health, so far as the public health is concerned? *A.* — I should say not. A number of woollen and cotton mills, I should say, would discharge a certain amount of dirt in the river, those that do any cleansing.

Q. (By Dr. WILSON.) — It has been said that certain machinery, wheels, etc., had become corroded by the water. Now, is that due, in your opinion, to sewage properly so called, or due to the discharge from some manufacturing establishment or establishments? *A.* — A certain amount of that would be due to sewage. The general process of oxidization, if brought in contact with the iron, would produce a certain amount of oxidization, but not great. I should say it was more likely to come from chemicals in the water.

TESTIMONY OF COL. GEORGE E. WARING, JUN.

Q. (By Mr. MORSE.) — Col. Waring, what is your profession? *A.* — Sanitary engineer.

Q. — Will you state to the Committee what attention you have given to sanitary matters? *A.* — My attention has been devoted very largely to works of sanitary improvement for the last sixteen or seventeen years, and in a general way for fifteen years before that.

Q. — Well, without asking for too many details, will you state generally to the Committee what works you have been connected with, what opportunities you have had for observing what is necessary, and what can be done for sanitary improvements? *A.* — The first public work with which I was connected was the drainage of Central Park in New York. From there I went into the army; and since the war, or rather since 1867, I have been more or less occupied with giving advice as to the improvement of houses and towns, making plans for sewerage, and investigating with reference to the disposal of sewage where there was no good natural means for the inoffensive delivery, or where that means of disposal was an alternative to be considered.

Q. — In the course of your study and work, have you gone abroad and studied the systems in use there? *A.* — Yes: several times.

Q. — How extensive an observation have you had there? *A.* — I have seen the works at a number of towns in England, at Paris, Berlin, and at Dantzic on the Baltic.

Q. — Now, coming down to our case here, will you state to the Committee when your attention was first called to the Blackstone River? *A.* — The subject was first brought to my notice, I think, about the first of August last; and soon after that time I made an examination of the ground.

Q. — That was at the request of the town of Millbury? *A.* — That was at the request of the town of Millbury. I was about leaving for Europe, and was therefore obliged to leave the collection of engineering details to an assistant.

Q. — What personal examination did you make of the river? *A.* — I went from Worcester, from the outlet of the main sewer, as far south as Burling's mill-dam. My attention was directed chiefly to the character of land that might be used for the purification of the sewage.

Q. — You went to the factory, did you? Burling's Mill? *A.* — I went to a mill; but whether it was Burling's Mill, I don't know.

Q. — You may continue, if you please, and state further in regard to this matter. *A.* — My attention was given chiefly to the character of the land available, with or without pumping, for the purpose of sewage. Incidentally, in the course of the examination, I saw the condition of the Blackstone River at different points. I was made cognizant of the fact that it seemed to be in a polluted condition; but my professional attention was not called or directed specially to the degree of pollution. The question submitted to me was one which assumed that the river was polluted, and that means were to be adopted to remedy the pollution. It was only with reference to that, I visited the ground.

Q. — Have you, Col. Waring, with a view to stating concisely your views here this morning, prepared a statement? *A.* — Yes: I was requested to put in form my general views on the subject.

Q. — Won't you read it, or make such use of it as you please, and then state to the Committee the results of that examination? *A.* — It will perhaps state what I should say discursively in a much more concise manner.

So far as I understand the scope of the question to be considered by your Committee, it relates to the Act of the Legislature of 1867, authorizing the city of Worcester to change the channel of Mill Brook, and convert it into a common sewer, as giving an implied right to use that sewer in such a manner as to deliver foul and offensive matter into the natural water-course of the Blackstone River. Also to determine to what extent it is practicable for the city to use the drainage rights thus conferred without such pollution of the Blackstone River. I assume that it is with reference to these points that I have been called upon to testify, not as to the general question of the right of a riparian occupier to pollute a natural water-course in such a manner as to affect the health or the comfort of those past whose lands the later course of the river flows. This portion of the general question either has been or must be decided as a question of proprietary right, in which, as I take it, legislation must be subject to well-established principles.

Under all ancient practice, a sewer is only a drain, a channel for the removal of waters which the proper enjoyment of territory requires to be removed. Until well into the present century, this was probably the only meaning of the term; and up to that time the office of a sewer was simply to furnish a safe outlet for rain-water, for soil-water, for the overflow or backing up of streams, etc. The use of these sewers for the removal of excrementitious and other refuse matters is very recent. In Boston, according to Mr. Elliott C. Clark, as late as 1833, and in England much later, the admission of foul matters was prohibited. The use of common sewers for foul drainage is an assumption of recent date, which has grown up largely through neglect, and with no well-determined conception of the ultimate effect to be produced. It is not at all in my province to consider the degree to which the Legislature, in giving this right to the city of Worcester, contemplated the use of the Mill Brook sewer for the removal of organic wastes. Certainly, under the circumstances of the case, the removal of rain-water and of subsoil water, so far as this was necessary, was contemplated; and such removal, in conformity with the long-recognized office of sewers, was proper, and could not be objected to by the inhabitants of Millbury or by other residents along the Blackstone River. The ground of their complaint relates, not at all to such legitimate use of an artificial channel of drainage, but entirely to what, under the circumstances, must be regarded as an improper use of the right of the city. It relates solely to there being added to the effluent water the varied off-scourings of a large industrial community. Neither is it within my province to consider the degree to which the Blackstone River has been polluted by these off-scourings: that is a question of fact, to be decided according to the testimony of those who are personally cognizant of it to a greater extent than I am. Probably, if the very existence of the Worcester community depended on the discharge of foul water into the Mill Brook sewer, and equally on the discharge of the unpurified effluent of the Mill Brook sewer into the Blackstone River, there would be a justification for the continuance of the present state of things. Even then that community should be allowed to enjoy this vital privilege only on the condition of making full compensation to all whose rights had suffered from its act. That the people of Millbury and others living along the Blackstone River do so suffer, I assume after a personal examination and from analogy. The leading question, therefore, is, Is the present condition necessary to the existence of the community? This question, I think, would be answered in the negative by any person who had paid more than casual attention to such subjects. There are two means of relief which it is open for the authorities of Worcester to adopt, either one of which would keep out of the Blackstone River the impurities complained of. The first would be to withhold from this common sewer, and, incidentally, from all of the sewers of the city, all manner of foul substances. The streets may be submitted to such a system of scavenging as will prevent the accumulation of horse-manure and other offensive matters on the surface, to be washed into the sewers by the rainfall; every house may be disconnected from the sewer; its liquid wastes may be collected in tight cesspools, to be emptied by pneumatic process, and its garbage may be cared for by separate removal; every factory which produces foul wastes may, without oppression, be compelled to adopt some of the known suitable methods for purifying the water which its processes have made impure. This being done, the water discharged by the Mill Brook sewer, which will still serve its legitimate purpose of a sewer, will be substantially purified; and all cause for complaint on the part of the residents along the Blackstone River will cease.

Lest this be considered an unprecedented condition, I beg to refer to the report of Sir John Hawkshaw on the purification of the River Clyde. In his general recommendations he says, "I have not attempted to enumerate all the various kinds of work and manufactories which contribute to the pollution of the streams and rivers of the Clyde district, nor all the mechanical or chemical processes which have been tried or recommended as means for purifying the liquids hitherto run to waste in such works and manufactories. The works, and many of the processes in question, are described in the valuable reports of the River-Pollution Commission, to which I beg to refer. I have made myself acquainted with previous inquiries, and have obtained information enough, I think, to justify me in believing my recommendations to be practicable of application, and that if they are adopted they will secure the end in view. Those who own and manage works and manufactories have been wholly under control, and can therefore provide and enforce the necessary provisions and regulations for removing the nuisances which they so often create. . . . It should be enacted that solid and dry refuse, not including under that term fæcal matters, should immediately cease to be conveyed or thrown or placed so as to fall or be carried into the Clyde or its tributaries. It should be enacted, that, after a definite period, which might be fixed at eighteen months, no fæcal matter or urine from manufactories or public works should, except under special license from sanitary commissioners, be passed into the Clyde or its tributaries. Within eighteen months it should be practicable, in most cases, to make provision for keeping back these offensive matters."

He would throw upon manufacturers the duty of purifying their manufacturing waste, and he recommends earth-closets for the domestic use of the people.

In Paris, until very recently, the water-closet matter, and even the chamber-slops of houses, was by law delivered into tight cesspools, to be emptied from time to time. Even now there is only an insignificant exception in the case of persons who adopt a certain prescribed straining apparatus, which allows the liquid portions thus produced to pass into the sewers. In many towns in England, generally as the result of judicial or legislative restrictions, the devices above indicated, or their equivalent, have been adopted, and are systematically carried out, with the direct purpose of preventing the pollution of rivers. In nearly every city on the continent of Europe, sometimes with this object, but more often with the view of preserving a valuable manure, there is and always has been an entire withholding of such wastes from the sewers, which are constructed to remove storm-water only. In fact, more precedent by far can be found for the above-prescribed course than for any other method of treating domestic and industrial wastes. Please understand that I do not make this suggestion as a recommendation. I realize very fully that for this restriction to be placed upon a community like that of Worcester would be nothing less than an economical and sanitary calamity. It would inevitably lay a cumbersome tax on all its people, and would lead to serious injury to the public health. I suggest it only as a possible means by which that community may, without sacrificing its existence, and without destroying its property, concentrate upon itself the disadvantages which it seems not averse to inflict upon others.

Fortunately no such radical and retrograde action is necessary. There are other means, well known, long tried, and fully demonstrated to be successful, by which the trouble complained of can be averted at small expense, leaving the city in the full enjoyment of its costly sewerage works. There can, of course, be no objection, at least, no legal objection, to the delivery of all the foul drainage of Worcester into the Mill Brook sewer. That whole sewer lies within its boundaries, and is under its exclusive jurisdiction. The people of Millbury have no right, and, as I understand it, they have no desire, to interfere in any manner with the use of the sewer. They do claim, and it seems to me that they have the right to claim, — certainly other peoples have made the claim, and have been sustained in it, — that before the water of the sewer is restored to the Blackstone River, a river originally pure, but now polluted, they shall take from it again that which, for their own convenience but without real necessity, they have added to it. There are many ways in which these objectionable matters may be withdrawn from the water of the sewer before it is returned to the river. The experts of the State Board of Health have indicated one means, I have indicated another; and there are still others, some of which competent engineers might consider preferable to either of these.

As the authority of the Committee and of the Legislature over this matter probably stops short of the prescription of the particular means to be adopted, it is hardly worth while here to discuss the details of any of these plans, or to do any thing more than to show that one, or several of them, is capable of affording the relief sought. Fortunately this is an easy task. The entire sewage of Dantzic on the Baltic Sea, where the climate is quite as severe as anywhere in New England, has the entire effluent of its very complete system of sewerage well purified, winter and summer, by surface irrigation. About one-eighth of the sewage of Paris, made very foul by the removal of street-dirt in a putrid condition by the sewers, and by the very considerable contamination coming from public urinals, and other sources, is perfectly purified by agricultural processes on the plain of Gennevilliers. A large portion of Berlin now sends all of its sewage to the irrigation-fields at Osdorf, where it is completely purified. Croydon in England, which is a larger city than Worcester, has most successful purification-works close to its border. The great health resort, Malvern, purifies its sewage by intermittent filtration. So does Kendall in the North, Leamington, and Rugby use broad irrigation. Over fifty other towns in England purify their sewage in a similar manner. The places named I have visited personally, and have made a careful examination of their purification-works. I might cite other towns where satisfactory purification is effected by chemical processes; but these seem to me so unsuited to the conditions we are considering, that the discussion of chemical purification is hardly worth while. Suitable works on either of the plans submitted can, with entire safety, be adopted for Worcester. Mr. Rawlinson, the Chief Engineering Inspector of the Local Government Board, says, in his "Suggestion as to the Preparation of District Maps, and of Plans for Improved Sewage," etc., "It is persistently urged by some parties that fluid sewage corrupts the soil over which it is spread, and produces malaria in the atmosphere over a sewage-farm. . . . The facts are, that continued irrigation, with foul, corrupt sewage in excessive volume, for very many years, as at Cragintinney Meadows, near Edinburgh, has failed to produce a sewage-swamp to corrupt the soil or to produce malaria injurious to health." That there is any peculiarity in the climate or in the soil of Worcester which indicates a special difficulty in the adoption of the processes of purification by agricultural treatment, is clearly disproved by the long and satisfactory experience in this very manner in connection with the insane hospital located there. I think all engineers who have given attention to the subject will agree with me in the broad proposition that there is no reason why the effluent of the Mill Brook sewer may not be made practically pure for an outlay much less serious than would be that required for the only permissible alternative, — the withholding of all the foul wastes of the population of Worcester from the sewers of that city. We probably should not agree as to the best means for accomplishing the desired result; but on the main question there can be no difference of opinion. Probably, also, there would be no difference of opinion as to the proposition that this purification may be accomplished for a sum much less than the amount of the damage inflicted, and hereafter to be inflicted, on the population of the Blackstone Valley, within the State of Massachusetts, leaving out of the consideration the suit for injunction which is quite sure, if relief is not found, to be instituted by the adjoining State of Rhode Island.

Q. — That last sentence leads me to ask you whether, as a resident of Rhode Island, you have any knowledge of the feeling of the people of that State? *A.* — The subject has been discussed among those interested in sanitary matters; and, as I understand it, the people in the north part of the State are apprehending serious trouble from the pollution of the Blackstone River, ascribing that pollution, of course, not only to the city of Worcester, but to the population along its whole course.

Cross-Examination.

Q. (By Mr. GOULDING.) — You are retained as engineer by the town of Millbury? *A.* — Yes, sir.

Q. — Have you ever studied law? *A.* — No, sir.

Q. — Have you given any attention to equity practice at all? *A.*

— I have never been a law student. I have had necessarily to study the legal bearings of questions involved in cases like this.

Q. — What works upon that subject have you studied? A. — I can't give the titles of them, sir.

Q. — You hardly would want to claim that you are a legal expert? A. — Decidedly not.

Q. — You were employed to report a plan for the purification of this sewage, which has been published in the Report of the State Board of Health and Lunacy and Charity, I think? A. — Yes, sir.

Q. — That expresses your view pretty fully upon the subject, does it not? A. — I can't say it expresses my view very fully. I necessarily made that report in great haste; and, if I were to re-write it more deliberately, I should add very much to it.

Q. — Do you desire to modify any opinions you expressed there? A. — I do not.

Q. — You speak about an injunction. Do you mean to say that proceedings are being instituted, or about to be instituted, to enjoin? A. — Not at all. Only in the natural course of events, as the population of the valley of the Blackstone River in Massachusetts increases, if it continues to throw its off-scourings into the river, the people will do something to prevent it. I only spoke of it as an injunction quite sure to be asked for.

Q. — You have no idea of the court it would be issued from? A. — From the United States court, I suppose.

Q. — You speak about the use of sewers for the purpose of carrying off foul matter as a recent thing, if I understand it? A. — Yes, sir.

Q. — Do you know when the Fleet-street sewer was built? A. — I don't know what year the Fleet-street sewer was built; but I know that long after it was built it was part of the system that household waste and excrementitious matter was excluded from.

Q. — Fleet Brook was a pretty clean brook originally? A. — It was a brook running through a dense population, and there was no sewage in the city.

Q. — When was Ludgate-hill sewer built? A. — That I couldn't tell you.

Q. — Wasn't it built more than two hundred years ago? and didn't it carry off the excrementitious matters of the city from the beginning? A. — No, sir: it only carried off so much as washed from the surface of the ground.

Q. — Mr. Rudolph Hering is a pretty good and reliable sanitary engineer? A. — So far as I know, he is a very careful man.

Q. — He gave a report recently to the National Board of Health upon this whole subject? A. — Yes, sir.

Q. — You are familiar with it, I suppose? *A.* — I haven't read it. I have run over it, and read portions of it here and there. Strictly, I have not read it; but I am not entirely without knowledge of it.

Q. — The Croydon farm you have seen? *A.* — Yes, sir.

Q. — That is a good specimen of a favorable result, is it? *A.* — Yes, sir.

Q. — How large was the stream into which the Croydon sewage is emptied, after treatment, or before treatment, prior to the construction of the works? *A.* — At the point where I crossed it, I should say that the footbridge was about fifty feet long, and that the river occupied pretty nearly that width, perhaps forty feet, and was a clear, rippling stream; what we should call here a large brook.

Q. — What did they call it? *A.* — The Wandle.

Q. (By Mr. MORSE.) — How large a place is Dantzic? *A.* — Dantzic has a population of a hundred thousand, I think.

Q. — You speak of the climate in your report: how does that compare with the climate of Worcester? *A.* — I think that they have colder weather at times in every winter than they have ordinarily at Worcester. I think their extremes are somewhat colder. They have a long-continued and very severe winter. Their temperatures are kept by a scale that I never have encountered before, and I have never made any comparison between that and the Fahrenheit scale except as to the extremes.

Q. — In cold weather when the ground is frozen, is it practicable to use the irrigating process? *A.* — Yes, sir; or rather the irrigating process very largely prevents the freezing of the ground.

Q. — So that you would consider in our ordinary winter weather it would be possible to use this same mode of purifying the sewage of Worcester? *A.* — Perfectly so, sir.

Q. (By Mr. GOULDING.) — At Berlin don't they have large storing reservoirs to retain the sewage during the winter? *A.* — No, sir. The sewage of the principal works which I saw, which serve about one-fifth of the city, have a pumping-well, I should say, of about the capacity of this room. The pumping is emptied directly into pipes which discharge into the distributing channels of the sewage-irrigation farm.

Q. — The question I asked was, whether they did not have a large reservoir to store, basins to store the sewage during the winter? *A.* — No, sir.

Q. — This statement with regard to the sewage at Berlin, is that correct? — " All of the sewage is to be pumped and distributed over two farms for irrigation and filtration, — one to the north-east for the part of the city north of the Spree; the other to the south for the southern portion. The former is at Falkenberg, six miles from the centre of

Berlin, with an elevation of a hundred and two feet above the Spree, and an area of two thousand and fifty acres. The level portions are used for filtration and winter storage, the gently sloping areas for the furrow system, and greater slopes for broad irrigation. A novel feature are the winter basins, into which the sewage is turned, after vegetation ceases, to a depth of two feet, and allowed to soak away."
A. — He is not describing the sewage: he is describing the downward filtration system.

Q. — These basins, then, are for the downward filtration system?
A. — Yes, sir. While I have read parts of Mr. Hering's report, I have no knowledge of that part of it to which you have referred.

TESTIMONY OF DR. HENRY P. WALCOTT.

Q. (By Mr. MORSE.) — Doctor, you have come in at our request, have you not, to state to the Committee the result of your observations of the Blackstone River? *A.* — Yes, sir. I have come in to answer any questions the Committee may desire to ask me with regard to the substance of this river.

Q. — How long have you been connected with the State Board of Health? *A.* — Nearly two years.

Q. — During that time what personal observation have you had of the river? *A.* — I have had a good deal. I examined the river previous to my appointment on this commission; and upon the commission I visited the river frequently, and saw it under many different conditions.

Q. — Will you state to the Committee whether or not, in your judgment, the condition of things consequent upon the turning of the Worcester sewage into the river is such now, and promises to be such hereafter, that it is likely to affect the public health of that valley? *A.* — I think it is.

Q. — Whether or not, in your judgment, it is an evil that requires some attention? *A.* — Yes, sir: I think it does.

Q. — Now, doctor, I assume that you can hardly add any thing to what you have said in your report as to the best practicable plan; but I wish to know, in the beginning, whether or not you are of opinion that a method can be adopted at a reasonable cost which will prevent the evil? *A.* — I think there is no question that some system of irrigation can dispose of the evil entirely.

Q. — What is your judgment as to the best system?

Mr. GOULDING. Is it expected the doctor is going to change his views?

Mr. MORSE. Not at all; but still, for the purpose of completing the report, I would like to have him state.

WITNESS. I have seen no reason to alter the recommendation of the report I signed.

Q. — In brief, what did you there recommend as the best system? A. — A system of downward filtration.

Q. — The one described by Dr. Folsom? A. — Yes, sir.

Q. — Would you agree with Dr. Folsom in the statement he made? A. — Entirely.

Q. — I want to ask in regard to one member of your commission, Mr. Davis, whether or not he is a civil engineer of great experience and high standing? A. — Yes, sir: I think there can be no question of his very high standing.

Q. — What is his present position? A. — It is not that of an engineer. He has some connection with a New York telephone company.

Q. — He was engineer of the Boston system of sewage? A. — Yes, sir.

Q. — This large system in process of construction now? A. — Yes, sir; and preparatory to that he was sent abroad, and made a very thorough examination of the most approved modes of sewage disposal.

Q. (By Dr. CHAMBERLAIN.) — Is it your opinion that the sewerage turned into that river now, — I don't mean twenty-five years hence, — is it your opinion the sewerage to-day is detrimental to public health? A. — Yes, I think it is.

Q. — To any considerable extent? A. — Not yet to a considerable extent.

Mr. MORSE. I ought to say, in reference to your suggestion about touching the matter gingerly, I only did so for the purpose of saving time. The report of the doctor was put in. I do not want to go through the details of it.

Cross-Examination.

Q. (By Mr. GOULDING.) — I suppose that you intended (it is hardly necessary to ask you the question), but you intended to express your views fully and clearly, not including any matters that were not properly relevant in your report? A. — In as few words as possible: yes, sir.

Q. — You haven't changed your views about it in any way? A. — No, sir, not at all.

Q. — Whether you didn't understand from the terms of the resolution that you were bound to report some plan or other, if there was any, to the Legislature? A. — I didn't understand that it was any part of the business of the commission to determine the fact as to the existence of a nuisance.

Q. — You assumed that some plan was to be reported to dispose of the sewage of Worcester without following the Blackstone River? *A.* — Yes, sir.

Q. — And you assumed the pollution? *A.* — I think we were not asked to report upon the fact of the pollution.

Q. — But to report a plan for purifying that sewage? *A.* — Yes, sir.

Q. (By Mr. MORSE.) — Is there any question about the fact that the State Board of Health for several years, including your own term, has reported that the river is polluted? *A.* — They have always used very strong language on the subject, and represent it as the most polluted stream in Massachusetts.

Q. (By the CHAIRMAN.) — What do you understand the question submitted to you was? *A.* — To devise a plan for the purification of the sewerage of the city of Worcester, and relieve Blackstone River. We were not asked, as I understand it, to determine the question whether there was pollution or not, as a preliminary to that report.

Mr. MORSE. But the doctor states that the Board of Health have reported for several years it is the most polluted river in Massachusetts.

Q. (By Mr. GOULDING.) — Their reports will show for themselves what they did report? *A.* — Yes, sir.

Q. — And when they have reported as to the pollution of that river, they have stated their proposition in the terms in which they intended to state their proposition, as far as you know? *A.* — Yes, sir.

Q. — And their reports will, therefore, show what they meant? *A.* — Yes, sir.

Mr. GOULDING. Then, we do not need any glossary upon it at present.

The CHAIRMAN. I think it was in the mind of the Committee (of course, we may not have looked over the whole subject) that your commission was to report to the Legislature what, in the judgment of the commission, it was advisable to do, taking into consideration the practicability of doing any thing, and also the amount of damage that was being done. I think the Legislature intended to cover the whole matter.

WITNESS. Mr. Chairman, you must remember this commission is the creature of the State Board of Health, and not of the Legislature. What the Legislature intended to do in the matter, I do not know. I have answered the questions with regard to the action of this commission.

Q. (By Mr. MORSE.) — Doctor, whether or not, in your judgment, from such observation as you have made of the river, it is in such condition that some system of purification ought to be adopted, with proper regard to the health of the inhabitants of the valley? *A.* — Yes, sir: I think it has got to that point when something must be done.

Q. (By Dr. WILSON.) — Do you feel pretty sure, from the investigations of your commission, that to purify the river, so far as relates to Worcester sewage, will relieve the nuisance? *A.* — Well, I think that any measure of that sort will have to be accompanied by further measures for the reduction of manufacturing waste. If the river is to be purified, it has got to be free from manufacturing waste as well as sewage.

Q. — If you should purify the city of Worcester, would the Blackstone River be a free river below the Singletary Stream? *A.* — If they don't purify the Singletary Stream, which is a very small stream, and the Blackstone is a large one, it would not be clean. The Singletary Stream is a very impure stream, — no question about that.

Q. (By Committee.) — Is the factory-waste a detriment to the public health, — this coloring matter? *A.* — The coloring matter I don't think is. I think the matters from washing wool are unquestionably detrimental. I don't think the logwood amounts to any thing.

TESTIMONY OF CHARLES E. WHITIN.

Q. (By Mr. FLAGG.) — You live where? *A.* — I live in Whitinsville, town of Northbridge.

Q. — You are engaged in business at Rockdale? *A.* — Yes, sir.

Q. — How far below the mouth of the sewer is Rockdale on the river? *A.* — Well, by road it is twelve miles from Rockdale to Worcester. I suppose the sewer is about two miles below. I should suppose, by the running of the river, probably fifteen miles, as it is a circuitous stream.

Q. — You have heard the testimony of those who have testified as to the condition of the river, — either heard it or read it. Does your experience in a measure agree with theirs? *A.* — Yes, it does. Of course the river is not as impure below, at Rockdale or Northbridge, as it is at Millbury; because at Farnumsville just below Fisherville we get the water from Quinsigamond, which is probably half the stream, and purifies the water as much as that amount more of water would purify it.

Q. — Have you any hesitation in saying even at Rockdale the contamination from the sewage is such as to be a nuisance to you in your business there? *A.* — No, sir. In 1873 I had occasion to build a bulkhead for the water to drain down; and last year we drew the water down from the Providence and Worcester Railroad some ten days in order to give them an opportunity to put in a bridge, and the condition of the pond was very much more impure; that is, before it was mud, more of a clean mud. Last year it was slimy, sticky mud, and offensive. It affected our boilers more or less. In 1857, when the mill started, we used to blow the boilers off once a month. Now we do so every day; not that it is absolutely necessary to blow it off every day the whole year, but there are seasons of the year when we have freshets that it is necessary to blow it off, and therefore we make it a rule. Certainly it is necessary to do it as often as once in three days. Some five or six years ago I had occasion to clean the river out below me: it got filled up with gravel, and I found under the banks of the river it was very offensive. You who know the Blackstone River know it is a sandy river on the banks, and a sediment collected under there which was very offensive. Until last year cattle or horses never refused to drink the water when the mill was stopped. Last year, when the mill was stopped, they refused to drink the water: there was a bad odor to it after a freshet. For instance, in February of last year, the water had run comparatively pure, because we had a very dry season. We used to have fall freshets, but for some reason they don't come as early. Last year, gentlemen on the Committee will remember, we were short of water until the first part of February. Then we had a freshet, and the water ran very impure there for a week or ten days. It was yellowish water; and there was so much of it, it ran a long ways down the stream, coloring the river: then it ran pure.

Q. — The dam at Rockdale is an old dam? *A.* — Yes, sir. There has been a dam at Rockdale for seventy years. There was a dam there when Worcester was but a village.

Q. — How about the Riverdale dam? *A.* — Longer than that.

Q. — Farnumsville? *A.* — Sixty years.

Q. — Fisherville? *A.* — I don't know about that. They take their water some ways below Saundersville. I don't know in regard to that dam.

Q. — Whether or no the people in Northbridge other than the manufacturers have a general interest in this agitation? Has any action been taken by the town in appointing committees? *A.* — Well, what is for the interest of a part is for the interest of the whole. We had an article in the warrant last year upon appointing a committee to secure some action about taking the sewerage out of Blackstone River. Of course it wouldn't affect the westerly part of the town as it would the easterly part, except that what is an injury or benefit to one portion is to the other.

Q. — As one familiar with the manufacturing of Blackstone Valley, the effect of taking down all the dams would be what upon the industries of that valley? *A.* — Well, the effect to the valley of taking down the dams would be to deprive the property of so much power that it would produce, in my judgment, such a condition of things that it would be utterly impossible for anybody to live there.

Q. — How about the river becoming contaminated as Worcester increases in size?

Mr. GOULDING. Is Mr. Whitin supposed to know any thing about that?

Mr. FLAGG. He is supposed to know about the Blackstone-river Valley for a number of years. He is an expert on that very point.

A. — If the dams were all taken down and the water should be shut back of a Saturday, or any time when the mills stopped, and there was nothing to collect this impurity, instead of having the water purified from the sewage in it that comes from the city of Worcester, the water would cease running at Cherryville, perhaps at six o'clock in the afternoon, and there would be nothing after that water got down to dilute the sewage that comes from the city of Worcester, and it would be then exposed to the rays of the sun, with nothing to dilute it or purify it until the time the water reached it again. And I have no question from the condition of things when we have been obliged to draw our pond off, the valley would be in such a condition that no one could live there.

Q. (By Mr. GOULDING.) — Wouldn't it be better to keep the ponds up all the way there, and not draw them down at all? *A.* — Will you allow me to ask a question in answer to that?

Q. — No, sir.

Mr. FLAGG. You may answer it as you see fit.

A. — Yes, sir: I think it would be better to keep the ponds up, provided you give us water enough to keep them up. But, as I suppose you know very well, the Blackstone is a bottled-up stream: six months we have a plenty of water, and six months we are supplied by reservoirs. The water has got to be stored up, unless you buy all the property to keep the reservoirs full.

Q. — If the ponds were full, they wouldn't get lower unless you drew them down, would they? *A.* — Yes: I suppose they would.

Q. — They would evaporate? *A.* — Yes, sir: unless you have got a different law in regard to evaporation from what I have ever seen.

Q. — Wouldn't it be possible to keep those ponds full if you didn't draw them down through the summer? *A.* — Yes, it would, with a system of reservoirs above, and a plenty of water; but the people of the valley think as the Legislature, in their wisdom, saw fit to take away one of the sources of the water-supply of the Blackstone River, from which they have always drawn in the dry months, and put in the hands of Worcester, it is worse than before; and now, of course, you use that water as you please, and you wouldn't be very apt to draw that water entirely during the season; and then we should have less water running down the river than now.

Q. — I didn't ask you any thing about the mills. A. — I didn't say any thing about the mills.

Q. — What I wanted to know was, not whether you could run the mills and keep the ponds full, but whether you could keep the ponds full? A. — Yes, if you had plenty of water back of it.

Q. — Isn't there water enough in the Blackstone River to keep it full, if you didn't use your mill? A. — No, sir: not if you didn't have reservoirs back. If you give the Blackstone River the amount of water you now get from freshets, the water would be comparatively pure.

Q. — Do you know how much the evaporation is on the pond? A. — No, I don't.

Q. — About one-eighth of an inch a day, isn't it? A. — As I said, I don't know: I shouldn't want to answer I do.

Q. — You don't think you would lose an eighth of an inch a day from the Blackstone River? A. — I have seen the time when there was no water coming into the Blackstone River except what came from reservoirs.

Mr. FLAGG. And sewers.

WITNESS. And sewerage, of course.

Mr. FLAGG. I have two letters from witnesses who could not be here, which I will read: —

MILLBURY, March 12, 1882.

G. A. FLAGG, Esq.

I am unable to be present at the sewage hearing, but you may read the following: —

This is to certify that I have been a resident of Millbury, living within twenty-five or thirty rods of the Blackstone River, for nearly two years past. During this time I have had more sickness than in all my life before. No member of my family, which consisted of six persons, has been exempted; and my youngest child, after a brief illness, *died* suddenly more than a year ago. From carefully studying the subject, I am convinced that the principal cause of all this has been the proximity of the Blackstone River, which, most evidently, pollutes the atmosphere near it, and from which, at times, the odor is very offensive.

B. J. JOHNSTON,
Pastor of M. E. Church.

MILLBURY, MASS., March 1, 1882.

Mr. FLAGG. I should say that the death of that child was from scarlet fever.

To G. A. FLAGG, Esq.

Dear Sir, — I am unable to be present at the hearing as you desire.

About the first of August, 1881, I went down to see my lot in the cemetery which is near C. D. Morse's pond. While there, my attention was attracted to the rapid filling up of the pond. On my way home I had to pass near the sash and blind shop of Mr. C. D. Morse. He was standing out in the yard in front of his shop. I said to him I was much surprised to see how fast his pond was filling up; if it continued to fill up for three years to come as fast as it had for the last three years, there would not be much space left for the water to run. We were standing near the water on the east side of his shop: he asked me if I did not smell the water. I said, no, I had catarrh in my head so bad I could not smell any thing. He said the water smelt like rotten eggs. I said, if so, it was probably sulphuretted hydrogen gas. I had a few matches in my pocket. I split the end of a stick, and inserted a match, and ignited it, and applied it to the bubbles that came floating down on the surface of the water: they ignited at once. I could see the flash; did not hear the report, being rather deaf. Mr. Morse said he distinctly heard it. That satisfied me the odor complained of by Mr. Morse was sulphuretted hydrogen gas or hydrogen sulphide: both have the same putrid odor, but very different in their poisonous effects.

Sulphuretted hydrogen is instantly fatal to animal life when pure, and even when diluted with fifteen or twenty times its bulk of air has been found so poisonous as to destroy life in a few minutes. One hot day last summer the odor from the river was so offensive my wife had to put down all the windows. My house stands about seventy rods from the river, — may be a few rods more.

Respectfully,

ELIJAH THOMSON, *Millbury, Mass.*

MILLBURY, March 11, 1882.

Mr. FLAGG. As to Elijah Thomson, I have a slip from a newspaper here which will describe who he is. I will read but a sentence from it: —

"Mr. Elijah Thomson, an aged and respected citizen of Millbury, furnishes the following account of his invention of friction matches."

He is the inventor of friction matches, and is an old gentleman who has lived for a long time in Millbury.

Mr. GOULDING. Do you mean to say he is the original inventor?

Mr. FLAGG. Yes, sir.

CHARLES G. MORSE. *Recalled.*

Q. (By Mr. FLAGG.) — You were explaining the sprinklers when you left the stand yesterday. Have you any thing more to say to the Committee about them? *A.* — Perhaps I might state, in the first place, as to the size of the hole: it may be better to give the definite size, — one-twelfth of an inch.

Q. (By the CHAIRMAN.) — That is the diameter? *A.* — Yes, sir. I stated yesterday that the holes were probably about ten inches apart. I found, by measuring them, they are between five and six: they vary a little.

Q. (By Mr. FLAGG.) — These sprinklers are in common use in manufactories? *A.* — Yes, sir. I put this sprinkler in under the direction of the secretary of the Manufacturers' Mutual Insurance Company.

Q. — In other places they have been effective? *A.* — So far as I know, always. They were recommended very highly to me.

Q. — You heard this letter that I read from Mr. Thomson? *A.* — Yes, sir.

Q. — Did you see what he describes in that? *A.* — I did.

Q. — You saw the flash of the bubbles? *A.* — Yes, sir.

Q. (By the CHAIRMAN.) — How long did you say this sprinkler had been in? *A.* — Perhaps seven or eight years. I couldn't state positively.

Q. — You have tried them frequently in the mean time? *A.* — Never tried them inside the building, except in the basement.

Q. — Ever have water turned into them? *A.* — Never turned it in, except in the case of the fire. I have tried them on the outside of the adjoining building. I think I called your attention to a little gable. I have also tried it in the basement, and it worked to perfection. It protected the basement all the way, except the farther end, at the time of the fire.

Q. (By Mr. FLAGG.) — I want to ask you as to the age of the dams, including all those from the mouth of the sewer to the Blackstone line. You have looked into that matter? *A.* — Yes, sir.

Q. — What can you say? *A.* — The original Burling dam was built at the time the Blackstone canal was built. The exact date I couldn't tell you, — in the 30's, I think.

Q. — Next is your dam? *A.* — Next is my dam. My dam, as near as I can ascertain, was built in 1827.

Q. — Next is the Atlanta dam? *A.* — That was built a hundred and fifteen years ago, as near as I can ascertain.

Q. — Millbury Cotton-Mills dam? *A.* — That was part and parcel of that same privilege, as I understood.

Q. — Cordis Mill dams? *A.* — About seventy years ago.

Q. — Simpson's dam. *A.* — Simpson's dam soon after.

Q. — Wilkinsonville dam was built by Asa Waters? *A.* — Yes, sir.

Q. — About what time? *A.* — I looked that up; but, really, I think —

Q. — Asa Waters died about 1834, did he not? *A.* — As long ago as that, I am told: it was before I came to town. I think that is an older dam than the Cordis or Simpson's dam.

Q. — Saundersville dam has been testified to? *A.* — Yes, sir. I went no farther than the Wilkinsonville dam.

Q. — You pay taxes, I understand, on property in Worcester? *A.* — Yes, sir: and have, for the last ten years, paid about two hundred dollars a year.

Q. — What can you say as to the general opinion of the public in Millbury and adjoining towns? *A.* — As has been stated, we were appointed a committee a number of years ago; and we reported, perhaps, twice or three times to the town, and asked for more time. Perhaps three years ago an article appeared in the Millbury paper, in regard to the pollution of the Blackstone River, rather criticising the committee for not acting. The feeling there is very strong that something should be done. It is very injurious to health. We have had trouble from it in our own family, and I have experienced trouble personally; and among the laboring men it is not only a matter of talk, but it is spoken of on the street. It has been a subject of prayer at the church. It is often spoken of going to and from church, and in almost any gathering we are at. Last spring the matter was up. Of course we were incurring expense on the sewerage matter; and I think one man expressed the opinion of the town when he said not to limit the committee in the amount of money, but let them go to the extent of the valuation of the town if it was necessary. The committee have been very careful in spending money. We were asked last summer, by the State Board of Health, to bring forward some plan, and present our views in regard to doing away with the pollution of the river. We immediately, within ten days, I think within less than a week, contracted with Col. Waring to bring forward his plans. At the same time the city of Worcester, a rich city, was asked to do the same thing, and brought forward none.

Q.— You have informed yourself as to the number of operatives in the mills at these dams we have spoken of? *A.* — I think, in the aggregate, there are some thirty-two hundred or thirty-three hundred.

Q. — Formerly your pond was used by the Baptists for their ceremony of baptism, was it not? *A.* — In May, 1868, and July, there were twenty-four persons baptized.

Q. — It was customary to use that pond very often, but is not now? *A.* — Yes, sir: my pond above the dam had a beach for bathing.

Q. — And very much ice formerly came from that pond? *A.* — Yes, sir. Two ice-houses were upon that pond. I think the first one was abandoned in 1870 or 1869 for family use, and for market use the following year.

Cross-Examination.

Q. (By Mr. GOULDING.) — What did you have these pictures of the ruins of your factory made for? *A.* — The first I knew about them, I found the artist there taking them.

Q. — You didn't have them taken, then, yourself? *A.* — I did not order them taken, but I bought some.

Q. — Have you not used this picture with your friends for the purpose of showing how effective this sprinkler was in saving your property? *A.* — I never showed them outside the Committee, at my place, and here.

Q. — Did you ever have any experience in the use of those sprinklers, except on this occasion? *A.* — Yes, sir.

Q. — When? *A.* — I stated that I tried it in my basement several times.

Q. — You never had occasion to use them in the case of a fire except this time? *A.* — That is the only fire I ever had.

Q. — Were you ever present at any fire before when there was an attempt to use them? *A.* — No, sir.

Q. — Your gear broke the first thing? *A.* — No, sir.

Q. — Didn't you have to supply new gear? *A.* — I did after running some time.

Q. — At the time of the fire, I mean? *A.* — At the time of the fire it ran about an hour. I got control of the fire before my gear broke. The fire-engines were there, and had full control of the fire before my gear broke.

Q. — Did you use your pump after you supplied the gear? *A.* — Yes, sir.

Q. — How long did you run it after you got a new one on? *A.* — Perhaps we run it half a day the next day. I was seven minutes putting the duplicate gear on. I done all I could to save my property. I had a duplicate gear, and I was just seven minutes putting it on by the time of a watch.

Adjourned to 10 o'clock A.M.

FIFTH HEARING.

THURSDAY, March 16, 1882.

THE hearing was resumed at 10 o'clock.

Mr. GOULDING. I want to call the attention of the Committee to the following State reports, and other documents, to which I shall refer in argument: —

Report of State Board of Health, 1876, pp. 96, 107, 122, 140, 73.

Report of State Board of Health, Lunacy, and Charity, 1881, pp. 11, 24, 122, 123.

Report of State Board of Health, 1874, pp. 70, 98, 99, 116, 117, 130, 135.

Report of State Board of Health, 1878, p. 66.

Supplement to No. 16 National Board of Health Bulletin, Dec. 24, 1881, p. 17.

Rudolph Herring's Report, p. 6: "Disposal of Sewage."

Then I desire to refer, *passim*, to that report, so far as it relates to Berlin, London, Dantzic, Brighton, Croydon; generally to that report, specially to those points.

I shall cite, besides the cases cited by the other side, *Merrifield* v. *Lombard*, 13 Allen, 116, and *Wheeler* v. *Worcester*, 10 Allen, 591.

TESTIMONY OF CHARLES F. ADAMS.

Q. (By Mr. GOULDING.) — Where do you live? *A.* — Worcester.

Q. — What is your business? *A.* — I am a teacher at the normal school.

Q. — What departments do you teach? *A.* — Natural philosophy and chemistry.

Q. — How long have you been there? *A.* — Seven years.

Q. — Have you compiled any statistics with reference to the death-rate from various causes in Millbury as compared with other places? *A.* — I have.

Q. — Have you a copy of your compilation? *A.* — I have.

[Witness submitted the following paper.]

STATISTICS OF MORTALITY IN MILLBURY, 1861 TO 1881.

To His Honor E. B. STODDARD, *Mayor of Worcester.*

It seems hardly worthy of this age of scientific investigation, that so grave a matter as the disposal of the sewage of a city should be so largely discussed upon a basis of assumption and guesswork. The question whether Worcester is poisoning a sister community is a question of fact, which ought not to be difficult to answer by the recognized methods of investigation. A death-rate showing a regular increase for a series of years corresponding with the increased amount of sewage, if otherwise unexplained, would indicate a serious matter, especially if accompanied by a higher percentage of deaths from those diseases more or less connected with filth. Assuming that the clamor for restrictive legislation grew out of a deep sense of injury, I was interested to make a little research in the official registration of deaths, hoping to measure to some extent the poisoning influence of sewage. The results are so at variance with much of the current talk as to be worth your attention.

At the outset, if we ask whether Millbury is an unhealthy town, we have to choose a standard of comparison. Evidently the sanitary conditions of a valley manufacturing town differ as much from those of the hill-farming towns as from those of the cities, and in the state and county the higher death-rate of the centres may or may not be balanced by the greater healthfulness of the more sparsely populated sections.

The location, industries, and population are factors which show themselves more or less in the death-rate, so that perhaps the fairest standard is the average of other towns of similar size and situation. Choosing, then, all the towns between 3,000 and 7,000 population in the valleys of the Nashua, Miller's River, the Chicopee, and the Quinnebaug, we shall include the towns of Clinton, Leominster, Winchendon, Gardner, Athol, Spencer, Warren, Palmer, Ware, and Southbridge, all manufacturing towns, and averaging, in 1875, a population of 4,677 against 4,529 in Millbury. These furnish conditions approximately similar.

The following table will enable us to compare the death-rate of Millbury with that of the ten towns, and it will also be seen in Table IV. that for twenty years Millbury has been apparently growing healthful: —

TABLE I. — Comparative Death Rate * in Different Localities.

	1871-75.	1871-80.	1876-80.	1880.
Massachusetts	21.05	20.02	18.98	19.80
Worcester County	19.17	18.23	17.29	18.92
Worcester	23.90	21.58	19.27	20.89
Ten Towns	18.79	18.22	17.65	19.71
Millbury	20.57	19.91	19.24	18.77
Millbury in 1881, 14.76.				

* Rate per 1,000, based upon average population.

If we continue our inquiries further, and ask whether Millbury has an undue amount of those diseases which are more or less associated with poisoned air and water, it still further appears that Millbury compares favorably with the other towns, and that on the whole such diseases seem to be growing less, as is shown by Tables II., III., and IV.

It will be observed that the figures for the filth diseases fluctuate through a wide range, though generally diminishing. The infantile death-rate also appears to be decreasing, and the school attendance increasing. Thus the statistics, examined from many points of view, seem to show little ground for legislation or experiment.

Hoping that the enclosed tables may be of interest, I remain

Respectfully yours,

CHARLES F. ADAMS.

WORCESTER, March 4, 1882.

187

TABLE II. — The Comparative Death-Rate * from Filth Diseases in Different Localities.

	1871-1875. Average.	1871-1880. Average.	1876-1880. Average.	1881. Total.
Massachusetts	5.13	4.81	4.49	4.27
Worcester County	4.79	4.35	3.91	4.46
Worcester	6.56	5.55	4.54	3.33 (1880)
Ten Towns	4.89	4.66	4.45	†4.12 (1880)
Millbury	5.11	4.52	3.93	4.01

TABLE III. — The Comparative Death-Rate * from the Separate Filth Diseases.

	1871-1875.						1876-1880.						1881.					
	Scarlatina.	Diphtheria and Croup.	Typhoid.	Dysentery, Diarrhœa, Cholera Morb.	Cholera Infantum.	Cer. Sp. Men., Wh. Cough, Erysip., Met.	Scarlatina.	Diphtheria and Croup.	Typhoid.	Dysentery, Diarrhœa, Cholera Morb.	Cholera Infantum.	Cer. Sp. Men., Wh. Cough, Erysip., Met.	Scarlatina.	Diphtheria and Croup.	Typhoid.	Dysentery, Diarrhœa, Cholera Morb.	Cholera Infantum.	Cer. Sp. Men., Wh. Cough, Erysip., Met.
Massachusetts	.87	.65	.83	.60	1.60	.58	.41	1.60	.46	.55	1.06	.41	†.32	†1.34	†.49	†.56	†1.19	†.37
Worcester County	.98	.59	.82	.49	1.39	.52	.44	1.23	.53	.42	.90	.39	†.64	†1.11	†.60	†.55	†1.16	†.40
Worcester	1.22	1.05	.78	.55	2.23	.73	.60	1.08	.49	.63	1.26	.48	.79	.58	.31	.14	.82	.69
Ten Towns	.96	.55	.95	.56	1.44	.43	.49	1.70	.61	.37	.93	.35	.38	†1.17	†.73	†.22	†1.07	†.55
Millbury	.76	.67	1.17	.49	1.62	.40	.99	.95	.52	.35	.86	.26	1.90	.00	.42	.00	1.48	.21

* Deaths per 1,000, average population.   The average of 1871-75 and of 1876-80 is taken for 1871-81.   † 1880.   ‡ Nine towns, the record of Palmer not received.

TABLE IV.—Deaths in Millbury, 1861-1871-1881. Population, 3,296-4,397-4,741.

Increase in Population, 1861-71, 33 per cent.; 1871-81, 44 per cent.

	1861.	1862.	1863.	1864.	1865.	61-70.	1866.	1867.	1868.	1869.	1870.	1871.	1872.	1873.	1874.	1875.	71-80.	1876.	1877.	1878.	1879.	1880.	1881.
Scarlatina	0	1	0	0	3	1.0	0	0	1	4	0	10	7	0	0	0	4.0	2	2	3	0	15	9
Diphtheria and Croup,	1	12	11	4	3	3.6	1	1	2	0	1	8	2	3	1	1	4.7	8	3	1	3	7	0
Typhoid and Fever	4	1	3	7	2	3.2	4	3	0	1	7	5	11	5	4	1	3.8	2	5	2	0	3	0
Dys., Diar., C. Morb.	0	2	2	5	6	3.6	5	4	1	10	1	2	4	1	2	2	1.9	0	4	0	0	4	2
Cholera Infantum	2	3	1	7	5	3.9	2	4	5	3	7	1	10	11	7	7	5.6	0	7	5	3	5	7
Met., W.C., Ery., C.S.M.	1	0	1	1	1	.8	2	0	0	1	1	1	0	3	2	3	1.5	0	3	1	0	1	1
Total filth diseases	8	19	18	24	20	16.1	14	13	9	19	17	27	34	23	16	14	20.5	12	25	12	7	35	19
" "	-	-	89	-	-	161	-	-	72	-	-	-	-	114	-	-	205	-	-	91	-	-	-
Pr. ct. filth dis. above or below av. 10 yrs.	—56	+18	+12	+49	+24	-	—13	—19	—44	+18	+06	+80	+64	+11	—23	—32	-	—39	+34	—39	—66	+69	—08
Av. death-rate filth d.:																							
Scarlatina	-	-	.23	-	-	.26	-	-	.29	-	-	-	-	.76	-	-	.68	-	-	.99	-	-	1.90
Diphth and Croup	-	-	1.75	-	-	1.00	-	-	.24	-	-	-	-	.67	-	-	.81	-	-	.95	-	-	.00
Typhoid and Fever	-	-	.96	-	-	.85	-	-	.75	-	-	-	-	1.17	-	-	.84	-	-	.52	-	-	.42
Dys., Diar., C. Morb.	-	-	.84	-	-	.93	-	-	1.02	-	-	-	-	.49	-	-	.42	-	-	.35	-	-	.00
Cholera Infantum	-	-	1.02	-	-	1.02	-	-	1.02	-	-	-	-	1.62	-	-	1.24	-	-	.86	-	-	1.48
Met., W.C., Er., C.S.M.	-	-	.23	-	-	.22	-	-	.20	-	-	-	-	.40	-	-	.33	-	-	.26	-	-	.21
Av. pr. 1000 of av. pop.	-	-	5.03	-	-	4.28	-	-	3.52	-	-	-	-	5.11	-	-	4.52	-	-	3.93	-	-	4.01
Deaths from all causes,	58	70	66	82	82	73	75	69	63	78	86	97	115	91	79	77	90	85	113	91	68	89	70
†Per 1000, all causes	17.6	21.2	20.0	24.9	24.9	21.4	19.8	18.3	16.7	26.0	19.6	22.0	26.2	20.7	17.4	17.0	20.2	18.8	25.0	20.6	15.0	18.8	14.8
‡Av. pr. 1000, all causes,	-	-	21.7	-	-	20.4	-	-	19.0	-	-	-	-	20.7	-	-	20.2	-	-	19.6	-	-	14.8
Av. school att. bet. 5-15,	-	-	-	-	-	-	-	-	-	-	-	.59	.69	.50	.63	.63	.63	.64	*.42	*.66	*.74	*.84	§.85
" "	-	-	-	-	-	-	-	-	-	-	-	-	-	.61	-	-	.63	-	-	.66	-	-	.69
" of No. belong.	-	-	-	-	-	-	-	-	-	-	-	.88	.66	.66	.56	.81	.72	.66	.51	.79	.73	.88	†.92
Av. dths. und. 5 to bths.	-	-	.26	-	-	.24	-	-	.22	-	-	-	-	.29	-	-	.27½	-	-	.26	-	-	.21

* Increase in school attendance exceeded by only three towns in the State. † Based upon preceding census. ‡ School attendance in 1881, exceeded by only four towns in the county.
§ Based upon census of 1875. Death-rate in 1881 lowest for 20 years—two-thirds that of 1861-65—and lower than that of eight-ninths of the State (1880).

Q. — From what sources were those statistics compiled? *A.* — From the reports of registration for the State for the successive years, and also from the original returns in the document-room.

Q. — When did you make this compilation? *A.* — Since the first of March.

Q. — I will not stop to read this paper, but I will ask you what the compilation undertakes, in the first instance? *A.* — It undertakes to compare the mortality of Millbury with that of ten towns similarly situated and having similar industries; and also with the city of Worcester, the county of Worcester, and the whole State. It also attempts to compare the mortality of Millbury in the various years, that is, from 1860 up to 1880, so that a person may see at a glance how it compares; and it also attempts to compare the mortality from the several diseases that are commonly regarded as filth diseases; that is, the filth diseases from 1870 to '75 and '80 are compared with those of 1860 to 1865, and so on.

Q. — Now I will ask about the tables. Table I. is on the fourth, or last, page. "Comparative death-rate in different localities." What is that rate, — the rate per thousand? *A.* — Yes, — rate per thousand of the average population in the periods; that is, for 1871 to '75, the average of the census of the year 1870 and the census of 1875 is taken.

Q. — And you take how many periods there for comparison? *A.* — Three periods and two years, making five.

Q. — In the last year, you do not give the other towns in Worcester, Worcester County, and Massachusetts, do you? *A.* — I did not have time to get the data, and the data from the State are not in. Some of the towns in the county have not yet been returned.

Q. — Do you think of any thing in that table that needs explanation beyond what you have given? *A.* — I think not: it is intended to be plain.

Q. — The towns that you have chosen are given at the bottom of the first page, I believe, — the towns of Clinton, Leominster, Winchendon, Gardner, Athol, Spencer, Warren, Palmer, Ware, and Southbridge. I will now turn to the second table, which is on the third page, — "Comparative death-rate from filth diseases in different localities." You take the same localities, do you? *A.* — Yes.

Q. — And the same periods? *A.* — Yes, sir.

Q. — Now, what do you undertake to do in that table? *A.* — To compare the rate of mortality in those different localities and in those different years.

Q. — And upon the same basis of deaths per one thousand of average population? *A.* — Of average population: yes, sir.

Q. — Now, is there any further explanation needed of that table, that you think of? *A.* — Nothing, except that all the filth diseases that are included in this are named in the following table, Table III., which, I believe, contains more than are sometimes given.

Q. — Now, state what Table III. undertakes to show. *A.* — It undertakes to show the rate of mortality, and the comparative rate of mortality, from each of those separate diseases; that is, so that a person can compare the death-rate of Millbury, of typhoid, with that of the ten towns, with that of the county, and with that of the State, for the successive periods, or can compare the death-rate from any of these diseases from one period to another.

Q. — Take, for instance, the first column. Opposite the word "Massachusetts" I find the decimal .87 under the column "Scarlatina." What does that mean? *A.* — It means, that out of every thousand average population, between 1871 and 1875, the average death-rate per year was 0.87; or, if you make it for every 100,000 people in the State, 87 people died a year, on the average, during that period of five years.

Q. — That is, the average, during that year, of deaths from scarlatina, was .87 of one? *A.* — Per thousand average population: yes, sir.

Q. (By the CHAIRMAN.) — That is, if the population was one thousand, eighty-seven died? *A.* — No, sir: less than one died.

Q. (By Mr. GOULDING.) — Now, Table IV. is a little more elaborate. There are some things which I may require you to explain. What does that table undertake to show? *A.* — It undertakes to show, first, the number of deaths per year from each of the several causes known as filth diseases; to show the total amounts for the successive years, and for the periods; and to show the average death-rate, so that they may be easily compared; and the death-rate from the diseases during different periods, so that you may compare the last period with the first, or with the middle, or any other, as you may determine.

Q. — Just go into a little detail, so as to be sure we understand it. I find here several columns headed 1861'–62'–'63, up to 1865; then a division by a heavier line; and then, between that heavy vertical line and another which follows, "1861 to 1870." Does that mean a summary of the whole ten years? *A.* — What is under that means the average for that period of ten years. These tables, of course, are to be considered with reference to the population.

Q. — That you had provided for? *A.* — Yes, sir.

Q. — "Scarlatina," for instance. In that column I find the figure "1." What does that mean? *A.* — It means that there was, on the average, one death per year, for each year, from that disease, from 1861 to 1870 inclusive, in Millbury. All this table relates to Millbury.

Q. — Then, the column at the head of which is "1871 to 1880" is the same thing with reference to that period, isn't it? A. — Yes, sir.

Q. — Now, opposite the line "Total filth diseases," I find, under the heading "1861," "8." What does that mean? A. — That means that in the whole town there were eight deaths during that year from these several causes.

Q. — And the same as to the other years following? A. — Yes, sir.

Q. — Now, the figures "89," in heavier type, below "1863," mean what? A. — They mean that during those five years, from 1861 to 1865, there were 89 deaths. It is really the sum of these totals just added. It is the sum of 8, 19, 18, 24, and 20; showing that, during those five years, in the whole town, there were 89 deaths from these causes. I wanted to place it in that five years, and so put it in the middle of the five years. The printer wanted to run the line right down there.

Q. — Now, the figures "161," in the same kind of type, under the "1861 to 1870" column, mean what? A. — That, in the ten years from 1861 to 1870, there were 161 deaths from those several causes. It is really adding up the 8, 19, 18, 24, and 20; and then skipping over to the 14, 13, 9, 19, and 17.

Q. — Now, the "72" is what? A. — For the five years, the total deaths from 1866 to 1870, the same as the "89" in the other case.

Q. — Now, is the "114," in the same line, the sum total for the five years from 1871 to 1875 inclusive? A. — It is.

Q. — And the "205" is the total from 1871 to 1880 inclusive? A. — Yes, sir.

Q. — And the "91" for the last five years, from 1876 to 1880? A. — Yes, sir.

Q. — Now, I find this abbreviation "met.:" that means "*metria*"? A. — *Metria:* yes, sir.

Q. — Next I find here "per cent filth diseases above or below average ten years," and, under the separate columns, some figures with plus and minus marks before them. Explain what that signifies. A. — The first one, "—.56," under "'61," means that during the year 1861 there was fifty-six per cent less of deaths from these causes than there were on the average of those ten years of the whole number of deaths under the death-rate. Now, the whole number of deaths in the ten years is shown just above, "161," in bold type; that is, an average for the ten years of sixteen deaths per year.

Now, this first year there were only eight; so there was really fifty-six per cent less than the average for the ten years.

Q. (By Dr. WILSON.) — May I ask you, if you add all the pluses together and subtract the minuses, would not that leave just one per cent? Would not that be what it would amount to? *A.* — No: it would be just nothing, would it not?

Dr. WILSON. No, sir: it would be just one per cent.

WITNESS. Adding the plus above the average, and subtracting the minus below the average, I should think would leave just zero.

Q. — Now, below that line we find "Average death-rate filth diseases;" then follow these several diseases; and under the middle columns of the separate five years, and the middle columns of the separate ten years, we find rows of figures which seem to be added, in bolder type? *A.* — They are added up.

Q. — Now explain that. *A.* — The upper part of the table that we have been speaking about refers to the whole number of deaths in the town, and this part refers to the rate per thousand, and the average rate per thousand per year; so that, under the first year, ".23" means that during that five years the average rate per thousand every year of deaths from scarlatina was .23 of one. Or, putting it the other way, out of one hundred thousand people in the town, there were, on the average, twenty-three deaths per year, in that period of five years, from scarlatina.

Q. — You mean, if there had been one hundred thousand inhabitants there? *A.* — Yes, sir.

Q. — Now, these figures below, opposite the line "Average per 1,000 of average population," the line of figures through the table, beginning with "5.03," in similar type, means what, sir? *A.* — Those are the sums of these separate death-rates that we have been speaking of: the sum of .23, 1.75, and so on. That is, the "5.03," for the five years from 1861 to 1865, shows that, on the average, for every thousand people in the town, there were five deaths from all these diseases put together.

Q. — Then the next line, "Deaths from all causes," that needs no explanation, I suppose? *A.* — That is the entire number of deaths, and not any relative number.

Q. — Next, "Per 1,000 from all causes"? That is simple, I take it. Then there is, "Average per 1,000 from all causes," in heavier type. *A.* — That is derived from the line above, grouped in five-year periods and in ten-year periods, and the average of the whole taken.

Q. — The next line I see here is, "Average school attendance between five and fifteen," and these columns begin later, in 1871. What is the explanation of that? *A.* — The returns previously to

1871 were divided into summer and winter attendance, and I did not take the time to add them up, and balance them, to fill in from 1860 to 1870. The number "59," under "1871," means, that, if there were in the town one hundred pupils between five and fifteen, there were, on an average, all through the school year, fifty-nine of them present at school every day.

Q. — The "61" in the next column but one is the average of those five years? *A.* — Yes, sir: on the same basis.

Q. — The "63" between the heavy lines is the average of the ten years? *A.* — Yes, sir.

Q. — "66" is the average of the five years between 1876 and 1880 inclusive? *A.* — Yes, sir.

Q. — The next line gives the average number belonging to the school? *A.* — The whole number whose names are enrolled on the registers. Some of the pupils are entirely out of school, but of those whose names are enrolled on the registers, out of every hundred that are enrolled, eighty-eight of them were there, on the average, every day.

Q. (By Mr. CHAMBERLAIN.) — What do you mean by "pupils entirely out of school"? *A.* — I mean, a boy of ten years may be in a factory, and, if he is, he is not considered in this account; but only those who have been at school a week or two, or something of that kind.

Q. (By Mr. GOULDING.) — Then, the next is "Average deaths under five to births." Does that need any explanation? If so, please to state exactly what it means. *A.* — That first one, ".26," means that for every hundred children born for the first five years, on the average, there were twenty-six died before reaching the age of five.

Q. — I see that is a decimal. Does that mean that, on the average, twenty-six out of every hundred children that are born die before they are five years old? *A.* — If there are a hundred births in the town in a year, the table shows that, deducting all the children living up to the age of five years, and including all the infantile diseases, the death-rate would be twenty-six.

Q. — Did you compare that with the other places? *A.* — I did not figure it out with other places. The rule is, that about a quarter part of the population of the State die before five years of age.

Q. — Now, I see some statements below this table, "Increase in school attendance exceeded by only three towns in the State." That means the increase from what time to what time? *A.* — Under the year 1877, "*.42" means .42 per cent of one. .66 in '78; .74 in '79; .84 in '80, — that that rate of increase was only exceeded by three towns in the State.

Q. — Then I see the next statement in the table is "based upon preceding census." What does that mean? *A.* — Among the list, about the fourth or fifth line from the bottom, you see "Deaths from all causes;" and then, under that, "Deaths per 1,000 from all causes," and that is copied directly from the reports of registration. The first one you see is 17.6 for 1861. It is the last number under 1861. That is based, of course, on the census of 1860. Now, 21.12 is based on the same census, and the 20.0 is based on the census of 1860 until you get down to 1866 ; then the rate is based on the census of 1865.

Q. — Do you mean 1860 or 1870? *A.* — The rate under 1866 is based on the census of 1865. We have five-year censuses.

Q. — " School attendance in 1881 exceeded only by four towns in the county." Did you find that to be so? *A.* — That is printed in the last report of the State Board of Education. All the towns in the county are arranged in line, and that appears from those statistics. With reference to the use of Table III., on the third page, if you follow down the line in the first part of it, "1871 to '75," until you get to Millbury, you find ".76 ;" that is, of course, to every thousand of population, on the average; that is scarlatina. You can compare it with the ten towns which had a higher rate ; with Worcester, which had a still higher rate ; and with Worcester County, a higher rate still ; and with Massachusetts. In the same way you can follow this along. You find diphtheria higher in Millbury than in the ten towns ; lower than Worcester, higher than Worcester County, and so on. Then you can skip along until you come to "1876 to '80." The number ".99," under scarlatina, shows that Millbury had a higher rate than the ten towns, or Worcester, or Worcester County ; but of diphtheria ".95," which was much lower. Of typhoid, it was below the average of the county, and the average of the ten towns ; of dysentery, it was much below ; cholera infantum, it was below, and so on. In 1881 you get a large increase in scarlatina as compared with the rest of the towns; no fatal cases of diphtheria and croup ; none of dysentery, and so on.

Q. — I will ask you as a statistician what is the general result of this tabulation in respect to the comparative death-rate of Millbury relative to these towns, the county, and the State, with which you have compared it? *A.* — I think it appears that the death-rate of Millbury is neither high nor low ; of the two, I should not know ; it runs along, as far as I can see, about even. It appears in Table IV., I think, very near the lower part of it. There you see the average per thousand, and then you get, next to the last figure in 1863, "21.7 ;" that is the rate for that five years ; and then, in the first ten years, 20.4 ; and then, if you look along to '80, it is 20.2 ; that is, for the last ten years, it was less than for the first ten. And still further, in 1876 to 1880, it is 19.6. In 1880 it is 18.8, and in 1881, 14.8. I think it appears that there is a diminishing death-rate.

Q. — Now, in these figures showing the average death-rate, have you provided for the increase of population? *A.* — The lower numbers, "17.6," "21.2," "20," etc., which are next above the "21.7," under 1863, are relative to the population; but the numbers in heavier type, about a third of the way from the top of the table, — the "89," "161," "72," "114," "205," and "91," — are the absolute numbers of deaths from an increasing population all the time.

Cross-Examination.

Q. (By Mr. MORSE.) — Mr. Adams, do you undertake to give an opinion as to any thing except the accuracy of your computations? *A.* — The drift of the computation is plain. As to the medical matters, that is out of my province, of course.

Q. — Well, have you made a study of vital statistics? *A.* — I have, as a teacher of physiology, read carefully all the reports of the State Board of Health, and refer to them in my classes, which contain a considerable amount of carefully tabulated statistical matter.

Q. — Do you think that, when you have prepared tables to show the comparative death-rates of Millbury, of the State, of the county, and of certain towns, that you have exhibited all the facts which are material to determine whether or not Millbury has been growing healthy or unhealthy? *A.* — I think that that, to me, is the leading fact, the most available fact, the prime fact.

Q. — Did you ever read any book on vital statistics? *A.* — No, sir.

Q. — Let me call your attention to a statement in an authority on vital statistics, — Dr. Carpenter of London, — in which he says, "The number of deaths in a given district bear no constant ratio to its healthiness or unhealthiness. It does not necessarily follow that the conclusions respecting the sanitary conditions of a town or country are correct, because the ratio of mortality is low." *A.* — In Dr. Stewart's book he states that it does bear a ratio, that the two are commensurate. I have read that.

Q. — Then you have proceeded upon the theory, have you, that the statistics of death are the most important statistics in determining the health of a community? *A.* — I proceeded upon the theory that those would throw important light upon the subject under discussion. It is a matter that I have nothing to do with, you understand.

Q. — No: I simply wished to ascertain to what extent you desired the Committee to accept the conclusions to which you come as authority here. Are you aware of the fact that the birth-rate is an important factor in determining the condition of the health of a town? *A.* — Yes, sir; and for that reason I put in ten similar towns in which the birth-rate is somewhat commensurate with that of Millbury.

Q. — Have you made any tables which show the birth-rate of Millbury as compared with other places? *A.* — I have not. This matter has been taken up within a fortnight, and there has been what work I could do in what is here.

Q. — Your tables, if I understand them and the conclusions to which you come in this communication, tend to prove, if they prove any thing, that the health of Millbury has been improved in consequence of the sewage of Worcester being turned into the river? *A.* — No, sir, nothing of the sort.

Q. — You state in your report, if I understand you, that "Millbury compares favorably with the other towns; and that, on the whole, such diseases [that is, filth diseases] seem to be growing less." *A.* — Yes, sir.

Q. — And you then go on and say that "for twenty years Millbury has been apparently growing healthful." Now, you certainly do draw inferences from these tables, and undertake to present them to the Committee. *A.* — I say nothing about sewage. I say that the facts are, so far as I can ascertain them, that Millbury shows a lower death-rate to-day than for twenty years previously.

Q. — You say you "say nothing about sewage;" but you begin your communication by saying, "It seems hardly worthy of this age of scientific investigation, that so grave a matter as the disposal of the sewage of a city should be so largely discussed upon a basis of assumption and guesswork." And you then go on and give statistics which you think are important, and at the conclusion of those statistics you draw the inference that the health of Millbury has been improving. Now, I ask you whether you did not intend to convey by that communication the idea that the disposal of the sewage of Worcester had been so judiciously managed that it had tended to improve the health of the town of Millbury? What is the connection in your mind between the disposal of the sewage of Worcester and the conclusion to which you come here? *A.* — I am unable to find in the general death-rate, or in the death-rate from those separate filth diseases, which appear to be diminishing, substantial ground for thinking that the sewage of Worcester is not so abundantly diluted and so thoroughly oxidized, burned up, that it produces a higher death-rate: that is the point. I have nothing to say beyond that.

Q. — Please understand, Mr. Adams, that I am not seeking in any way to disparage the accuracy of your tables. I have no doubt they were figured honestly, and I presume accurately: it is simply whether, when you go beyond your tables, and undertake to draw inferences from them, you desire the Committee to understand that your opinion upon this subject is of any particular value as due to any special investigation of such matters. *A.* — I think it is entitled to some little consideration, but not that of an expert or a physician at all.

Q. — Don't you think that you are in error in assuming tha the statistics that you have put into your tables are sufficient data from which to draw the inference that you do, — "that for twenty years Millbury has been apparently growing healthful"? *A.* — I think I tried to state, that, in so far as the tables contribute any thing, just so far Millbury appears to be growing healthful. Beyond that I have no data: I don't know any thing about it.

Q. — Pardon me, Mr. Adams: I think that your language in this communication is much broader than that. I use your exact expression: "The following table will enable us to compare the death-rate of Millbury with that of the ten towns, and it will also be seen in Table IV. that for twenty years Millbury has been apparently growing healthful." I understand that to be the expression of your opinion, that, as shown in Table IV., the town has been apparently growing healthful during that time. *A.* — So far as appears in Table IV., it certainly has.

Q. — Have you any opinion as to the causes which have led Millbury to improve in health during the twenty years? *A.* — I think that very likely the science of medicine is understood better to-day than it was twenty years ago, which may have been a contributing cause.

Q. — But I understand the tables to show that Millbury has improved more than other towns during that time in health. Do you mean that the physicians of Millbury are shown by these statistics to have advanced beyond the average physicians of the State? *A.* — I think not; I don't know that that is a necessary inference from it.

Q. — Well, is that your opinion? *A.* — No, sir.

Q. — Then I go back again to the question as to what you attribute the improved condition of Millbury to? *A.* — Millbury is compared with itself for twenty years; and, compared with itself, it seems to me that may be one of the contributing causes. I think, also, there filters down from competent people a better knowledge of the conditions of health than was common in the community twenty years ago, and I think that is one reason.

Q. — Do you think that any more of it has filtered down into Millbury than into any other place? *A.* — No.

Q. — Perhaps I don't understand your tables; but I suppose that they were intended to show, or that you claim that it is a fair inference from them, that Millbury has advanced in health in a larger ratio than other places. Isn't that your conclusion?

Mr. GOULDING. There is no statement of any such claim.

Q. — Perhaps I do not understand your statement. You say here, "For twenty years Millbury has been apparently growing healthful." Do you mean that Millbury has been growing more healthful than other places have been, or that there has been a general advance in health in the last twenty years? A. — I think that Millbury has been growing in health for the last twenty years, and it also appears in another table that ten towns have not grown healthful as fast as Millbury.

Q — I thought I stated correctly that you intended to draw the inference that Millbury has grown healthful with more marked rapidity than other towns. A. — It appears that, compared with those ten manufacturing towns, the improvement is on the side of Millbury.

Q. — Now I ask if you have any opinion as to what that improvement on the side of Millbury is to be attributed to? A. — I think it is due, perhaps, as I said before, to a better and more general knowledge of hygienic laws, for one thing.

Q. — I ask you to confine yourself to Millbury, as distinguished from these other towns. You have compared Millbury with ten other towns, and you say there has been a marked improvement in health in Millbury over the other towns. Have you any opinion as to what that marked improvement in Millbury is due to? A. — I don't know why it should turn out that way.

Q. — Do you think it possible that the sewerage of Worcester can have tended to improve the health of Millbury? A. — It don't seem likely.

Mr. GOULDING. It is hardly worth while to waste time on such a question as that. We don't claim any such thing.

The CHAIRMAN. I think the Committee understand that.

Q. — Upon what principle did you make your selection of the ten towns? A. — It appears on the first table. It seemed to me that to compare Millbury with the whole State was not fair. To compare it with the whole county, made up of cities and farming towns, I thought would not be fair; but if a large number of manufacturing towns, of about the same size, in river valleys, could be picked out, there would be a fair standard of comparison. So I looked over the map, and found the river valleys of the State, and picked out towns of similar size, between three thousand and seven thousand population, and made an average population practically about the same as that of Millbury.

Q. — Did you pick out rivers where there had been any complaint of pollution? A. — I picked out the rivers by the map, without the slightest reference to the complaints of pollution.

Q. — Did you know, in point of fact, whether any of those rivers were polluted? A. — I did not.

Q. — You did not investigate it? *A.* — No, sir.

Q. — You are not aware of the fact that one of the rivers which you took is one of those condemned by the Board of Health as a polluted stream? *A.* — I have since read that it was.

Q. — Which was that? *A.* — Miller's River, I think. I know there was a report with reference to that.

Q. — Isn't there another one of the rivers that has been condemned by the State Board of Health, — the Nashua? *A.* — Not as far as I now remember.

Q. — The report of this year, Mr. Adams, on p. lxv, states that complaints have been made for some years of the condition of the Nashua River below Fitchburg. *A.* — That report has not come to my hands.

Q. — I understood you to say that, in selecting the towns, you did not undertake to ascertain what the condition of the rivers was upon which the towns were situated? *A.* — No: I simply intended to take all the rivers of the State that had small towns on them.

Q. (By Dr. WILSON.) — In Table III., under the year 1881, in the columns relating to Millbury, I find a column headed "Dysentery, diarrhœa, cholera morbus." That means, I suppose, that there were no deaths in Millbury in 1881, from any of those causes? *A.* — Yes, sir; and the same is true in regard to diphtheria and croup.

TESTIMONY OF DR. ORAMEL MARTIN.

Q. (By Mr. GOULDING.) — You are a practising physician in Worcester? *A.* — I am.

Q. — How long have you been a practising physician? *A.* — About fifty years.

Q. — How long have you been in Worcester? *A.* — Thirty-two years.

Q. — I want to ask you, doctor, what is your opinion with regard to the question whether the death-rate of a town or county is any criterion of the health-rate? *A.* — I supposed that our effort to get the Commonwealth to give us the death-rate was to help us in some way to improve the general health, and therefore reduce the amount of death that would otherwise occur; that every thing that lessens the amount of health diminishes the chances of human life.

Q. — That answers the question, perhaps, indirectly; but what is your opinion upon that question, as to whether the death-rate is a criterion, to any extent, and to what extent, of the health-rate of a community? *A.* — I believe it is admitted by the profession, as a rule, that the death-rate shows comparatively the amount of sickness in the community; that the death-rate is in proportion, as a rule. There are exceptions to the rule.

Q. — Where does your practice extend, doctor? *A.* — Well, it extends about Worcester, and in the adjoining towns, in consultation.

Q. — Have you been called in consultation to Millbury at all? *A.* — Not much in late years: no, sir.

Q. — You, as a physician, have a knowledge of the general health of the community in the towns adjoining Worcester? *A.* — From general reports: yes, sir.

Q. — You belong to the Medical Society, I presume. *A.* — I do.

Q. — I want to know if you have information of any epidemic or diseases in Millbury, or on the Blackstone River in Worcester, attributable to the river or the sewage? *A.* — I have not known of any peculiarly attributable to the river.

Q. — I will ask you whether throat-diseases have prevailed to any extent within the past few years in your practice? *A.* — Yes, sir: there has been a good deal of throat-disease for the last year or two in my practice.

Q. — Ascribable to the river, or any river, or in any marked way confined to localities near rivers? *A.* — Those cases that I have seen were the result usually of changes of atmosphere or changes of weather.

Q. — I want to ask you a question in regard to typhoid fever: whether a change of climate by coming into a new country is likely to produce it in the patient? *A.* — When I first came into Worcester County, I came into a farming town that employed a large number of workmen through the summer that came down from Vermont. Those people that came down from the mountainous regions of Vermont, a great number of them, had typhoid fever in the course of the summer, a great deal more than the regular inhabitants there had it.

Q. — Do you mean to say that is the rule? *A.* — That was the rule there. The people that come from healthy neighborhoods where there has not been typhoid fever are very apt to have it. It is a rule, I suppose, that people do not have typhoid fever twice.

Q. — With regard to the contagiousness of typhoid fever, what do you say about that? *A.* — Well, the profession differ about it. I think that the profession now are settling into the belief that it is contagious or infectious. I have some doubts about it myself.

Q. — To what do you ascribe the fact, that, when one person takes it in a house, several are apt to, similar conditions or causes existing? *A.* — The theory at the present day is, that it is the result of the ejections that are passed from diseased surfaces, some particles of which are inhaled. The real fact is, we do not know much about it.

Q. — Do you know any thing about the increased prevalence of intermittent or malarial fever in New England recently? *A.* — Yes, sir: I know considerable about that. I know that in Worcester County, until within the last few years, I never had seen a case of intermittent fever that had not been brought in from abroad, from malarial neighborhoods; but it is not only a fact in regard to Worcester, but it is a fact that our medical literature takes notice of, that there has been a recurrence of malarial fever through Massachusetts, especially in the southern portion of it. In Springfield there has been a great deal more than usual, and I have been informed by medical men in the county that there has been more generally. We used to have it brought from abroad, and it did not occur in people who hadn't been away from home. We have it now among people who have not been away from home either West or South.

Q. — Whether diphtheria and diphtheretic sore throats are confined in your practice to river-courses, or whether you find them on hills and everywhere? A. — Diphtheria appears strangely everywhere and anywhere. It has appeared on the highest hills we have, and in the healthiest neighborhoods, without any apparent cause. It is not established at all what diphtheria is, in my mind, and I do not think it is generally with the profession. The term "diphtheretic sore throat" is a term we all use to satisfy people who want to call every thing "diphtheria." Real diphtheria is a rare disease, very rare indeed; but such a person has a little ulceration of the throat, and we call it "diphtheretic:" that means "like." Most of what we call "diphtheretic sore throat" is the result of a common cold. The first case of real diphtheria that I ever saw in my life was on a place called Ragged Hill, in West Brookfield. It was terrible. It was a high place, where you would suppose that a person couldn't help but be healthy. Children died off there in great numbers. That was a great many years ago.

Cross-Examination.

Q. (By Mr. MORSE.) — What are the principal diseases that are occasioned by sewage or cesspool effluvia? A. — I don't know any disease that is absolutely caused by cesspools. The negro that cleaned out most of the cesspools in Worcester is between eighty and ninety years old, I believe, and he has had rheumatism.

Q. — Do you think that is a fair answer to my question, doctor? A. — Yes, I think it is.

Q. — Don't you agree with other physicians in considering that sewage and cesspool effluvia are very efficient causes of disease? A. — I think bad smells, effluvia, injure the general health, and a person whose health is below the ordinary grade is more likely to have disease; but I really cannot say that I know any disease absolutely, any individual disease, that was the result of sewage.

Q. — You mean to say that the tendency of the presence of sewage or cesspool effluvia in the atmosphere is not to cause disease? *A.* — No, I don't say that.

Q. — Is it, or not, the tendency of sewage effluvia to cause disease? *A.* — I think the tendency is to injure health, and any thing that injures health tends to disease.

Q. — What are the diseases that are most likely to be caused by the presence of such effluvia? *A.* — I should say dysentery and bowel complaints.

Q. — Are the diseases that are ordinarily caused by the presence of sewage effluvia usually fatal diseases? *A.* — I should think they were as fatal as diseases in general, setting aside tuberculous diseases. I should think dysentery was as fatal a disease as typhoid fever.

Q. — Are you not aware of the fact that a very large number of diseases is caused by such effluvia when the diseases are not themselves fatal? *A.* — Oh, I don't think that all the ills that are the result of bad sewage are fatal, by any means. Bowel-complaints with adults are not very fatal. Dysentery is a pretty fatal disease when it is severe. A number of years ago we had a very fatal epidemic of dysentery. That was in 1852, I think.

Q. — Is it not a fact that the attention of medical men, and therefore, to some extent, of the community at large, has been specially attracted within the last few years to the necessity of preventing what are known as filth diseases? *A.* — Yes, I think so. I believe the profession have got attacked with a little epidemic themselves. I think we are like all other classes in the community: when our attention is brought to a specific thing, we run it into the ground a little, like other professions.

Q. — Do you think that physicians have gone too far in enjoining the necessity of cleanliness and provisions against filth? *A.* — No, sir, I do not think they have. I think that, like everybody else, when we want to accomplish an object, we state it as strongly as the real facts will warrant. We are like all other folks.

Q. — Don't you go as far as the Board of Health of this State go in their views as to the necessity of preventing those diseases? *A.* — I don't know exactly how far they go.

Q. — Have you ever read their reports on the subject? *A.* — To a certain extent. I have not read them very thoroughly.

Q. — Did you ever read a document which they circulated in their report for 1876, an article written by Dr. Simon? *A.* — I think I read it at the time. You see I am somewhat along in years, and I don't remember as distinctly as I did thirty or forty years ago.

Q. — May I ask you whether you have yourself given any special attention to this class of diseases? *A.* — I have given attention to sewerage, so far as to endeavor to keep clear of bad smells myself, and to induce my neighbors to do so; and I have tried to see that my clients were not seriously injured by the sewerage in the south part of the city. It is only a short time ago that it was carried below Quinsigamond. I don't go as far now as I used to from home.

Q. — You were content, I presume, when the sewage was carried beyond Quinsigamond? *A.* — Well, I mean to say that I don't practise below there so as to know the effect of it. That is all I mean to say.

Q. (By Mr. CHAMBERLAIN.) — You have had an extensive practice in your time; when you have had a case of typhoid fever, have you inquired where the sink-drain was, to ascertain whether that was the cause of it or not? *A.* — When I have a case of typhoid fever, I inquire all about it.

Q. — Then you must have thought that might have been the cause? *A.* — I think a bad stink has a tendency to lower health; and, when the health is lowered, I think a person is a great deal more liable to have an attack of fever than when he is vigorous. I think that any thing that lowers the health renders a person more liable to disease.

Q. (By Dr. HARRIS.) — I understood you to say that you did not think sewage produced typhoid, except in the way you have stated, by lowering the general condition of health, and in that case the disease might be developed? *A.* — If I said that, it didn't exactly give my ideas. I said that I didn't know of a case where the direct cause was sewage; that I couldn't give a disease that I knew was in the habit of being produced by sewage effluvia, etc. That is what I meant to say.

Q. (By the CHAIRMAN.) — That is, it might be produced by that, but you don't feel sure of it? *A.* — Yes, sir.

Q. (By Dr. HARRIS.) — Perhaps you have a case where every member of the household is down with typhoid fever. There can be no doubt of that. You look around to see if you can discover the cause, and you find that the drain from the sink runs into the privy, and from the privy it has got into the well, through a coarse, gravelly soil, and the family have been drinking the water from that well. In such a case as that, would you be led to suppose that the well-water had any thing to do with the disease? *A.* — If I had one case of typhoid fever that was brought down from Vermont into a farmhouse, and the excreta were carried out, and put into a privy, and from there went into the well, and the rest of the family had the disease, I should suppose that was the cause.

Q. — That is not my question at all. The case I put was an actual one. There was a family who lived on a gravelly knoll, and almost at the same time, within two or three days of each other, the whole family came down with typhoid fever. Of course we looked around to find the cause, and we found that the sink-drain was turned into the privy, and from there we traced it into the well; and they had been drinking that water. My question was, whether, in your opinion, that water had any thing to do with producing typhoid fever in that family? *A.* — I should strongly suspect it did; and I should strongly suspect that there was typhoid matter that got in of some kind. And it is generally admitted, I think, by the profession, that typhoid fever is propagated quite largely from the excreta getting into wells. I have seen lots and lots of instances of that.

Q. (By Dr. WILSON.) — Don't you think there is a material difference between the case of the drainage from a water-closet or privy getting into a well, and people drinking the water, and the case of sewage flowing into a river, and people smelling the sewage, in the way of producing typhoid fever? *A.* — I should think there was a vast difference. I can imagine that smelling bad material might reduce nervous energy, and produce nausea, vomiting, and diarrhœa, especially with a sensitive person. But I have not known (I don't claim to know every thing, but I have seen a great deal of disease), I have not known any case of disease that I supposed was directly produced by effluvia from sewage like the sewage that you are contemplating. I always look after the sewage when I have a case of typhoid fever; always look the house over, and, if I find there is any thing bad in Worcester, I call on the Board of Health to take care of it.

TESTIMONY OF DR. J. MARCUS RICE.

Q. (By Mr. GOULDING.) — You have been a practising physician in Worcester for how long? *A.* — Since 1855.

Q. — You were in the army for a while? *A.* — Yes, sir. I was absent from Worcester for four years, nearly, and also absent about a year afterwards, when I was abroad.

Q. — Where does your practice extend generally? *A.* — Through the city of Worcester.

Q. — Into the country towns at all? *A.* — Somewhat.

Q. — Do you go to Millbury? *A.* — I have not been in Millbury much. I have not practised in Millbury. I have been there occasionally.

Q. — Have you known or heard of any diseases, ascribed by the profession to the river, as existing in Millbury? *A.* — Nothing more than the general reports which are made to the Board of Health, and the report which has been made here to-day by Mr. Adams.

Q. — That, perhaps, is not exactly an answer to my question. I asked you whether you, as a doctor, have heard in your profession of the prevalence of any diseases on that river, ascribable to that cause? *A.* — No, sir: not of my own personal knowledge.

Q. — Is there any such report in the profession, that has come to your knowledge as a physician, that there is any epidemic prevailing down there? *A.* — No, sir.

Q. — I want to know if you know any thing about an increase of malaria in this region within a few years? *A.* — There has been a marked increase of malaria within a few years, whereas previously there was none. I know in Worcester I have seen some cases within the last two or three years that I was unable to trace to any place outside of the city, and I had no doubt that they originated in the city. Also, the reports of other towns show that malaria has increased in other regions in Massachusetts, and in the adjoining towns in Connecticut. The malarial disease which I have seen (and I have seen, during previous years, a good deal of it) was all imported with us; that is, they were cases of farming men who came from the West or South.

Q. — In regard to typhoid fever, what is your opinion as to the contagiousness of that disease? *A.* — I suppose that typhoid fever may be conveyed in the excretions of a typhoidal patient. I am not aware that any case is communicated directly from the person, in the sense that measles and scarlatina are communicated. That is perhaps a mooted point.

Q. — What do you say in regard to the death-rate of a community furnishing a standard of its health-rate? *A.* — It undoubtedly does, in my judgment, furnish a standard for comparison, and the death-rate will be largely influenced by the health-rate of a city or a community, although we have, so far as I know, no sufficient statistics to establish that point entirely.

Q. — You agree in general with Dr. Martin's testimony? If there is any point that you think of, where you desire to express a difference of opinion from him, please to state it. *A.* — I believe I have not any thing to offer.

Cross-Examination.

Q. (By Mr. FLAGG.) — You are familiar with the works of Dr. Simon, are you not? *A.* — I have read some of them.

Q. — You know that he is directly against you when you say that the death-rate is a criterion of the health-rate? *A.* — I understand that Dr. Simon says that it does not bear a constant ratio; but he does not say that it does not have an effect upon it. It seems to me, as a medical man (and I presume it is the same with others), that the health of a community must have something to do with the death-rate.

Q. — You say you do not practise in Millbury? *A.* — No, sir.

Q. — You are not familiar with the health-rate in Millbury as distinguished from the death-rate? *A.* — No more than the returns show. The statistics I am familiar with.

Q. — The statistics of death-rate, you mean: there are no statistics of health-rate, and, in the nature of things, there cannot be? *A.* — I don't agree with you there: in the nature of things, there could not be statistics of the health-rate.

Q. — And there are none such? *A.* — There are none such, so far as I know.

Q. — As a medical man, knowing Millbury and the Blackstone, if you had heard reports (as you say you have not) of zymotic diseases in Millbury, would you have been surprised? *A.* — I don't think I said that.

Q. — You said you had not heard reports of diseases. *A.* — *Epidemics*, I said.

Q. — You have heard, then, of diseases? *A.* — Yes, and I have seen them.

Q. (By Dr. WILSON.) — Would you have been surprised if you had heard of the prevalence of epidemic or zymotic diseases in any locality about Worcester? *A.* — Only this: that we have not often had in Worcester, or in the adjoining towns, any epidemic disease. There has been a long series of years since we have had any severe epidemic disease in Worcester.

Q. — So you would have been somewhat surprised? *A.* — I should have been surprised only in that sense. It is always a matter of surprise. We don't know how epidemic diseases come.

Q. (By Mr. CHAMBERLAIN.) — Whether or not you think the sewage of Worcester draining into Blackstone River is injurious to public health? *A.* — I should suppose that it would not be beneficial to the inhabitants living along the line. The facts as they are shown to us by statistics do not prove it. It is very difficult, sometimes, to sustain our theories by facts, and that is the difficulty in this case. The facts are against the theory.

Q. — If you go by the statistics, it rather promotes the public health? *A.* — I didn't say that.

Q. — But you can say that it does not have a sufficiently detrimental effect to produce any increase in the death-rate? *A.* — I don't say that it improves it.

Q. (By Mr. FLAGG.) — Suppose the facts are, that the health-rate of Millbury and the towns below has decreased, is that fact against the theory? *A.* — I said the facts were against the theory. I said this: that I should suppose that the admission of sewage into the Blackstone River would be unhealthy, but that the facts, as developed at this hearing, and the facts which appear in the reports of the Board of Health, do not bear out that theory.

Q. (By Mr. SMITH.) — Let me ask you a question in regard to your theory. Is your theory based upon general observation and knowledge, or is it merely imagination? *A.* — It is based upon general observation; but, when I come down to details, I am not able to state.

Q. — But these special cases don't sustain the general observation? *A.* — No, sir. There might be contaminating matter put into sewage, and it would depend very much upon the distance which it had to traverse as to the effect. For instance, if typhoid fever is communicated by the excretions of the patient, how long a time those excretions must remain in the water, and exposed to the air and water, before they become inert, is a matter which I am unable to state.

Q. (By Mr. HAMLIN.) — Did you ever know a case of typhoid fever where those excrements were absent, or could not be ascertained? *A.* — I have seen cases of typhoid fever where one person was taken down in a family, and I was unable to determine that the matter emanated from any other patient. I have seen others in the same family contract the disease; and those cases, I supposed, were the result of contamination from the first patient. But, then, there are things which modify it. If the people who are in contact with the first typhoid case have already had typhoid fever, they are not likely to have it. So that that is eliminated.

Q. (By Mr. FLAGG.) — You stated that there was a theory that the pollution of the river was injurious to the public health. Now, if the fact should prove to be that the health-rate in Millbury and adjoining towns on the river had decreased, would not that support that theory? *A.* — In so far as that was the fact.

Q. — And you have already said that you do not know any thing about the health-rate of those towns? *A.* — I don't know any thing about the statistics.

Q. — So that you don't know that the facts are against the theory? *A.* — I have stated that I do not know any thing about the statistics.

TESTIMONY OF CHARLES D. PRATT.

Q. (By Mr. GOULDING.) — You live in Worcester? *A.* — Yes, sir.

Q. — And have lived there how long? *A.* — Forty-two years.

Q. — You were mayor of the city for two or three years? *A.* — I was mayor of the city three years.

Q. — How long ago? *A.* — I was mayor in 1877, '78, and '79.

Q. — Have you property in Millbury? *A.* — Yes, sir.

Q. — Where is your property? *A.* — In Bramansville.

Q. — Do you have occasion to go there frequently? *A.* — Yes, sir.

Q. — How do you usually go there? *A.* — I drive, almost always.

Q. — Have you been troubled with any smell in going down the highway there? *A.* — No, sir.

Q. — Do you know where this rendering establishment is? *A.* — Yes, sir, — Jeffard & Darling's.

Q. — Ever smelt any thing from that on the Millbury road? *A.* — In warm weather I have, when the streams were low.

Q. — Were you present one Sunday when there was a collection of dead fish in the river in Millbury? *A.* — I was.

Q. — Won't you tell us about it? *A.* — I drove down to Millbury one Sunday, just before noon, and my attention was called to the fact that there was a lot of dead fish just below Mr. Morse's factory, and they wished I would go down and look at them. I stopped there, and found a good many dead fish in the water, just below the factory.

Q. — What was the apparent cause of that collection of dead fish? *A.* — Well, there were three or four of us there: we talked the matter over, and came to the conclusion (I believe we all agreed), that as the ponds were very low, and the water was very low there, and it was very hot, — the sun shining in there very hot, — the water was so heated and so impure that the fish died from that cause. I took pains to go below this place, and I found no dead fish there; and on my way home I stopped at several places along to see if I could find any between Worcester and Mr. Morse's factory, and I did not find any. I would say, that when I came back from my son's place in Bramansville, I got out and put my hand in the water, to see what its condition was; and it was very warm, very hot water. The water was low, and the sun poured in there so that I should think it would be difficult for fish to live there anyway.

Q. — Was there any water flowing into that pond that day? *A.* — I think not: the water was very low. There was no water flowing into this little pond.

Q. — Where the fish were? *A.* — Where the fish were, — no, sir.

Q. — You have known Mill Brook and the Blackstone River ever since you have been in Worcester? *A.* — Yes, sir, I have.

Q. — Whether manufacturers have not always been on the river, and always put their filth into it, and the inhabitants along the line of Mill Brook, ever since you have known it? *A.* — Yes, sir, they have.

Q. — What kind of a stream was Mill Brook, as long ago as you first knew it, in respect to purity? *A.* — It was never very pure since I can remember, — forty years. There have always been these manufactories in operation there, and they have always emptied all their filth into Mill Brook.

Cross-Examination.

Q. (By Mr. FLAGG.) — Since the sewage of Worcester has been poured into Mill Brook by its present system of sewerage, Mill Brook is in a filthier condition than before, is it not? A. — I should think, when the water is low, that it would be.

TESTIMONY OF JOHN McCLELLAN.

Q. (By Mr. GOULDING.) — You live where? A. — I live on the Blackstone River, just below Saundersville, near Saundersville depot, on the Providence road.

Q. — In what town? A. — The town of Grafton.

Q. — How near to the river do you live? A. — My house is about five rods from the river, I think. The public highway goes between.

Q. — How long have you lived there? A. — Twenty-seven years.

Q. — What is your business? A. — Farming.

Q. — I want to ask you whether there is any trouble with regard to your cattle drinking the water of the river? A. — No: my cattle drink the water just as readily as they ever did, when they have occasion to drink it. To qualify that a little, I would say, in the common run of water. In high water, like what there was two weeks ago, there is a good deal of impurity in the water. It is very roily and colored; but probably eleven months in the year they would drink the water as freely as any water.

Q. — How does that river generally compare, in point of apparent purity, with the condition in which it was the day the Committee were down there? A. — Well, very different, indeed. I dipped up a pail of water that day, as I have done several times, to test the water in the river, and it was very dark, very muddy, and left quite a sediment. I let it stand over night, and let the sediment settle in the bottom of the pail. The next morning, although the water in the river had not abated but little, there was not a quarter as much sediment as there was the day before.

Q. — How is it generally with reference to sediment in the river at your place? A. — Usually there is no roil in the water. Last fall I tried it in a tin pail, and I could see the bottom perfectly, in the common run of water, and there was no sediment. There was some color in the water. The water has a yellowish color at low water.

Q. — Did you do this in consequence of any conversation with Mr. Esek Saunders at any time? A. — I do not wish to answer that question. Mr. Saunders and I are neighbors. I am not disposed to question the testimony of any one here. I only state in regard to the condition of the river at my place. I had heard it stated that the water was very roily, which led me to investigate it somewhat.

Q. — Now, what was the result of your investigation? *A.* — In the common run of water, as I have said, there was no roil in it. It had a yellowish color, about such as I see in the jar standing there, just about the sewage color: but in my water there is a great deal more impurity now than there was twenty years ago; and, as proof of this, where the water overflows its banks, and spreads on to the mowing land, it increases the production of the mowing land perceptibly.

Q. — More than it used to? *A.* — Yes, sir: I have not any doubt. I have said repeatedly, that there is a great amount of impurity emptied into the stream by the sewerage of Worcester; but my theory is, that, there being about nine or ten dams between my place and Worcester, nearly all of it settles in those ponds before it reaches my place.

Q. — Have you ever had any difficulty in getting your cattle to drink this water except during low water, one month in the year, as you say? *A.* — During high water. No, sir: our cattle, as they are driven from the pasture, where there is plenty of pure spring water, will frequently stop at the river, and drink, before going into the yard, almost always some of them; and we have aqueduct water in the yard generally from the hills.

Q. — They prefer the river-water apparently? *A.* — Well, I don't know that they prefer the river: I mean to say, that, when they have occasion to drink, they go there to drink, just as freely and just as readily as they do any water that we have.

Q. (By the CHAIRMAN.) — That is the regular drink that you intend to provide for your animals? *A.* — Yes, sir, in the summer: we do not wish to let them out of the yard in the winter time.

Q. — You are speaking now of their drinking it of their own accord? *A.* — Yes, sir: when they are passing by, they go to it of their own accord.

Q. (By Mr. GOULDING.) — Do you ever pasture them at any time on the banks of the river after mowing? *A.* — They sometimes go to the intervale.

Q. — In such cases do you provide any other water? *A.* — Nothing but river-water.

Q. — Are your cattle healthy? *A.* — Yes, sir: I don't know but they are.

Q. — Have you heard any complaint about it? *A.* — I have not heard any complaint at all. There is a neighbor of mine in Saundersville who drives his cattle down to my barn now twice a day to drink.

Q. — To drink from your aqueduct? *A.* — No: at the river.

Q. — Do you know of any farmers on the river, in your vicinity, whose cattle will not drink the river-water? *A.* — I do not know of any cases, sir.

Q. (By Mr. CHAMBERLAIN.) — Do I understand you that those cattle will drink this water just as readily as they will pond or rain water, or brook-water? *A.* — Yes, sir, just the same, if they have occasion to drink. When they are passing the river, they go in and drink. If any of them have not drank what they want in the pasture, they stop at the river and drink before going into the yard.

Cross-Examination.

Q. (By Mr. FLAGG.) — I understood you to say that there were times in the year when they did not like the water apparently? *A.* — Yes, sir: when the water is very roily.

Q. — I understand your answer to be, that there are times in the year when they do not like the water? *A.* — Yes, sir.

Q. — If people living in Saundersville, Mr. Chase, Mr. Saunders, Dr. Wilmot, and perhaps half a dozen others, have testified that they have noticed quite a smell from the river, and that cattle would not drink the water, would you have any doubt that their testimony was true? *A.* — Well, all I can say in regard to that is, that we have not noticed it at our house, — none of the family.

Q. — How near is your house to the river? *A.* — About five rods. There is considerable current along by my buildings.

Q. — Do you know whether there are places on the river where it would be likely to be more noticeable than at your house? *A.* — I have no doubt that there are above me. I have said repeatedly, that I did not doubt that Millbury people were suffering from the effects of the sewage.

Q. — And at the dams below? *A.* — There are four dams in Millbury below Mr. Morse's. If I mistake not, one complaint has been, of damming the water and letting the impurities settle; and it seems from my experience that most of the impurities settle in those ponds above the dams.

Q. — And that would be true of the dams below you? *A.* — Yes, sir.

Q. (By the CHAIRMAN.) — Is this subject a matter of common talk in your neighborhood, as to the injury that is done? *A.* — Yes, sir: I think it is.

Q. — Is there any division of sentiment about it? *A.* — I should think not much division of opinion. Most of the opinion is that it is an injury, and I do not say that it is not an injury to us.

Q. — Is it the general opinion that it is an injury to the public health? *A.* — I have never discovered it, and have never supposed it was.

Q. — What I mean is, whether it is a subject of general talk? *A.* — I have heard it spoken of as being unhealthy in Millbury.

Q. — Is there any question between the mill-owners and farmers? For instance, is there a set there who say that it is injurious to the manufacturing interests, but, on the whole, beneficial to the farming interests? *A.* — I do not know that there has been any division. The fact that the mowing lands where the river overflows are more productive than they were formerly proves to my mind that there is more impurity in the water.

Q. (By Mr. SMITH.) — So that there is some little advantage to the farmers from the overflow of the sewage? *A.* — Yes, sir: I think the lands are more productive.

Q. (By Mr. HAMLIN.) — Your family has always been well? *A.* — Yes, sir, we enjoy pretty fair health.

Q. (By Mr. GOULDING.) — What is your age? *A.* — Seventy-five.

Q. (By Dr. WILSON.) — Have you always lived there? *A.* — I have been there twenty-seven years.

Q. — Do you live near the river? *A.* — Five rods from the river: about twenty rods from the depot.

Q. — Do you own real estate in Worcester? *A.* — No, sir.

Q. (By Mr. FLAGG.) — How wide is the river at your place in summer-time through the dry season? *A.* — It is about sixty feet. I have a bridge crossing the river opposite my barn which is sixty feet long.

Q. — And about how deep in the deepest part? *A.* — Well, in low water there is very little running there. For instance, at low water, upon the Sabbath, when they shut down above, there is sometimes very little running: you can walk across, without going over shoes, on the stones. Then there is an offensive odor, as there always is where water is drawn off. There was just as much twenty-five years ago as there is now.

Q. — What sort of odor? What kind of smell is there from the water? *A.* — Well, it is a smell which probably you have noticed when you have drawn off water, — any water that has been standing. It is said, by those who claim that there is a bad odor from this river, that it is a very different odor from that where the water is drawn off.

Q. — Well, is it a very bad odor — very offensive? *A.* — Yes.

Q. — Have you smelt such smells near cesspools? *A.* — No, not at all.

Q. — You don't mean to say that wherever you draw off water that has been standing you necessarily have a very bad smell? *A.* — Well, it is a different smell from a privy or cesspool.

Q. — You say that the current at your place is quite swift? *A.* — Well, there is some current: not swift, but the water is moving. It is between the Saundersville factory and the grist-mill privilege.

Q. — What is the nearest dam below you? *A.* — It is the grist-mill.

Q. — How far is that from your place? *A.* — Perhaps a quarter of a mile. That is owned by the Saunders cotton-mill.

Q. — How much fall is there between your place, should you say, and that dam? *A.* — The grist-mill pond sets up very near to my barn.

Q. — Do you know what the amount of fall is? What is the difference in the height of that dam and the height of your land bordering on the river? *A.* — Well, my land is some two feet above the level of the water in that pond. As I said, the pond sets up very near to my barn, — the grist-mill pond.

Q. — Well, there is fall enough to make a current? *A.* — Yes, sir, against the house. The house is nearer the factory. Perhaps for ten rods there will be a current below the house.

Q. (By Mr. GOULDING.) — Is there a difference, deacon, between this smell which you discover when the water is low, and the smell which you discover when any pond is drawn down, or any river, over an extensive area which is usually covered with water? *A.* — It is the same.

TESTIMONY OF STEPHEN HARRINGTON.

Q. (By Mr. GOULDING.) — You live in Worcester? *A.* — Yes, sir.

Q. — How long have you lived in Worcester? *A.* — Thirty-two years.

Q. — Where were you born? *A.* — Westborough.

Q. — How long have you known the Blackstone River? *A.* — I went to Millbury to live fifty-six years ago this month.

Q. — Where did you live in Millbury? *A.* — I lived in Armory Village.

Q. — Whereabouts, with reference to the river? *A.* — My house was on what is called Canal Street.

Q. — How near to the river? *A.* — It is, perhaps, fifty rods.

Q. — Were you familiar with the river from that time? *A.* — Yes, sir.

Q. — For how many years? *A.* — Fifty-six years. I lived in Millbury about twenty-six years. I lived in Grafton about two years and a half. The rest of the time — fifty-six years — I have lived in Millbury and Worcester.

Q. — After you went to Worcester, did you continue to be familiar with Millbury and Millbury people, and the Blackstone River? *A.* — I occasionally worked down the stream.

Q. — On what? *A.* — I worked on Whiting's machine-shop one year. I worked most of the season down there, and built the Whiting factory, just below the Whitingville depot.

Q. — Did you work on any other mills on the stream? *A.* — I built the one at Farnumville twice.

Q. — Any other places on the stream? *A.* — And the Millbury Cotton-Mill. The Wheeler Mill I built three times on Singletary Brook. I have worked in all the factories on the streams, setting dye-kettles, and any kettles that they had.

Q. — Have you any relatives living in Millbury? *A.* — I have a brother who lives on the road to Worcester, where the bend of the river is, just before you get into the village.

Q. — What is his Christian name? *A.* — David B.

Q. — Is it in his family that the aged lady died recently? *A.* — Yes, sir.

Q. — How old was she? *A.* — Ninety-two years and four months.

Q. — Who was she? *A.* — She was the aunt of his wife.

Q. — How long had she lived there? *A.* — She had lived in his family fifty years. She lived there thirty-two years.

Q. — How old is your brother? *A.* — He was eighty-one last month.

Q. — Have you known Mill Brook from an early period? *A.* — Well, I have known it from the Washburn & Moen Works down below.

Q. — For how long? *A.* — Since 1835, when the factory was built there. I was at work in Worcester at the same time.

Q. — Have you ever worked on the brook? *A.* — I spent two and a half years there, at the Washburn & Moen Works.

Q. — What can you say with regard to the purity of that stream before it was used for the drainage and sewage of the city? *A.* — Well, all the factories standing on Blackstone River, and Singletary Brook, and Mill Brook, always had their privies standing over the stream; and all the vitriol they used went into the stream.

Q. — Can you say any thing about the condition of Mill Brook when you came to Worcester thirty or thirty-two years ago, or later? *A.* — Well, at times when there has been no rain, the stream is a great deal fouler than it is when there is a great deal of surface-water emptying in, as there is in the spring of the year.

Q. — Have you known about any other deaths in Millbury recently? *A.* — There is a Mrs. Bixby, whom I used to know when I lived there. She died this winter at the age of ninety-seven.

Q. — Where did she live? *A.* — I don't know where she died.

When I lived in Millbury she lived on the road to the Old Common, we call it.

Q. — Not near the river? *A.* — No, sir. Mr. Greenwood died right on the bank of the river. One of the witnesses, I understood, testified to a case of typhoid fever, — one of the doctors, I think it was. That man lived near the cotton-mill when I was there, the next house to the boarding-house. I noticed a well on the lower side of the house, and there was a pump in it, and a heap of manure about eight feet high near by, and a sink-drain emptying outside of the house, about twenty-five feet off. Underneath is a ledge pitching towards the well, and I thought perhaps the impurity might get into the well, as I have known several cases in my experience.

Q. (By Mr. HAMLIN.) — How old was Mr. Greenwood? *A.* — He lived to a considerable age: I don't recollect exactly; upwards of eighty, I should judge.

Q. (By Mr. GOULDING.) — Have you been engaged in digging drains? *A.* — I put in the first sewer, probably, that was laid in Worcester, from the Old Exchange down to Thomas-street Brook. That was in 1851, I think, or 1852.

Q. — The Bay State House, when that was built, what did that drain into? *A.* — That drained into Mill Brook. I did not put that in.

Q. — Is the Thomas-street Brook you speak of, Mill Brook? *A.* — It was originally Mill Brook. They have been changing the course of Mill Brook, and left that portion off.

Q. — That is, in more recent years? *A.* — Yes, sir: within a few years.

Q. — But at the time you laid the sewer, that was the natural channel through which Mill Brook flowed? *A.* — Yes, sir. Then, two years after, I put one in Front Street, from Chestnut Street down to where the viaduct crosses.

Q. — When was that laid? *A.* — I think that was in 1853 or '54. It took part of two years; that is, I did part in one year, and the other afterwards.

Q. — Now, have you observed the water in the sewers at any time? *A.* — I have frequently had occasion to go into the sewers in the street for the purpose of entering the side drains, and I have noticed that, when there had been no rain for a week or two, the water seemed to be as pure — that is, clear — as spring-water. There is an odor about it.

Q. — Do you see any floating refuse matter or suspended matter in the water on such occasions? *A.* — No, sir: there will be a little sediment at the bottom.

Cross-Examination.

Q. (By Mr. FLAGG.) — What is the condition of the Blackstone River now compared with what it was when you were living in Millbury? *A.* — Well, I was at Millbury just before this last heavy rain —

Q. — Are you familiar with the general condition of the river now? *A.* — Yes, sir.

Q. — What is its general condition now, as compared with what it was when you lived in Millbury? *A.* — I do not see any marked difference.

Q. — Were you familiar with Armory Village in Millbury? *A.* — Yes, sir: I worked there a great many days.

Q. — A great many workmen worked there at the forge-shop, triphammers, etc.? *A.* — Yes, sir.

Q. — You knew then that they used the river-water for drinking, didn't you? *A.* — I heard the testimony here the other day: that is all I know. I didn't suppose they drank it: I didn't drink it.

Q. — You did not know that they did? *A.* — No, sir.

Q. — You do not know they did not? *A.* — I have seen the workmen go to wells very frequently, or send boys.

Q. — That was in the summer-time, when they wanted cold water? *A.* — I never knew that they drank it.

Q. — You were not one of the workmen? *A.* — No, sir: I did not work in the mill. My business was mason business.

Q. — Those old people who died lived the greater part of their lives before the system of sewerage was put in — that was put in in 1868, '69, and '70? *A.* — Yes, sir.

Q. — And died soon after? *A.* — It is thirty years since I put the first sewer in.

Q. — You spoke of the sewers that you put in or worked on in Worcester in 1851, and the Bay State sewer in 1856 or '57? *A.* — I did not put in the Bay State sewer.

Q. — Are you familiar with the present system of sewers? *A.* — Yes, sir: I work about them all the time, more or less.

Q. — Do you know that there are forty-four different ones? *A.* — I don't know how many sewers there are.

Q. — And that the system drains eight square miles? *A.* — I should not think it would drain so much. There is a good deal more draining to be done in Worcester. I should think about two-thirds of the inhabitants were accommodated with sewers.

Q. — Two-thirds of about sixty thousand? *A.* — Yes, sir.

Q. — The sewer that you speak of as having put in in 1851, and the Bay State sewer, that was put in in 1856, and one or two others, drained nothing in comparison with the present system? *A.* — No, sir.

Q. — Can you tell how many houses were connected with those sewers? I am now referring to the one you worked on in 1851, and the Bay State House sewer. *A.* — In 1850 or 1851 there were not a great many; perhaps a dozen. That was a short sewer: it did not go low enough to benefit the people on the lower side of the street, but those on the upper side. The one on Front Street, there was quite a number, — some shoe-shops, — and as fast as they built buildings, they entered them all, and a good many of the old ones. It took all the surface sewage.

Q. — Was not the purpose of those sewers mainly to take care of the surface drainage? If it had not been for that, would they have been put in? *A.* — I suppose that that was one purpose, and the other was to enable people to get rid of the sewage of their estates.

Q. — How many water-closets on Front Street were connected with that sewer that you put in in 1851? *A.* — Well, there was the Harrington Block and Piper's Block.

Q. — Were those connected with the sewer at the time you put it in? *A.* — Those were large blocks.

Q. — Were they connected with the sewer when you put it in, in 1851? *A.* — Very soon after.

Q. — How soon after? *A.* — I should say within a year. I won't be positive whether the Harrington Block was built just before or just after the sewer was put in. The City Hall was connected, and Dr. Kelley's Block, and S. R. Leland's, and the new blocks all along down Front Street, as they were built.

Q. — As they were built? *A.* — Yes, sir.

Q. — But not in 1851? *A.* — No: they have not all stood so long as that.

Q. — Now, can you state about the Bay State sewer? What emptied into that when it was built in 1856? *A.* — The sewage of the house. I don't know of any thing else, — the house and the washing-department.

Q. — Now, you speak of one other sewer, — the one on Thomas Street. How many houses emptied into that? *A.* — Well, there was Hobbs's Block — that was a block of stores and houses — and several smaller houses.

Q. — How large were these sewers that you speak of? *A.* — That on Main Street, Thomas-street Brook, was about twenty-six inches high, two feet horizontal. The one on Front Street, if I recollect right, is thirty by thirty-nine inches, — egg-form.

Q. — And the Thomas-street sewer? *A.* — The Thomas-street sewer is a continuation of Main.

Q. — You have named, in fact, about all the houses that emptied into those sewers when they were built, have you not? *A.* — Yes, sir.

Q. — And the rest since they have been built, from time to time? *A.* — I have named them as they were put in, a good part of them, as we went along; that is, within a year or two: and then others have entered from that time to this.

Q. — Are those sewers that you speak of in use now? *A.* — Yes, sir.

Q. — They have put in no new ones in place of them? *A.* — They have built one up Front Street, on the north side, because, when the first one was put in, Fox's Pond prevented their putting it as low as they have put the new one, and they are both in use to-day. That is about four feet deeper. Since they drew the pond down, it enables them to put their sewer lower, so as to drain the cellars, and avoid the back water which came up in the old sewer. There was not capacity enough to convey it off, and it flowed into the cellars; but, by having this new sewer four feet lower, and connecting with that, they avoid that difficulty.

Q. — Is the Bay State sewer used in the same manner? *A.* — The old Mill Brook, you understand, came up very near, almost under the Bay State stable; I don't know but it did come up under the corner of the stable, and the sewer was simply to go right into that valley, without connection with any other sewer. It goes right into the brook.

Q. — Is that the present drainage from the Bay State House? *A.* — I suppose they have continued it, because there is a sewer goes up now to meet their case in Central Street.

Q. — It was originally used only for the Bay State House? *A.* — Yes, sir. The sewage of the Bay State House now goes probably into the continuation and into the Central-street sewer.

Q. — Made for all the inhabitants between the Bay State and Mill Brook at present? *A.* — There are no inhabitants but horses there between the Bay State and the old Mill Brook. There are people living between the Bay State and the present Mill Brook.

Q. — About the Thomas-street sewer, has that been replaced? *A.* — They have built a new sewer. When they changed the course of Mill Brook, they had to continue it farther east, and started a new sewer, and came up to Main Street. I cannot say whether it is larger or not.

Q. (By Mr. SMITH.) — You said that when you had occasion to make an entrance into the main sewer for the purpose of connecting the side sewers, you found the water to be clear, with very little sediment? *A.* — Yes, sir.

Q. — What inference would you have us make from that, — that there is no polluting matter? *A.* — There is an odor attached to it.

Q. — Nothing that apparently pollutes the water, and no heavy sewage, apparently, passes through those drains? *A.* — It is diluted so that it is not perceptible.

Q. — You would not know there was any thing being passed through there, unless it was from the odor? *A.* — No, sir: if I saw it in the spring, and did not smell any odor.

Q. — Do you think that is the rule in regard to sewers in Worcestes? *A.* — That has been my experience.

Q. — Therefore you would not think there was any thing objectionable that went out of those sewers into the Blackstone River? *A.* — I have never experienced any thing offensive, and I have been engaged putting in these side sewers more or less for thirty years. I have not called a physician but once for six or eight years for sickness. I have had accidents.

Q. (By Dr. HODGKINS.) — Is that the condition in which you find them at the present day, or are you speaking of their condition some years ago? *A.* — It has been so for years.

Q. — At the present time you find it so, you think? *A.* — I think so.

Q. (By the CHAIRMAN.) — You work there about all the time, as I understand you? *A.* — Yes, sir. I don't cut so many holes now in the sewers as I used to. I have given that up; but I have access to it, and see them cut, and sometimes cut them myself.

Q. — You work in all parts of the city? *A.* — Yes, sir.

Q. — Then your remark does not apply to any one section of the city more than another? *A.* — No, sir, only as regards the sewers that I have put in. The main sewers, the principal sewers, have been put in by other parties. I am engaged in constructing private drains all over the city, wherever they want them.

Q. (By Mr. CHAMBERLAIN.) — You are engaged in putting in those drains for the city, or for private parties? *A.* — The main drains are put in by the city: the private drains, by the owners of the estates.

Q. — Then you are employed now, and have been during these past years, mainly by the city of Worcester? *A.* — No, sir.

Q. — You are principally in the employment of private parties? *A.* — The city of Worcester have a gang of hands of their own, and men to manage, and do it themselves. There is very little done outside of what they do themselves now.

Q. (By Mr. GOULDING.) — Was not a sewer built up Pleasant Street as early as 1851 to Dr. Gordon's house? *A.* — There was a stone sewer put up Pleasant Street. I think that went up nearly to Oxford Street, — a square stone sewer.

Q. — With what did that connect at the lower end? *A.* — It connected with the sewer that I put into Main Street.

Q. (By Mr. FLAGG.) — Mr. Harrington, twenty years ago, you were familiar with Mill Brook? *A.* — Yes, sir.

Q. — It then was noticeably dirty? *A.* — There were times when it would be considerably clear, and others when it was foul.

Q. — Looking at it, you would say, irrespective of the odor, that it was dirty? *A.* — After heavy rains, the surface water would always make it look roily.

Q. — Now, do I understand you that, looking at it to-day, but for the odor, you would not notice any difference? *A.* — I should not know any difference between its condition now and twenty years ago, as far as that is concerned.

Q. — Twenty years ago, what did you notice? *A.* — There is always more or less color in Mill Brook, and most all streams about any manufacturing city of forty thousand inhabitants. We get this appearance anywhere. There is a good deal of the Piedmont sewer that don't strike the river until it gets down below Quinsigamond. Mill Brook goes down in the channel of the old canal. I notice that it is frequently colored by the matter that comes from the shops and dye-houses.

Q. (By Mr. HAMLIN.) — It is not the sewage that makes that color wholly? *A.* — No, sir.

Q. (By Mr. TIRRELL.) — I would like to ask you what proportion of the population of the city of Worcester has Mill Brook for its natural sewer? That is, supposing there was no sewer in Worcester, and the people dumped their water and so forth right into the streets, or anywhere, what proportion would find its way naturally into that stream? *A.* — Two-thirds, I think.

Q. — And, taking the other third, where would it find its way naturally? *A.* — Well, it would ultimately go around through New Worcester, as we call it, and come in just above Quinsigamond Pond, before you get to Quinsigamond Pond. All west of that would come into Quinsigamond Pond either from Mill Brook or the stream the other way.

Q. — Now, by what theory do you explain what you have stated, that the brook is no more filthy at the present time than it was twenty or twenty-five years ago? *A.* — The quantity of water that comes in contact with the simple sewage of the city is so large that it would be hardly perceptible, I think.

Q. — How does the quantity of water that is poured into that brook now compare with the quantity of water there was in the brook before the sewerage system was introduced? *A.* — I was not aware that there was very much difference, take the year through.

The freshets are not so high as they used to be before they got so many dams. For instance, I have seen the water at Millbury, before they raised the road west of Gowan's Bridge, as it is called, go up over the top of the wall, in an average year, because there were no dams to keep it back. Now they have raised the road a little higher than the wall, and I have not seen it go over there.

Q. (By the CHAIRMAN.) — How long has Worcester had aqueduct water from Bell Pond? *A.* — They had it about thirty-five years ago, I think.

Q. (By Dr. HODGKINS.) — In consequence of the city having introduced water, there is more water running into Mill Brook now than formerly? *A.* — Of course: all the city water runs into Mill Brook.

Q. — And for that reason, you suppose the sewage is diluted in proportion to the increased quantity of water? *A.* — Yes, sir. For instance, a person using a water-closet will let on from one to five pails of water.

Q. (By Mr. GOULDING.) — You said that one-third of the drainage of Worcester would go into Quinsigamond Pond. You don't mean Quinsigamond Lake, but you mean Quinsigamond River? *A.* — No, I don't mean Quinsigamond Lake; I mean Washburn & Moen's pond, and the pond at Quinsigamond village.

Q. (By Mr. FLAGG.) — On p. 118 of the Report of the State Board of Health, Lunacy, and Charity, the sewerage area of the city of Worcester is stated to be about 20½ square miles. Is that about right? *A.* — Well, I should think it was very nearly correct.

Q. — Now, do you wish the Committee to understand, that if the inhabitants of this 20½ square miles used privies, and threw all their slops and disposed of their sewage as they would have to without a system of sewerage, two-thirds of that would go into Mill Brook? *A.* — Why, I don't know where it would go to, unless it would be absorbed into the ground.

Q. — Would not the contents of the privies remain there until they were removed? *A.* — Yes, sir.

Q. — Didn't it cost the city of Worcester $1,500,000 to make this stuff go into Mill Brook, because it would not naturally go there? *A.* — Why, if you take the whole expense of introducing water, and building all the sewers, and stoning up Mill Brook, and every thing, perhaps it has.

Q. — So that it is not a fair inference from your answer that two-thirds of this would have gone into Mill Brook any way? *A.* — I don't know where it would have gone, only it would have been absorbed into the ground. The water-sheds all pitch into the brook.

Q. — That is all you mean to say: that the natural water-shed of part of the city of Worcester drains into Mill Brook? *A.* — Yes, sir.

Q. (By Dr. WILSON.) — You have been familiar with this river thirty or forty years, more or less? *A.* — Yes, sir.

Q. — How does its condition as to odor, looks, and so forth, to-day, compare with its condition twenty years ago? *A.* — I did not observe very much odor the other day when I was there. The difference in the appearance of the water was not marked to me. It is all the water that my brother's cattle have to drink.

Q. — How about looks? was it a clear stream twenty years ago? *A.* — I should think it would compare fairly with other rivers. It has rather a muddy bottom in many places.

Q. (By Mr. GOULDING.) — You speak about your brother's cattle. You mean, your brother's cattle drink that water there? *A.* — Yes, sir; and have done so ever since he lived there.

Q. — And do so still? *A.* — Yes, sir.

Q. — Do you know of any trouble about their drinking the river-water? *A.* — I never heard of any.

Q. (By Mr. HAMLIN.) — Where does your brother live? *A.* — His barn is about twenty-five feet from Morse's Pond. It is the place where one of the witnesses from Millbury testified there was the most odor in the street. It is what was a bend of the original river.

Q. (By Mr. CHAMBERLAIN.) — Do you know that they drink the water now? *A.* — I was there since the examination in February, and talked with him about it, and saw him turn out his cattle.

Q. — In regard to that particular point, do you know that cattle drink the water? *A.* — Just as readily as they do any other water. He said nothing to the contrary.

Q. — Has he any other water to give them? *A.* — He has a well that he could draw from, but he does not do it.

Q. — Instead of drinking the water from the well, they go to the river? *A.* — Yes, sir.

Q. (By Dr. HODGKINS.) — You saw them drink while you were there? *A.* — No: he was unwell, and didn't turn them out that day; but his yard runs right down to the stream. I have seen them drink repeatedly.

Q. (By Mr. CHAMBERLAIN.) — Instead of pumping the water, he lets them go to the river? *A.* — Yes, sir.

Q. (By Mr. MORSE.) — What is your brother's name? *A.* — David B.

Q. — You say he was unwell a short time ago? *A.* — Yes, sir.

Q. — What was the matter with him? *A.* — He thought it was indigestion. He didn't have a physician. He said he suffered pain across the chest.

Q. — Didn't he call it cholera-morbus? *A.* — No, sir.

Q. — Haven't you so stated? *A.* — No, sir: "indigestion" is what he said it was.

Q. — Didn't you tell Mr. C. D. Morse that he thought it was cholera-morbus? *A.* — Mr. C. D. Morse said something about cholera-morbus: I didn't.

Q. — Didn't you say it was cholera-morbus? *A.* — No, sir: I didn't consider it so.

Q. — Didn't your brother tell you he had never suffered so in his life? *A.* — He said he never suffered so much severe pain in six hours as he did the day before I was there. I didn't know that he was sick.

Q. (By Mr. GOULDING.) — When was this? *A.* — I think it was either Tuesday or Wednesday after the first hearing.

Q. — Your brother is how old? *A.* — He was eighty-one last month.

Q. (By Mr. MORSE.) — Do you desire to have the Committee understand that you personally saw cattle drink out of the river? *A.* — I have seen them drink.

Q. — When? *A.* — Well, I haven't been there very often for the last four or five years.

Q. — What is the last time that you will say positively that you saw cattle drink out of the river? *A.* — I should think it was two or three years ago last August.

Q. — You feel positive of that time? *A.* — Yes, sir.

Q. — Whose cattle were they? *A.* — David B. Harrington's.

Q. — You mean, at your brother's place? *A.* — Yes, sir. I was stopping there with him a few days, and saw them.

Q. — Have you any special reason for remembering the time? *A.* — Yes, sir.

Q. — Why? *A.* — I was doing a job of work for him.

Q. — Why should you take special notice of the fact that his cattle drank at the river? *A.* — I went out to the barn with him when he was doing his chores, and saw them go down and drink.

Q. — Did you have any special reason for noticing that cattle drank water at the river? *A.* — I did not make any report of it then.

Q. — Ordinarily speaking, I don't suppose you, or anybody else, would notice particularly where cattle drink. I want to know whether there was any thing remarkable in the fact that cattle should drink the water of this particular river, that you should remember it? *A.* — No, sir: I should think it was a very natural case.

Q. — Then you have no special reason for remembering it? *A.* — Yes, sir.

Q. — You fix that one time? *A.* — Yes, sir: that one time that I was there. I spent several days there.

Q. — Which was it, two or three years ago? *A.* — I won't be positive whether it was two or three years.

Q. — Have you been there since? *A.* — Yes, sir.

Q. — How many times? *A.* — I have been there two or three times since.

Q. — Have you ever seen cattle drink at the river since? *A.* — I don't know as I have been out to the barn.

Q. — Have you noticed any other person's cattle drink from the river? *A.* — I don't know that I have.

Q. — Did you ever make any inquiry whether they did or not? *A.* — Yes, sir.

Q. — When? *A.* — The last time I was there.

Q. — Of whom? *A.* — Of him and his wife.

Q. — What did they say? *A.* — They said they had never provided any other place for cattle to drink.

Q. — Did your brother state that his cattle drink the water from the river now? *A.* — Yes, sir.

Q. (By the CHAIRMAN.) — Was it a subject of discussion, whether cattle would drink water from the river, or not? *A.* — Well, I don't always tell all the secrets.

Q. — Is it a question that is discussed in the town, whether the cattle will drink the water from the river, or not? Do you know of any question arising down there, whether cattle will drink that water, or not? *A.* — I don't know any thing in regard to other places.

Q. — Was that the reason why you happened to speak of it? *A.* — Just as I was starting off he and his wife both alluded to the hearing down here, and they both remarked that they did not see any occasion for making such statements. They never had made the discovery themselves. They spoke of this: that they had not had a physician for sickness, himself and wife and two daughters, I think, for over thirty years; and they did not attribute any of their ill-feelings to the water.

Q. — That is what I want to get at. Do you know any more talk of that kind down there amongst the people? Have you talked with other citizens down there? *A.* — I have not talked much with the citizens down there. I haven't been down there very much. I have been at his house once or twice a year.

Q. — Did you hear him say there was any difference of opinion on that subject down there? *A.* — Well, he said some folks were complaining, but he didn't discover it in its effects upon himself or his family or cattle.

Q. (By Mr. SMITH.) — How about the smell? Did he complain about it? *A.* — He did not complain of it; never has complained of it.

Q. — Did he say that he or his family had observed those bad odors? *A.* — He did not say any thing about his daughters saying any thing about it one way or the other; but himself and his wife both spoke of not making the discovery of this offensive smell except, when the water was low, there might be a little odor, but it didn't trouble them.

Q. (By Mr. HAMLIN.) — Did they mention it as being any different from what it was several years ago? *A.* — No, sir, they did not say thing about any difference.

Q. (By Mr. CHAMBERLAIN.) — I don't quite understand of what the people of Millbury were complaining. You say they had heard complaints from others. I want to know whether those complaints related to drinking the water, or whether there was any complaint except what you have heard since this hearing? *A.* — No, sir.

Q. — When they were speaking of it did they say that other people complained that their cattle did not drink the water? *A.* — I think they did not say any thing about cattle.

TESTIMONY OF JOSEPH S. PERRY.

Q. (By Mr. GOULDING.) — You are a resident of Worcester? *A.* — Yes, sir.

Q. — One of the Highway Commissioners there? *A.* — Yes, sir.

Q. — Been so for a number of years? *A.* — Yes, sir.

Q. — Have you always lived in Worcester? *A.* — With the exception of some twelve years. I was at Auburn for some years, but I was in the city every day.

Q. — Do you own any real estate on this Mill Brook or Blackstone River? *A.* — I own some twelve or fifteen houses below Cambridge Street.

Q. — How long have you owned them? *A.* — I have been building them for the last six or eight years.

Q. — You rent them? *A.* — Yes, sir.

Q. — Do you know of any trouble from any diseases in those houses arising from the river? *A.* — I never heard of any.

Q. — The tenements rent without any trouble? *A.* — Yes, sir.

Q. — I will ask you in regard to the scavenging of the streets, whether you clean the streets and carry off the filth, and to what extent? *A.* — Yes, sir: we aim to clean up the paved streets certainly twice a week; sometimes we clean up oftener. We calculate for the future to clean them up nearly every morning.

Q. — But you have in the past cleaned them twice a week? *A.* — Yes, sir.

Q. — And where do you carry this stuff? *A.* — We carry it off, and dump it for manure and filling.

Q. — Does it get into the Blackstone River? *A.* — No, sir.

Q. — How much, in round numbers, do you carry off a week? *A.* — I should think we average from fifty to a hundred loads a week, probably, from the paved streets.

Q. — How many of the streets are paved? How much in miles? *A.* — I think that we have some seven or eight miles of paving.

Q. — As to the gutters in the other streets, what do you do with them? *A.* — We clean them up every spring all in good shape, and then we clean them occasionally during the summer.

Q. — Have you known Mill Brook for a good many years? *A.* — Yes, sir.

Q. — And the Blackstone River? *A.* — Yes, sir.

Q. — With regard to the former condition of Mill Brook, as to its purity, how long has it been impure? *A.* — I should think, on account of the increase of manufacturing establishments, it is more impure than formerly; but, according to the statements made, it is not caused by the sewage. It is caused by the manufacturing more than the sewage, I should think.

Q. — But as to the fact of the impurity, how long has Mill Brook been an impure stream, to your knowledge, and to what extent? *A.* — Well, more or less, always.

Q. — How was it twenty or thirty years ago? *A.* — Well, it was somewhat impure, but not quite so much, perhaps, as at the present time.

Q. — When Fox's Pond existed, what kind of a hole was that? *A.* — A pretty nasty hole: a great deal of sediment used to settle in there. Fish used to grow pretty large there.

Q. — Your department takes care of the catch-basins. *A.* — Yes, sir.

Q. — You have some friends who live at Ludlow Pond, where Springfield gets its source of supply? *A.* — My wife's father owns the farm next to that place.

Q. — Do you know any thing about any dead fish appearing in that reservoir? *A.* — Last season a gentleman by the name of Graves, who has the care of the place, said he gathered about five hundred pailfuls of dead fish.

Q. — Was any cause assigned for that? *A.* — No, sir: there is no sewage away up in the country.

Q. — Do you know any thing about the rendering establishment down there in Millbury, or near Millbury? *A.* — Jeffard & Darling's: yes, sir.

Q. — What is it? *A.* — It is where the dead horses, refuse, bones, and every thing that is gathered up of that kind, is carried.

Q. — How near to the river is it? *A.* — The railroad is between that and the river; I should think, something like twenty-five or forty rods.

Q. — Did you ever notice any smell from it? A. — Yes, sir: in warm weather I do very much.

Q. — Where? A. — As you go down to Millbury, not from where I am.

Q. — On the Millbury road you have noticed it? A. — Yes, sir.

Q. — Might not that be the smell that people would smell from the road? A. — It is the only smell that I have ever smelt.

Q. — Ever discovered any smell from the river as you were passing along the road? A. — No, sir.

Cross-Examination.

Q. (By Mr. FLAGG.) — Do you work on the sewers? A. — I have charge of the highways.

Q. — The effect of this scavenging is to keep from the sewers all the refuse of the streets? A. — Yes: we keep every thing as neat and nice as we can.

Q. — Do you know the sewage-flow into Mill Brook, the number of gallons per day? A. — No, sir, I do not.

Q. — The State Board of Health state it to be three million gallons. Do you think, if you keep out the refuse of the streets, that that three million gallons is mostly water-closet sewage and house refuse? A. — We clean up the streets as well as we can.

Q. — If that went in, the sewage would be still worse than it is? A. — I should think it would.

Q. — What is done with this stuff? A. — It is carried off: the farmers come and get it where we dump it. We sell it to them.

Q. — Where do you dump it? A. — We carry it on to Summer Street at the present time.

Q. — Is there not a rendering establishment on the old road to Millbury, the road leading down towards Dorothea Pond? A. — There is a place there where they take in dead horses.

Q. — Is not that the only place where dead horses are taken? A. — No, sir.

Q. — Isn't it a fact that for two years they have not taken any dead horses to this place of Jeffard & Darling's? A. — I did not know it was a fact.

Q. — Do you know of anybody taking any there? A. — Yes, sir: I have taken them there myself, I think, within less than three years.

Q. — Then you have no knowledge of that being a rendering establishment within three years? A. — I did not know but they carried horses there the same as usual. I know they carry any amount of bones there that smell pretty strong.

Q. — So that you now change your testimony, and say you don't know that dead horses have been rendered there? *A.* — I know they have been.

Q. — But not within three years? *A.* — I can't say as to that.

Q. — So that if there is no rendering there, and has not been for three years, and there is an odor, it cannot come from dead horses? *A.* — There are thousands of loads of bones go there that are pretty strong.

Q. — In working over the sewers, has your sense of smell been blunted? *A.* — Not that I know of: I can't say.

Mr. GOULDING. There is one false impression that might be created by a question put by Mr. Flagg, in which he said that the State Board of Health reported the sewerage area of Mill Brook as $20\frac{1}{2}$ square miles. We are not quite as large as that comes to. The drainage area of Mill Brook is $12\frac{1}{2}$ miles. The $20\frac{1}{2}$ miles is the entire drainage area of the city.

Adjourned to Friday at ten o'clock A.M.

SIXTH HEARING.

FRIDAY, March 17, 1882.

THE hearing was resumed at 10 o'clock.

TESTIMONY OF A. B. LOVELL.

Q. (By Mr. GOULDING.) — You reside in Worcester? *A.* — Yes, sir.

Q. — How long have you lived in Worcester? *A.* — Sixty years.

Q. — That is about your age, I suppose? *A.* — A little above that.

Q. — How long have you been acquainted with Mill Brook? *A.* — I have lived close by it all my life; within three-quarters of a mile, and sometimes bordered on it.

Q. — As a boy, you were accustomed to sail on it, fish in it, etc.? *A.* — Yes, sir.

Q. — Won't you tell us a little about the history of Mill Brook as briefly as possible? *A.* — I lived at one time near the old jail, at a public house, close by the brook. The brook there is stoned up on both sides.

Q. — That is some distance above Lincoln Square, in the northerly part of the city? *A.* — Yes, sir.

Q. — Was it stoned up on both sides when you first knew it? *A.* — Yes, sir. Then it went down just below the square to a factory or machine-shop. The old jail stood there on the corner. The stream ran down there, and took the sewage from the privies and the public house and the old jail. Then it ran down below the factory, stoned up, to old Market Street, and entered a pond near where there was a machine-shop and blacksmiths' shops; and there were privies all along the border of the brook, on both sides. Then it left there, and went down from School Street to Thomas Street, stoned up on both sides; and the buildings on each street bordered on the brook, and the privies used to stand over the brook all along down. Before you got there, about midway between School Street and Thomas Street, J. P. Kettell used to have a hat manufacturing-shop, and used to drain his dyestuff into the brook, and darken the water; and, when we wanted to fish in the brook, we had to wait until this dyestuff cleared away so that we could see the bottom of the brook. At that time, after we left Thomas Street, there was no other street from Thomas Street all through the meadow; but the brook was stoned up below Thomas Street, about half-way between Thomas Street and Central Street (as it is now): the rest of the way it was an open brook through the meadow. Then it went down through the meadow near Rice, Barton, & Fales's factory, and then it entered the brook, — that is, the old natural brook. But, after the canal was made, there was a gate built in it, and the water filled the upper basin. There were two entrances from this brook into the canal.

Q. — When was the canal made? *A.* — I think in 1827 or '28. That is, the first boats came up then.

Q. — Were there any other manufactories or dwellings on the brook below Rice, Barton, & Fales's, before it reached the Blackstone River? *A.* — Oh, yes, sir: all along down the river.

Q. — How many, and what were the principal ones? *A.* — A good many of them that bordered on the brook had their privies over the brook. There was a basin right there by Rice, Barton, & Fales's, where the boats used to land. Then, there was another basin up near the square where the boats used to come up. It was in 1828 that the first boat came up. The sewer was before that.

Q. (By Mr. FLAGG.) — The sewer? *A.* — You might call it a sewer. Stone sewers went into the brook. They have been discontinued since.

Q. (By Mr. GOULDING.) — How many stone sewers went into it as long ago as 1828? *A.* — There was one in Thomas Street, that lies there now: that lies there dormant. There was another in School Street. I guess the one on School Street has mostly been taken up. It went down there below the basin into the old mill-pond, where the old red grist-mill stood. The first gate was there; and the water flowed back to the upper basin. That old mill-pond was a kind of catch-all for every thing, at that time, and all along for years. Dead animals of all kinds were thrown in there. I have seen dead hogs, dogs, cats, and every thing else of that kind, in the pond. That is not done now, because the city don't allow it. At that time it was a very common thing to see dead animals, and one thing and another, floating around in the pond.

Q. — How did the purity of the stream forty years ago compare with its present condition, so far as is apparent to the eye? *A.* — After the canal was built, we couldn't see the bottom of the canal: it was muddy. The black meadow mud used to wash in there, and keep it roily, so that we could hardly ever see the bottom of the canal after the boats began to run. In fishing for suckers, we have got to go to the bottom; and it was always dark. Once or twice they dug out where the brook enters the basin, because the mud got in there.

Q. — What brook do you refer to? *A.* — I refer to this brook that runs through the city.

Q. — You mean Mill Brook? *A.* — Yes, sir.

Q. — Mill Brook ran into the canal? *A.* — Yes, sir: supplied it.

Q. — After the canal was constructed, the whole of Mill Brook flowed through the canal, I suppose? *A.* — Yes, sir.

Q. — Have you known the Blackstone River ever since you were young? *A.* — Yes, sir.

Q. — What have you observed about the ancient purity of that stream, if any thing, as compared with its present purity? *A.* — I don't think it is any better now. Of course, there is more empties into it now than then, as far as that is concerned.

Q. — Do you remember any thing about its old condition, — whether it has been for a good many years polluted, or whether the pollution is a recent thing? *A.* — Ever since I can remember, there has been more or less drainage into it from the estates all along its borders.

Q. — Your business is that of a manufacturer of sewer-pipe, is it not? *A.* — Yes, sir.

Q. — Have you done a good deal of work on the sewers of Worcester? *A.* — I have, considerable.

Q. — And observed the water that flows through the sewers? *A.* — Yes, sir.

Q. — To what extent, and on what occasions? *A.* — Oh, well, I am tapping them all the time: more than one a week on an average, taking the year round, I think.

Q. — What do you say about the the appearance of the water that flows through the sewers? *A.* — If there has been no rain for a few days, it is very clear in some localities.

Q. — How does the water running in the lateral sewers usually look? *A.* — If there has been no surface-water running into them, in some portions of the city it is very clear. In other portions, where there are manufacturing establishments which turn in their dyestuffs, and shoe-shops that turn in their blacking, etc., of course it is colored.

Q. — Where there is only sewage from houses emptied into them, what is the condition of the water? *A.* — It is very clear indeed. In some localities, if you stood at a man-hole and looked down, you would think it was clean enough to drink, as far as clearness is concerned.

Q. (By Mr. CHAMBERLAIN.) — Any scent to it? *A.* — I presume there is.

Q. — Don't you know whether there is or not? *A.* — We get a smell from the sewers worse than the water, — the gases from the sewer. In some localities we do not get but very little of that. They ventilate the sewer in Main Street, and in some other streets.

Q. (By the CHAIRMAN.) — Is the water clear throughout the city? *A.* — Some of it is dark-colored.

Q. — When the drains carry mainly house-sewage, how is it? *A.* — Some of it is perfectly clear.

Q. — And where it is otherwise, it is where mills empty in? *A.* — Yes, sir. There is a felt-factory where they use blacking, and, I suppose, other stuff, the same as shoe-shops.

Q. (By Mr. CHAMBERLAIN.) — How do you account for the fact that the water in the sewers is clear? *A.* — I don't understand it myself, unless it is because there is so much water running in from the springs. There are a number of springs on the hills above Main Street that drain into the sewer, and that water makes it clear.

Q. — You say, that on looking down through a man-hole, the water appears perfectly clear? *A.* — It is, if there has not been any surface-water running in for a few days.

Cross-Examination.

Q. (By Mr. FLAGG.) — Looks clear enough to drink, you say? *A.* — Yes, sir: you look down ten or twelve feet, and it is perfectly clear.

Q. — Have you ever made a mistake, and drank it? *A.* — No, sir.

Q. — How do you avoid that mistake? *A.* — I ain't dry about that time.

Q. — There is an odor about it, you say? *A.* — I presume so: I don't know whether the odor comes from the sewer or the water. There are gases in the sewer.

Q. — Do you know that fresh sewage is not so offensive as sewage that is older? That it decomposes in a few days? *A.* — I don't know what difference there is. Perhaps it is more concentrated.

Q. — The description you gave of Mill Brook started somewhere about 1828, I understand? *A.* — Before that time. That was the time the first canal-boat came up, if I am not mistaken.

Q. — You spoke of some sewers that emptied into Mill Brook from Thomas Street? *A.* — Yes, sir.

Q. — Do you mean to say they emptied in there in 1828? *A.* — Yes, sir.

Q. — Mr. Harrington testified yesterday that he built the Thomas-street sewer in 1853 or '54? *A.* — Yes, sir: that was built after the stone sewer had been in a good many years.

Q. — Describe that. *A.* — It was a common stone sewer, two feet square, stoned up at the sides, and covered over with flat stones.

Q. — What was its purpose? *A.* — To take off the water from Main Street.

Q. — Not to take off the sewage of water-closets and house-refuse? *A.* — I don't know what there was in it. I have been in that sewer a good many times.

Q. — What was the purpose of the School-street sewer? *A.* — That was to take the water from the street.

Q. — But not the sewage from privies? *A.* — I don't know: I never have been into that.

Q. — Do you know how long Worcester made a business of carting off cesspool matter? *A.* — I can't tell you how many years ago it was begun.

Q. — Is it not true, that, until a few years, that has been done on a large scale? *A.* — It has been done more or less ever since the catch-basins were built. Of course they fill up, and have to be cleaned out.

Q. — I mean cesspool matter from the different houses? *A.* — No, sir: from the street, — the wash from the street.

Q. — How have they got rid of the cesspool matter from the different houses in Worcester? *A.* — Oh, there are cesspools built all over the city now. Some of them enter the sewer, of course. There are underground cesspools in some localities. There are three cesspools connected with some houses, one after another, at the present time.

Q. — Is it not true, that, within a few years, that has been done away with? *A.* — It has been a number of years since they began. Since the sewer was put in, all have entered it that could; but of course there are a good many that have not entered it yet.

Q. — Do you know of any sewers that were constructed to 1851? *A.* — I do.

Q. — Mr. Harrington put the first sewers into the streets, didn't he? *A.* — I don't know.

Q. — Before that, whatever sewers were constructed were built for the purpose of taking merely the storm-water, — surface-water, — were they not? *A.* — There were brick sewers put in before that. There was one connected with the Lincoln House that went down into the meadow, and of course the drainage found its way into Mill Brook.

Q. — You spoke of fish in Mill Brook. Are there any fish in it now? *A.* — I don't know; I have not fished there of late.

Q. — Do you know the quantity of sewage flowing into Mill Brook? *A.* — No, I don't know.

Q. — The State Board of Health, on p. 119 of their Report of this year, state it to be 3,000,000 gallons a day. Do you know the ordinary dry-weather flow of Mill Brook? *A.* — I do not.

Q. — The State Board of Health state its ordinary flow at 3,500,000 gallons. Now if 3,000,000 gallons of sewage are put into 3,500,000 gallons of water, its effect would be to pollute the water, would it not?

Mr. GOULDING. The State Board of Health do not make any such statement, I think.

A. — Of course it would not be so good as pure water. There is no question about that.

Q. — What is the condition of Mill Brook in its ordinary flow now as compared with what it was in those early days to which you have referred, with reference to the purity of the water? *A.* — Of course the water is colored by reason of the fact that there are so many more manufacturing establishments: double, treble.

Q. — There is more sewage, is there not? *A.* — Yes, sir.

Q. — You spoke of Mill Brook draining all the estates on its border, and of Blackstone River draining all the estates on its border. What did you mean by that? *A.* — The privies used to stand over the brook all along down, and the drainage from the houses used to run down into the brook. All the estates that bordered on the brook could get in there with their sewage, and did so years ago all through that region.

TESTIMONY OF E. B. STODDARD, *Mayor of Worcester.*

Q. (By Mr. GOULDING.) — How long have you lived in Worcester? *A.* — I went to Worcester to live in September, 1847.

Q. — Where were you born? *A.* — I was born in Upton, about four miles from Farnum's Village, which is on this stream.

Q. — You have always been familiar with the Blackstone River, even before you went to Worcester? *A.* — I have.

Q. — After you went to Worcester, did you become familiar with Mill Brook, and what special reasons, if you have any special reasons, have you had to know about it, from that time to the present? *A.* — I studied law in Worcester, was admitted to the bar in 1849, and commenced practice there, and had more or less to do with the estates on Mill Brook. I had more or less occasion to go to Millbury. Having a sister who resided in Millbury, I used to go down there quite frequently. I have a brother-in-law who resides there now; some of my family residing there, and I go there now, and am about the stream in a measure. In 1852 I built a house on Pleasant Street, and that is a street where there is a sewer. At that time, on the top of the hill, or near the top of the hill, was a blind drain. After we got down about half-ways, was the stone drain, as I have seen it open opposite Dr. Martin's house, who testified here. I don't know when that was put in; but, at any rate, Dr. Martin drained his cellar into that, because he built his house the same year I did. That drain ran down somewhere. Whether it ran down Front Street, or not, I cannot tell: I don't recollect about that. Along about that time, I think, they had drains laid down in some of the streets. In 1856 the Bay State House was built. I know that that drain ran into Mill Brook, because I had an interest in the Bay State House, and that drain was laid at that time, and there was more or less of the sewage got into Mill Brook. Of course, after I went to Worcester, the old canal was abandoned. I recollect the brook when I went there: it was a dirty brook, and went round a good deal in the meadow. In 1863 I was alderman of the city of Worcester, and then, I think, water was introduced into the city. After the water was introduced, we began to talk about the matter of sewage, and the drains more especially. About that time, in 1867, the Act was passed by the Legislature giving the city of Worcester a right to drain into Mill Brook sewer, giving them all of its rights. Mr. Merrifield, who owned a large machine-shop in the valley there, found that his boilers were injured by the dirty water of Mill Brook. He didn't complain at that time of the sewage, but he came to consult me about it; and a suit was brought against Mr. Lombard, who owned the machine-shop above him, for putting vitriol and stuff into the brook. Mr. Merrifield claimed that it killed the fish, and that was one evidence that the chemicals were injurious to his boilers. That suit was carried into court, and was decided, and you have the report of the decision, which has been referred to by counsel here.

Q. — You brought that suit? *A.* — Yes, sir. I was in that case with Mr. William Brigham of Boston, who was a brother-in-law of Mr. Merrifield.

Q. — Did you have occasion at that time to investigate the character of Mill Brook? *A.* — Yes, sir. It was a dirty stream, from the manufacturing establishments that emptied into it, and also from the sewage that was run into the stream at that time. Then afterwards I had particular occasion to examine into the matter in 1872 or '73, when we built the viaduct, with the Nashua and Worcester Railroad. I was on a committee with the railroad to build that, with Mr. Allen the engineer. We had to take up a portion of the sewer between Foster Street and Franklin Street, and go down there and rebuild it. The water was very impure at that time. I have known the stream ever since, and of course the water has grown more and more impure, I think, from year to year, not entirely from the fault of the sewage, but because manufacturing establishments have increased to such an extent. For instance, when Mr. Washburn started in Worcester, he began with a small blacksmith's shop on that stream, and now the establishment of Washburn & Moen, the wire-mill at the head of Salisbury's Pond, employs to-day, I think, two thousand men. The Committee saw that stream the other day, and the water, as I looked at it, appeared to me in about the same condition, roily and polluted, as it did where the sewer enters the Blackstone River at Quinsigamond. At any rate, I had specimens of the water taken the next day after the Committee were there at Washburn & Moen's mills, and you can see those specimens. But before the Committee came to Worcester, I directed that specimens of the water should be taken, all along down the brook, and they were taken; that is, before the recent storm came on. We have them here to show to you. Specimens were taken also, I think, from Singletary Brook, and from the sewage below. Of course, at the time the Committee were there, as we had had a storm, there was an immense quantity of water flowing over. I think it was against us, that it was unfortunate for us, that the Committee saw the water when it was at its highest stage, because it was turbid. Everybody knows that, with such an immense body of water as was flowing at that time, the water would look very roily, the same, as everybody knows, even in a summer storm, that from an hour's flow into a trout-brook, or any thing of that kind, the water becomes so roily that you cannot fish: you have to wait for a day or two before you can do any thing; and the Blackstone River, at that time, looked as much disturbed as it would at any time during the year. I was glad that the Committee saw it at its height at that time, because it was a good time, as it seemed to me, to compare with it Singletary Brook, and what was coming in there.

Now, after 1872 or '73, when the viaduct was built, the stream was used by everybody for manufacturing purposes, and they put in what they chose to. I do not know that there is the slightest restriction in that respect. Starting with Washburn & Moen, who put in their vitriol and other chemicals that are used in washing their iron, and their acids, if they are asked about it, they claim that what they put in helps purify the sewage of the stream. It is the same way with some of the other manufacturers, who put in dyestuffs. At Washburn & Moen's there is a large dyeing establishment, I think. That starts with the stream; and so it goes down. There is a large dyeing establishment on Foster Street, where they dye a good deal of felting, and things of that sort. There used to be at Fox's factory, when I first went there, a pond, and the sewage went into that, until the dam was taken away. The pond of Washburn & Moen at Quinsigamond Village has been substantially taken away, because the sewage goes under it.

We start with the proposition which I do not think the counsel on the other side can deny. You will find this statement in the case of Merrifield against the city of Worcester, and I think it will be borne out by any quantity of testimony that we can bring before you. Before the passage of the Act of 1867, chap. 406, the city and town of Worcester had laid out and built sewers in several of the streets, which sewers terminated and discharged into the brook, at a point above Green Street; and from time immemorial the stream has been used by the city and town, and by the inhabitants, for sewage purposes, and numerous private drains have discharged into it. That is on p. 509, vol. xii., of Massachusetts Reports. That is what we claim to be the condition of affairs now, as a claim of law. We say that the manufacturers have a right, an immemorial right, and the people of Worcester have a right, to use that stream for business purposes, fairly, and may enjoy the free use of it, and that that right extends far beyond the right of the mill-owners, who got their rights in 1795 by statute law. That is one of our claims.

Then, to come down to the sewage question, and the city entering the brook, we claim, under this Act, that we have a right to enter Mill Brook, and use it as a sewer, and nobody can deprive us of that right; and even this Legislature ought not to deprive us of it, and we do not think they have the right to.

Then we come to the point of the remedy. I was led into a slight investigation of this subject last November, and I took it up from my own stand-point, and made up my mind with regard to it; and before I had seen any report from the Commissioners appointed by the State Board of Health, Lunacy, and Charity, I made up my mind that perhaps some experiments might be tried by the city of Worcester, provided that the people of Millbury didn't object to trying some experiments, and I stated fairly what I thought in regard to it.

Q. — Inasmuch as the views of former mayors have been put in, I will ask you if you have here a statement in writing or print? *A.* — Yes, sir.

Q. — That was a part of your Inaugural Address? *A.* — Yes, sir.

"Situated as this city is, where only a single large stream flows directly to the sea, it has, or ought to have, the right to a way of necessity for its sewage to flow.

"The present controversy in relation to sewage between the city of Worcester and some of the towns on the Blackstone River, is one of momentous interest and concern.

"That stream is like a common passage-way; and no one corporation has exclusive ownership in the purity of its waters for manufacturing or culinary purposes. Every mill or house situated near its banks has for years contributed more or less to defile its waters. The theory that the city of Worcester is responsible in damages as a wrong-doer, because it is obliged of necessity to flow its sewage into the only channel which nature has provided, and where, by the express terms of a special statute, it is authorized to have such outlet, is not reasonable, and can hardly be sustained as good law.

"The old law of fixtures, for instance, has from time to time received new breadths of construction in the decision of the courts, to suit the requirements of business. So the unavoidable pollution of such a stream, long used to receive the impurities of mills and manufactories before the system of sewage by this city was adopted, is a potent reason why the city should be allowed to empty its sewage into the only stream which nature has provided to receive and remove it.

"Perhaps a different rule of law will prevail when it is shown that the sewage is allowed to accumulate on lands lying upon the stream, thereby creating a nuisance injurious to public health. The General Court has control of questions affecting public health, and can enact laws to have specific nuisances abated.

"In the present case it is not a matter of fact, determined by full investigation, that any injury to health from effluvia exists in an unusual degree when the ponds between Worcester and Millbury are drawn down in the summer months; though there are complaints that the health of citizens of Millbury is affected when such a condition exists, and the usually flowed lands are exposed.

"To meet these complaints, and any such exigencies as exist more or less in other cities or towns of the State, I think a remedy could be applied with some reasonable hope of success.

"The question of how far the ponds situated near the centre of towns in this State should be controlled by their Boards of Health, so as to keep them full of water during warm weather in the interest of public health, is one which the Commonwealth should investigate at its expense, through the State Board of Health.

"Entertaining this view, and desiring to urge this consideration before the General Court, I go still farther, and recommend, if a general law cannot be obtained, that this City Council should petition for authority to so control the water in the ponds in Millbury, on or near the Blackstone River, that it shall not be drawn down below the raceway of the dams, in order that the low lands may be kept flowed from May to November. Such control to be regulated by a proper Board, who should order when the waters may be used, and who should cause the ponds to be refilled from the flow of the natural stream.

"My suggestion is to ask for an Act limited to two or three years, with the provision that any reasonable damage caused to parties should be paid by the State or by the city of Worcester as might seem just. By such an experiment the fact could be ascertained whether there are just grounds of complaint of a nuisance to public health which could not be remedied without serious expense. I make the above suggestions anticipating the fact that mill-owners will object, because they may want to use the water in the daytime which collects at night.

"I do not wish it understood that I think the city is in fact committing any nuisance, or is responsible, morally or otherwise, for a condition of things inseparable from the existence of a large community at this point. In other words, this city has a right to exist, and become, from its situation and by its enterprise, still larger; with an inalienable right to enjoy light, air, and water, with the privilege of drainage added, and that without being subject to pay tribute to any one. I only suggest, since the matter is in controversy, that the facts may perhaps be ascertained, and possibly a remedy found, if any be needed, by a simpler and less expensive method than has hitherto been proposed."

I know the land very well where the commission recommend that the city should try the experiment of downward filtration. It is on what is called Hull Brook. I have been over that a good many times in the spring and in the summer. It is very wet land. Of course, in order to do any thing, it would have to be very much drained, as they recommend the plan. My theory is this: After the dam at Fox's Pond was taken down, — where a good deal of the solid material in the sewers used to settle, — the water ran there, so that there is no place for it to settle, except as it goes along. The next pond is Washburn & Moen's, at Quinsigamond. That used to collect the sediment and fill up; and now the sewage, instead of going into that pond, is taken below in the drain, where you saw it comes up. Then the next settling basin is a little dam which has been built by the Burling Mills within a year or two,—a little pond. Below that is Mr. Morse's pond, where the trouble is. Now, I say, in those remarks that I have made, that if Mr. Morse's pond was kept up in the summer time, if the city of Worcester had the right to control that dam, so as to keep the pond up during the summer, and let it go down when the freshets come, or have a sluiceway in their dam, so that they could let the sediment out, if there is any, there would be no trouble from any bad odor or bad smells. The great trouble, if there is any, comes from the flats round Morse's Pond. When it is dry in summer, and those flats are exposed, of course there will be as much odor from those flats as there is around any mill-pond. I could take you to a dozen ponds around Worcester, in the summer-time, where you would smell the same odor. If any gentleman has been in the habit of fishing in a mill-pond, and standing on the borders, he will know that much effluvium will come up from the soil, where there is no sewage. Now, if Mr. Morse's pond was kept full during the summer-time, I think there would be no trouble from any odor from that; or if the dam was taken down, and the water allowed to run through there, then the sediment would go into the next pond. I have made a little plan, which I think might be adopted perhaps to advantage; and that would be to form a receiving basin on the north side of Burling Mills pond. There are thirty-five acres there which could be used for a basin, where the sediment could collect. If the city of Worcester had a chance to try an experiment of that kind, and see whether the sediment would settle, it would not be a very difficult thing, at certain times in the year, to take it out, so that it should not run below. There are two or three places where small experiments could be tried, at very limited expense, which I think for the time being ought to satisfy the inhabitants of Millbury. Sometimes there is a difficulty, in a city government which is changing from year to year, in carrying out any particular series of experiments. As one mayor comes in, another goes out; the aldermen who had charge of the matter go out, and another set come in. If I was going to propose any thing, I would have a general law applicable to all cities in the Commonwealth, that they might have a right to have a commission, who should hold office for two or three years, whose duty it should be to look after the sewage, and have the whole control of sewage and water in those various cities in the Commonwealth. That is what ought to be done by every city, instead of leaving it in the hands of the mayor and aldermen. Of course, they do not want all their rights taken away, but that could be arranged. If there was such a commission in Worcester, I think they could have a chance to try experiments which ought reasonably to satisfy the inhabitants of Millbury that the city of Worcester does not desire and does not mean to do any thing which would really be detrimental to the general health of the town of Millbury. I object to the general rumors that are put forth; as, for instance, it was testified here the other day in regard to a man by the name of Wilmarth, who died at Farnumsville, I think of typhoid pneumonia, that Dr. Gage of Worcester said that his death was caused by malaria resulting from the sewage of Worcester. That was the statement that was made. I took occasion to write to Dr. Gage, night before last, not having an opportunity to see him, and asked him if he had made any such statement. I would like to put in his letter, to show that such evidence may be cooked, and that it is not true. Dr. Gage writes to me in this way: —

WORCESTER, March 15, 1882.
Hon. E. B. STODDARD.

Dear Sir, — Your note of this morning is at hand.

I was called to Farnumsville last Sunday (12th inst.) to see Mr. T. W. Wilmarth, in consultation with Dr. Maxwell. Mr. Wilmarth was a man about sixty-four years of age. and by occupation was superintendent of the cotton-mills at Farnum's. He had been ill about one week, and died, as I was informed, early Monday morning. His disease was *pneumonia*, and was undoubtedly caused by an imprudent exposure after taking a warm bath.

Neither sewage nor malaria nor the polluted river was spoken of by myself or any one else during my visit at Farnum's; and I am confident that no reason exists for supposing them in any way responsible for Mr. Wilmarth's sickness and death.

Yours very truly,

THOMAS H. GAGE.

The other morning there was a fire in Millbury, and I think they telephoned for one of our steamers to go down; and it went down, and did some little service: at least, it played on the fire. I inquired of Mr. Brophy, the engineer, to find out what effect the water taken from the river had on the steamer, because a fire-steamer is very sensitive to any polluted water, or any thing that would injure it; and Mr. Brophy's statement is, that he couldn't see that it injured it, or had the slightest effect upon the machinery of the steamer in any way, shape, or form.

There is one thing which I would like to state here. Midway between Worcester and Millbury there is an establishment which has been brought to my attention and notice, which I think has a good deal to do with the pollution of the Blackstone River, and for which the city of Worcester certainly is not in the slightest degree responsible: it is a *rendering establishment*, where the refuse is carried from the different markets, dead horses, and a variety of things, and those substances are rendered there. It is quite a large establishment. They try to keep it clean, so far as odors are concerned; but I understand that they have from that establishment a pipe drain which runs directly into the Blackstone River, which has nothing to do with the sewage of Worcester. They run their refuse into that; and, if there had not been such a storm the other day, I think I should have taken the Committee over there. I understand that blood and meat and other stuff go into that drain, and then go into the Blackstone River. I do not think the city of Worcester is responsible for that.

Q. (By the CHAIRMAN.) — Is that in the town of Millbury? *A.* — My impression is that it is pretty near the line, — just across the line, or right on the border. But I think the putting of that refuse into the stream is something that should be looked after, either by Millbury or the city of Worcester, one way or the other, because I think it has a good deal of effect upon the stream.

Q. — (By Mr. CHAMBERLAIN.) — Is that steam rendering done in tight tanks? *A.* — They are not such tanks as are used in other establishments of the same kind, by any means. They carry all the steam into earth beds.

Q. — They do not burn the gas? *A.* — I think not. There are times, when the wind is in a particular direction, that that rendering establishment can be smelt for a mile; and I am not surprised that the people of Millbury and Burling Mills should smell something there which they think comes from that stream. I think we shall show you by evidence that it does come from the rendering establishment.

Q. (By the CHAIRMAN.) — Are you prepared, as the Mayor of Worcester, to submit to this Committee any bill asking for any power or authority to do any thing that you have not the power or authority to do now, in the way of experiment? *A.* — I have my individual opinion, but of course I should not want to make any proposition without consulting with the other officers of the city.

The CHAIRMAN. I understand that the power of the mayor of a city is limited.

WITNESS. Without consultation, I should not want to submit any thing. Still, I think, if there was no other way, rather than have this experimental system adopted — with the prospect of putting the city of Worcester to the expense of five hundred thousand or a million dollars, which I do not think for ten years would answer any purpose, — until we know something more about it, I should not think there would be any harm in giving the city of Worcester, or any town or city in the Commonwealth, if the Committee saw fit, after full and mature consideration, the right to take land for the purpose of improving the water of their rivers or their sewerage system; and also the right to have the control, through their Board of Health, of any dams or ponds situated in such town or city, for a year or two, as I have suggested in my address, for the purpose of making the experiment, inasmuch as you will find throughout the Commonwealth complaints of this kind. I am informed that the people of Clinton complain of their river. Near Wachusett Mountain the water is drawn down from a pond in the summer. A gentleman from Clinton told me last Friday, "We are in the same box." There should be a uniform act, or one giving the city of Worcester the right to take land and the right to control any dam between Worcester and Millbury for two years if they see fit to make an experiment in the matter of controlling their sewage. I cannot see what harm there would be in such an act; but I can see that it would be very unwise to ask us to adopt the plan proposed by the report of the commission to the State Board of Health, and within four months from the passage of the act to remedy this difficulty. We could not do it. There would be no use in the passage of such a bill: we could not turn round. I should have no objection to the appointment of a local commission, although a great many people in Worcester might object.

Cross-Examination.

Q. (By Mr. FLAGG.) — I assume that you are a man familiar with Worcester and the Blackstone, and look at this thing as a public-spirited man. And now I would like to ask you, if, from all the causes of which you speak combined, including sewage, there is not such a state of affairs in the river as that something ought to be done by somebody? I understood you to say there was trouble, and something ought to be done by somebody. *A.* — Yes, there is trouble; and I think that perhaps an inexpensive experiment might be tried, and perhaps ought to be tried. But, when the manufacturers of Millbury claim that we ought to take care, not only of the sewage, but of the pollution of the stream which is caused by the manufacturers, I object.

Q. — I don't understand that the manufacturers of Millbury do that: you have not heard from them. *A.* — That is what they talk; we get it from them. They say the water comes down there, and injures their steam-boilers, etc.; and that is the chief complaint that we hear. The chief trouble comes from them.

Q. — The chief trouble comes from this Committee now, does it not? *A.* — I am willing to meet the Committee of Millbury fairly; but when the Committee of Millbury say, "If you adopt this system of downward filtration, and spend five hundred thousand or a million dollars, you have got to pay us for the evaporation of the water," it don't seem to me to be very liberal, and I should not expect the people of Worcester to indorse any such statement as that. They think they have a right, if their dams give way, to use a little of the water above, without having everybody put in a little bit of a claim. They think they have a right to go into this brook and use it, without paying tribute to Cæsar every time they move.

Q. (By the CHAIRMAN.) — Do you think, from what you hear said by the citizens of Worcester, that there is a feeling among them that some experiments ought to be tried? Do you think there is a disposition favorable to such experiments? *A.* — I think they would leave it to the City Government. I have talked with some of the heavy tax-payers there; and I think there would be a disposition to have a reasonable experiment tried, if the city were allowed to try it; but to be forced into trying an experiment suggested by engineers, who do it simply for the purpose of trying an engineering scheme, without knowing what the results would be, — I think there would be opposition to it.

Q. — Do you think, as a matter of judgment, that an experiment would be tried, if this Committee should report some such bill? I don't want you to infer that they think of doing it, but I want to know how near we can get together. *A.* — I think the City Government would. I think that they would like to see if something could be done. I have no doubt about it; at least, I should try to help them. I want to be fair about this thing. I do not want to injure Mr. Morse, who is my friend, or any of the people of Millbury.

Q. (By Mr. SMITH.) — On the whole, you think the time has come when something ought to be done about it? *A.* — I think it should be looked to. For instance, I think we should see if we could not form another basin there, that would act as compensation for the two basins that have been taken away, Fox's Pond and Washburn & Moen's Pond; and I would see whether, if we allowed the sediment to settle in a basin, or in two basins, and then removed it, it would not afford a remedy. If it would, it seems to me that that would be a very reasonable remedy; in the same way as when they talked about filling up the Back Bay, the engineers all said, "If you fill up the Back Bay, you must make compensation, so as to give the water a chance to run somewhere else." I think something of that kind should be tried, but I don't think the city of Worcester should be forced into this scheme.

Q. — Then I understand your answer to be, that you think the time has really come when some movement should really be made in that direction? *A.* — I don't know why we could not begin just as well in this way as any. What I mean to say is, that I think the city of Worcester, standing upon its rights, is willing, if you can suggest any reasonable experiment, without too much expense, to take hold of it and try it. I think the city ought to do something, and I do not know but what the State ought to say, "We will try the experiment, and see what will be the effect of keeping up a pond full in the summer, where there is complaint of malaria and of odors when the pond is drained down."

Q. (By Mr. SMITH.) — It would be a pretty dangerous experiment for the State to undertake that with regard to one city, would it not? *A.* — Perhaps it would, but the State has got something to do.

Q. (By Mr. FLAGG.) — Assuming that something ought to be done by somebody, the State Board of Health have recommended certain plans. You speak of authority. Now, in order to carry out any of those plans, must you not have a bill something like the one proposed by us? *A.* — It has been so long since your bill was read that I have forgotten what its provisions are.

Q. — You must have a bill of some sort, must you not, to do the things you suggest? *A.* — I say *permission;* your bill *compels*. My idea is to have a bill framed, and see what that will do. I do not think that this legislature, when they say *compel*, know what we ought to do. I think we better have a chance to see what we can do.

Q. — In the case of Merrifield against Worcester, what was the practical result? *A.* — Well, so far as that question was concerned, I believe they decided that Mr. Lombard —

Q. — Not the case of Merrifield against Lombard, but the case of Merrifield against the city of Worcester. *A.* — I think the practical result was that the manufacturers, not the city, had the right to drain into the stream.

Q. — Did not the city of Worcester make some recompense for the damage it was causing him in the way of furnishing him some water? *A.* — They did not. Before my administration they took away his water-wheel, and had the right to take away his dam. In taking away that dam, they had to compensate him, give him other water for it; but I do not think it was on account of pollution of the stream.

Q. — The bill he brought was for the pollution of the stream? *A.* — Well, he would bring a bill large enough to cover every thing.

Q. (By Mr. GOULDING.) — Were there damages paid for polluting Mill Brook? *A.* — No, sir: no damages for polluting Mill Brook.

Q. (By Mr. FLAGG.) — The Act of 1867 was obtained to give the city of Worcester the right to use Mill Brook as a sewer, was it not? *A.* — I presume that it was. They did not want to go on with the system of sewerage, unless they had an act. It would have been very unsafe to stand upon their immemorial use of that stream, without an act by which they had a right to alter the brook and straighten it. I think there are some parts of the river between Mr. Morse's mill and Burling Mills where the stream should be improved.

Q. — There must be a bill to enable anybody to do that? A. — There ought to be a bill; yes, sir.

Q. — The manufacturers of Worcester discharge into the sewers, do they not? A. — Not entirely.

Q. — To a certain extent? A. — Yes, sir. I was thinking of that. I thought you might ask that question. I take the Washburn & Moen establishment. They are not on a lateral sewer; they are on the original stream that has been walled up; but if the sewerage system had been there, they would have the right to put it in. Then there are some few persons who have little shops on side streets where there are lateral sewers; but mostly they are on the streams.

Q. — Tell me about the standing of such physicians as Dr. Joseph Bates, and Dr. George Bates, and Dr. Sargent of Worcester. Don't they stand very high in the profession? A. — I think they do. I think they have a high reputation. I do not wish to make any distinction between them. I should make some difference in my opinion of the men, if it were given.

Q. — Do you know their opinions on this subject? A. — I don't know the opinions of the three men; but I have that of Dr. Sargent, because I went to him and showed him that bottle of water that was taken from the gas-works, of which Dr. Sargent is president. I told Dr. Sargent, "I want you to see what your gas company are putting into that stream; there is a bottle of water that was taken from it near by the gas-works." Said he, "I was not aware it looked like that, but," said he, "there is tar in it, which doesn't pollute the stream any, excepting that it colors it; the tar itself does the sewage good. It does not hurt it, as far as health is concerned, a particle." Said he, "They cannot make out any thing against you; I don't believe there is any danger." Said I, "Thank you, doctor." I asked him to come down here and testify. He said he was so busy that he couldn't. This goes into a brook, and then it purifies itself by running through the brook some little distance. There is another bottle here that shows it. I never talked with Dr. Joseph Bates, nor Dr. George A. Bates. I presume likely they might say there was a smell or something of that kind, as some of your witnesses have, and I don't know but other witnesses have.

Q. — Is it not true that the rendering establishment is now on the road in an opposite direction from the tripe factory down by the Dorothea Pond? Don't you know about it? A. — I thought Dorothea Pond was a different place. I did not know that it was near Dorothea Pond; I thought it was before you got to Hull Brook.

Q. — You don't know what I mean, then. The river, when the Committee were there, was remarkably high? *A.* — Very high, I should think.

Q. — And your being glad that they saw it at that height was not because we were sorry? *A.* — No: I heard you were glad.

Q. (By Mr. MORSE.) — Mr. Mayor, you spoke of the provisions of the bill that has been submitted here, and first in reference to the time that was fixed. Would it obviate one of your objections if the time, instead of being limited to four months, were enlarged to six months, or even a year? *A.* — Well, of course that would be better; but that is not my idea. I do not think any bill should be passed compelling us, under the investigations which have been made up to this time, to do any thing.

Q. — I wish to take up each one of these points in its order. First, in regard to the time. I wish to say to you that the particular limit is not regarded as essential by the petitioners here. Then, second, in regard to the general provisions of the Act, you do not understand that the bill undertakes to prescribe the mode in which this trouble shall be remedied, do you? *A.* — Not at all; only it says that we shall remedy it.

Q. — Precisely, but it leaves the city of Worcester to determine what mode shall be taken? *A.* — Oh, yes.

Q. — Now, on the other point, which I assume is the principal one of difference that remains, — as to whether or not the city should be *required* to do this, or should be *permitted* to do it: the very remark that you made in reference to the transitoriness of the City Government would be particularly applicable here, would it not? That is to say, supposing a permissive act were passed, the present City Government might be favorably inclined to it, but another government might take a very different view, might it not? *A.* — Yes, they might do it; but there is no City Government that has taken hold of this subject; they have not even made any preparations in this case at all, until this City Government came in, that I am aware of. The subject has not been considered: they have not broached the subject; they have not considered it; they have not done any thing; but if this City Government should go on and make experiments, and get information, it might have an effect upon another City Government.

Q. — You have no reason for assuming that the present City Government would be ready to incur any considerable expense for this, have you? *A.* — Well, when you speak of "considerable expense," that is a thing no man can answer, only I think that while the present City Government might not see their way clear to do any thing which would be of advantage to Mr. Morse and the town of Millbury and the city of Worcester, with regard to stopping these complaints, they would be very happy to take hold of it.

Q. — Has there been any indication by the City Government that the city of Worcester would make any considerable appropriations of their own accord, to remedy this difficulty? *A.* — There has been no action by the City Government. This matter is simply left to the Sewer Committee. So far as that is concerned, I have recommended in the appropriations which are to be made, something, I do not know how much, for sewer construction, quite a large amount; but there has been no other action taken. I saw the piece that was published the other day in the "Sunday Herald" or "Sunday Globe," which I understood was written by a man who lives in East Douglas —

Q. — Let me say that I saw the article, and have it here. I have no knowledge myself as to where it originated, but I want to read it, and then ask you a question in connection with it. This article, I may say, I cut from the "Globe." It is dated March 1, 1882 : —

To the Editor of The Globe: —

Among the matters of more than local interest likely to engage the attention of the present legislature, is that of the pollution of the Blackstone River by the sewage of the city of Worcester.

The joint committee on public health now have the subject under consideration for a second time, and a hearing is now pending before them. Your regular Worcester correspondent has from time to time presented to the public what purport to be the views taken by the City Government of the respective rights of the city and of the inhabitants of the Blackstone valley; and from the whole tenor of his letters it is inferable that the authorities having the matter in charge on the part of the city, after careful consideration and due inquiry, have concluded : —

First, that the city has the legal right to empty its sewage into the river, regardless of consequences;

Second, that no nuisance is thereby created;

Third, that if there is, or hereafter may be, the inhabitants of the valley have no remedy; and

Fourth, That the city will stand upon its legal rights, and pay no regard to the wants and wishes of the people below on the stream.

The first and fourth of these propositions are maintained for the purpose of quieting the fears of timid tax-payers, who otherwise, for prudential reasons only, might investigate for themselves, and compel the authorities to act before a great and additional expense was imposed upon them; the second and third, for the purpose of preventing any expression by a large class of conscientious and fair-minded people, who otherwise would exert a controlling influence in governing the action of the city, and for the purpose of quieting and suppressing the natural impulse of all good citizens to do justice and equity without regard to strict legal rights.

These propositions are undoubtedly inspired by the city authorities, and are indorsed by nearly all of the representatives of the city in the legislature, and were they advanced in good faith, and honestly entertained, would be fair matters for argument only; but, if not so advanced and entertained, are open to grave criticism. The people of Worcester undoubtedly believe them to be honestly entertained, and an effort is being made to satisfy the legislature of their truth, and of the sincerity of the city's representatives in their advocacy of them, and, unless the correspondent of "The Sunday Herald" has been misinformed, some measure of success has been attained; for he writes to that paper, under date of Feb. 19, "The subject has been talked up considerable already with the members of the legislature, and there isn't that fear there was last winter."

It is the belief of the writer that the people of Worcester, including your regular correspondent, by their authorities and most of their representatives, are being deceived; for it is almost an open secret here, that the opinion of the expert employed by the city coincides with the opinion of the experts employed by the State Board of Health, Lunacy, and Charity, both as to the practicability of the scheme presented by the Board, and also as to its necessity; and this opinion is well known to the city authorities, but has been concealed from the people, and their expert is now employed simply as a critic of the plan recommended, and is being held in readiness to criticise any other plan that may be suggested.

It is also reasonably certain that the City Government, in executive session, has arrived at the conclusion that there is grave doubt as to the right of the city to continue its nuisance, but also have had before them and have considered a plan presented by a citizen of Worcester, whereby the nuisance may be abated.

The plan alluded to has received the cordial indorsement of four at least of the aldermen; and its originator, having full faith in its efficacy, has presented it in detail, describing his process fully, explaining the details of the construction of his purification works, and the places in which they are to be erected, and their cost, both with reference to the works themselves and on account of land to be taken for them.

It certainly is to the credit of any citizen of Worcester, that he should recognize the evil, and devise a way to overcome it; and if it is creditable to the City Government to exhort him to secrecy, and command his silence, and to suppress all knowledge of it, it certainly is not commendable for them to deceive their own community with reference to it, and upon a scale of municipal magnificence play the unbecoming and deceptive game of bluff with the surrounding towns below them in their own county, and with the general public. *

Now I call your attention to that portion of this communication which refers to a plan, which, it is said, has been presented to the aldermen. Is there any foundation for that statement? *A.* — It is untrue, the statement that there has been one word said about this matter in executive session of the Board of Aldermen. I never heard of it.

Q. — That was hardly my question. My question is, whether it is true that any plan has been presented? *A.* — I am going to answer it. About a month ago, a man by the name of Fuller came into my office with a box about two feet long, with wire sieves in it. He said he had discovered a system by which the water of the Blackstone could be filtered, and wanted I should look at it. He was a man whom I knew very well, and I said, "I hope you have found a remedy for the pollution of the water in big rivers like the Blackstone." He said he thought it could be done. He wanted to know if I had any objection to his showing it to members of the City Government. I said, "Not the slightest, or anybody else you choose to." I have seen him once since, I think, three or four days ago, and he said, "Don't you think I had better come down to the committee, and show my box-filtering scheme?" I said I did not think it would do any good. That is all I have heard of it. If he showed it to any of the aldermen, he showed it to them as individuals. Although I am a member of the Sewer Committee, it has never been brought before us. I should not be surprised if he had shown it to two, or three, or four, aldermen, but there has been no action of any kind taken upon it.

Q. — Is he a citizen of Worcester? *A.* — He is a citizen of Worcester.

Q. — Is that the only plan that has been proposed by a citizen of Worcester, or any board? *A.* — Yes, sir: I don't know of any other.

Q. — Has not Mr. Ames, a representative in the legislature from Worcester, submitted some plan? *A.* — I have not seen a drawing or figure put on paper, in any way. Mr. Ames has talked about a plan, and said, I think, if he had thirty or forty acres of land where he could have a basin, and the water was allowed to run over stones, or something of the kind, it would purify itself.

Q. — That is the plan that I was told was referred to in this communication, a plan submitted by Mr. Ames? *A.* — There is nothing to it.

Q. — Will you state to the Committee what has been said by Mr. Ames to the Board, or any members of it? *A.* — Mr. Ames has not been before the Board.

Q. — What do you mean by saying, then, that Mr. Ames has had some talk about it? *A.* — He has talked with individuals, saying that he thought, if we were obliged to do any thing, if we could form a basin where this water could be aërated, it would remedy the difficulties complained of.

Q. — Did he express any opinion as to the expense of such a method? *A.* — Well, so far as I ever heard of Mr. Ames's talk, I think he said that it would not cost very much; not more than a few thousand dollars, or something like that.

Q. — About how many did he say? *A.* — I can't tell you any sum; but I should think that he represented that it would not cost over twenty or thirty thousand dollars, anyway. But that was mere talk or discussion, as I understand it: I don't understand it to have been any formal proposition.

Q. — Not before the City Government; but he has suggested in private conversation that such a plan as that would remedy the difficulty? *A.* — He has talked about it. I have heard him talk about it in the office, when they were discussing that question.

Q. — He thought that for twenty or thirty thousand dollars the trouble could be remedied? *A.* — Yes, sir. Now, I don't say that I have stated the sum correctly; but it was not a large sum. Mr. Ames is here: you can call him, and he will tell you what he said.

Q. — Was that plan or suggestion presented by him, to your knowledge, to different members of the Board? *A.* — Not at all. I have not known of its being presented. He may have talked with individual members of the Board, but he never came before the Board with any thing of that kind.

Q. — Would you be willing, Mr. Mayor, to have an act drawn in this form, authorizing the city of Worcester to appoint a commission on the subject of purifying or taking care of the sewage, and then give to that commission discretionary power as to what should be done? *A.* — No, sir, I should not. I think we can take care of our own business in the city of Worcester.

Q. — I understood you to advocate the appointment of a commission, on the ground that the City Council would be changing from year to year? *A.* — I thought you meant an outside commission.

Q. — No, I mean a commission of citizens of Worcester. *A.* — I have no objection to that, any more than to a police commission, or any thing of that kind.

Q. — Please see if you take in the scope of my question. My question would be this: whether or not you would be satisfied to have a bill passed, authorizing the City Government to appoint a commission, and then give the commission full power as to what should be done, — let them have authority to take land, and incur such expense as would be necessary. *A.* — I would rather, and I think it would be better, if an act was drawn in that way, to submit the question of its adoption to the voters of the city, to see if they would approve of the act; the same as, when Roxbury was annexed to Boston, the people of the two cities voted upon it. Then it would be a relief to the chief officers of the City Government.

Q. — Then it would appear that you do not want any act requiring the City Government of Worcester to do this? *A.* — No, sir: not in four months.

Q. — I have already said that the limit of time is not material. *A.* — I should object to it, if it was to compel us to do it in two years.

Q. — Do you think there is any indication, in any action that the City Government of Worcester has taken, of its own motion, without something imperative in the act, that they would assume the expense, which might be considerable, of any purifying of sewage? *A.* — What you might consider a considerable expense, might be more than we would be willing to incur. We might disagree about it. If you will name your sum —

Q. — Well, say two or three hundred thousand dollars. *A.* — I do not think that we ought to be compelled to assume that expense. I think that the City Government would be perfectly willing to assume an expense of thirty, forty, or fifty thousand dollars, if they thought they could provide any remedy which would satisfy the people of Millbury, and answer the purpose. For instance, there is Mr. Stockwell, a gentleman from Sutton, who was senator from the Millbury District two or three years ago. He says this thing could be remedied easily at an expense of twenty or twenty-five thousand dollars, so that the people of Millbury ought not to have any complaint to make.

Q. — Well, you do not understand, Mr. Mayor, that anybody wants the city to spend one cent more than is necessary? *A.* — No, I don't think they do.

Q. — But you would agree to this, that something should be done at once to remedy the evil? *A.* — I cannot tell. You start in by saying that you want Worcester to consent to spend one or two hundred thousand dollars. What are you going to do? The remedy I might propose to-day, so as to take the sewage of the manufacturing establishments, might not apply. You want Worcester to spend three or four hundred thousand dollars. I do not think the time has arrived yet when the city of Worcester ought to be compelled to go into a large experimental operation, attended with great expense, to divert its sewage.

Q. — When do you think the time will come? *A.* — Well, I should hope that we should do what was right about it.

Q. — I haven't any doubt that you would, personally, Mr. Mayor; but your remarks this morning are the first indication that we have had from anybody from Worcester that looks to a practical solution of this difficulty. *A.* — Well, my remarks have been open to the public since the first of January. It is not any new scheme; it is a thing that I have thought of myself, that is all. If all the engineers should come in, and say this is entirely impracticable, I should have to yield to what they say; but I should look at it as a common-sense matter.

Q. (By Mr. GOULDING.) — I do not want the Committee should misunderstand your relation to this matter in any way, and I want them to know how I stand. I will ask you whether you have come here as a representative of the city to request this Committee to consent to a bill imposing any obligation upon the city of Worcester to make any other disposition of its sewage than it does at present? *A.* — I have not.

Q. — When you speak of this scheme which you have suggested in your inaugural, I do not understand that the City Government of Worcester, or any organized opinion of Worcester, has at all approved of it? *A.* — No, sir: it is just the same as Mayor Chapin's and Mayor Verry's remarks, that were put into the case originally to start with.

The CHAIRMAN. I think the Committee understand all that. What I suppose the Committee would like to know is, how far there is a sentiment in Worcester that is willing to do any thing. I do not understand the petitioners to ask for any thing; they seem to want some reasonable assurance that the city will do something; I do not understand that they ask that it shall necessarily spend forty or fifty thousand dollars. What we want to understand is, what response they would be likely to get in that direction from the city of Worcester. We would like to have you agree to a bill of some sort.

Q. (By Dr. HARRIS.) — Is there, within your knowledge, any probability that the city of Worcester and the petitioners could come to any approximate arrangement in regard to this matter? If so, it would save the Committee some labor. *A.* — We cannot give the mill-owners pure water for their steam-boilers, or for dyeing their cloths; nobody can do it; and when this legislature undertakes to make the city of Worcester do it, I do not believe it can be done; I do not believe it is possible.

Q. (By Mr. MORSE.) — Has the City Government of Worcester, at any time, appointed any committee to confer with the representatives of Millbury and the other places, with reference to any plan for the purification of the water? *A.* — I have never heard that they have at all.

Q. — Have you seen any thing in the attitude of the Committee of Millbury, or of the other persons who represent themselves, who claim to be aggrieved here, which indicates any unwillingness on their part to confer with the representatives of Worcester? *A.* — Not the slightest.

Q. — You are on friendly terms? *A.* — Certainly. Mr. Morse is my friend. We are perfectly friendly to them. I think that Mr. Morse and I could talk this matter over, and, if he has not got it too much on his brain, I think we should not disagree very much.

Mr. SMITH. I wish you would get together, and put your agreement in writing.

The CHAIRMAN. It would help this Committee out.

WITNESS. A man at the head of a city cannot always do what he would like to do as an individual.

Mr. MORSE. Perhaps the mayor might be a little relieved by this Committee.

WITNESS. I do not want to mislead anybody. Perhaps I have said a little more than I ought to have said here to-day.

Q. — (By Dr. HARRIS.) — You don't suppose, I take it, that this

Committee have any thing to do with the manufacturers? There is no petition from the manufacturers here. *A.* — Not as manufacturers; but then you find the manufacturers coming in here and making complaint. For instance, Mr. Simpson came in here, saying that they have lost five per cent. on some of their woollen goods because of the impurity of the water. When I inquire of the manufacturers, they say that they lost five per cent. the same year, for some other reason. The vendees take advantage of the hard times, when goods go down. That is about the amount of that story, in my opinion.

TESTIMONY OF WILLIAM E. WORTHEN.

Q. (By Mr. GOULDING.) — Where do you live? *A.* — New-York City.

Q. — What is your business? *A.* — Civil engineer.

Q. — Have you ever paid any particular attention to engineering in connection with sewers and sewerage? If so, please state what your experience has been. *A.* — I have. I have been a good deal connected with the construction of sewers and with the application of plumbing to houses; and, when the Metropolitan Board of Health was organized in New York, I was appointed the engineer of that Metropolitan Board. I continued in that office of sanitary engineer of the Metropolitan Board of Health as long as there was a Metropolitan Board of Health, — four years. All that time all the complaints of sewerage, drainage, and ventilation were referred to me for orders for structural remedy; and at that time I probably issued many thousands of orders.

Q. — How extensive was the sewerage system that was under your charge? *A.* — We had the whole city of New York, the city of Brooklyn, Staten Island, Long Island, including Jamaica, and up the river to Peekskill, including Peekskill; taking the whole of what was called the Metropolitan District.

Q. — Are you familiar with the literature with regard to sewage and its disposal? *A.* — I am.

Q. — Have you ever been abroad? *A.* — I have.

Q. — Seen any of the sewerage works abroad? *A.* — I have been in the sewers of Paris; I have seen the sewers of London, and their system of utilizing sewage at Barking, on the Westminster side of the river.

Q. — Have you read Col. Waring's Report to the State Board of Health? *A.* — I have.

Q. — Have you been on the premises, and examined the river? *A.* — I have: yes, sir.

Q. — What is your opinion of this whole plan of disposing of sewage proposed by the State Board of Health, Lunacy, and Charity? *A.* — I do not agree with it.

Q. — I want to ask you whether, in your opinion, fæcal matter from Worcester would reach Millbury in the Blackstone River? *A.* — Not to pollute it: no, sir.

Q. — Now, won't you explain why? *A.* — This is the basis on which I work: I get from Mr. Allen, that in the lowest run of water about 40 gallons of water pass into the sewer per day, for each inhabitant; not counting any thing that goes into it from springs, as one gentleman testified here. I also understand from him that the lowest run of Mill Brook exceeds 3,000,000 gallons a day, which would be 75 gallons each for the population of 40,000 using the sewer. That would be 115 gallons of water for each person. A gallon of water weighs 8.3 pounds; but I call it 8 pounds. Eight times 115 = 920 pounds. The amount of ejections, taking urine and every thing, that passes through an average person, would not be over three pounds a day, of which, of hard matter there would not be perhaps more than four ounces. There are 900 pounds, say, of water, and 3 pounds of ejections. One-third of one per cent. is what goes through an individual; and there is only about one-sixth of that which is hard matter. That is all there is to it. That is taking Mill Brook alone, as it discharges down there at Quinsigamond. It there mixes with a larger proportion of water, I suppose, — I do not know how much more; but, anyway, it comes in such a diluted state that there can be nothing from the water-closets, or merely human ejections, that can be detected in any way down there. I doubt if even chloride of sodium would disclose the presence of urine.

Q. — Is there any such condition of that river, as far as you could discover from an inspection of it, and from the evidence of plants, animal life, etc., which in your opinion requires the adoption of any such scheme at Worcester as has been proposed? *A.* — At present, no.

Q. — What evidences did you see, when you were there, with regard to the pollution of the stream, or the purity of the river? *A.* — Dr. Folsom stated here yesterday, I think, that these analyses did not discover any thing. They are not reliable; but there is a pretty reliable test, which is adopted in France, and which I think is the standard, or should be the standard, and that is the quantity of oxygen in the water. It is shown very conclusively that the quantity of oxygen that is in the water is the standard of the purity of the water for these purposes. As the water first comes out of the sewer there is very little oxygen. The result is, there are few or no organisms of any sort. As soon as it is mixed with a little more air, the lower organisms, that show no chlorophyl, first begin to develop; then plants begin to appear; and you go on in that way, and as you go down stream you can test the amount of pollution by the flora and fauna of the stream. It has been stated here that the sewage that comes down there invigorates the plants in that pond, and that they grow to such an extent as to fill it up. If that is so, in my view of it, that is evidence that the sewage is not unhealthy, because it promotes the growth of plants. You will find that where sewage is old and decayed it is death all around it. The plants are dead; the animals are dead; there is no life in it. It is not until it becomes aërated that it begins to be good, until, at the end, when you get trout and water-cress, the water would be pretty safe to drink: then that is very pure water. But the other is not injurious to vegetable life, and would not be to us, up to a certain line; then we could come in very well. The evidence shows that fish died there, below the pond. They must have come from somewhere; and when I was down there, the other day, I saw a mink coming across, and a mink is pretty good evidence that there are fish. It seems to me that that pond is not at present injurious through its pollution.

Q. — Where did you see that mink? *A.* — Just above that woollen-mill.

Q. — Burling Mills? *A.* — Yes, sir, Burling Mills.

Q. — What would be the effect of taking this sewage out on those seventy-five acres, with reference to stench? *A.* — If they took that sewage and concentrated it on seventy-five acres, the result would be that the smell would be worse than where it comes out of the brook diluted. It would be localized. It would not go down to Millbury, but would be localized where this stuff is used. The more concentrated the sewage, the greater will be the smell. We talk about "intermittent downward filtration." I think it is better to omit the prefix, and say it is *filtration*. It is nothing but a filter. We make an earth filter, — that is the whole of it, — and when that filter gets clogged we move to another. A filter clogs, not from the quantity of water put upon it: it is from the amount of turbidness, and the foreign matter held in the water. I think that Dr. Folsom was wrong in estimating the capacity of a filter by the acre; it should be estimated upon the degree of the turbidness of the water. That sewer-water will take no more land for a filter with the water of Mill Brook in it than without it; not a bit.

Q. — Do you know how that stream, Mill Brook, compares with the outlets of the sewers in English cities? I see the report says that this sewage is twice as dilute as the average of fifty English cities. *A.* — I have not compared it; and, personally, I have never seen the outcome of those places in England.

Q. — Is there, in your judgment, any such settled condition of the science and high art of sewage disposal, as to render it to any degree certain that such a scheme as has been proposed by the experts of the State Board of Health, Lunacy, and Charity, would effectuate the result desired? *A.* — If they expect to get rid of the smell, no. They are going to localize more smell up there at Quinsigamond than there is now, because the sewage will be more concentrated. The first idea in purifying sewage is to mix it with water; if you can mix it enough, it is done. In New York, we throw it into the river, and have no more trouble with it. If you mix it with a certain amount of water, then it becomes suitable for the food of the lower orders of animals; and the more you put in, the nearer it comes to what we call good water.

Q. — There is another branch of the subject to which I will call your attention. What is the effect of drawing off a pond on the question of polluting the air? *A.* — I have had a great deal to do with it. I have heard the testimony in the case, that the smell was offensive. One doctor testified that it had the smell of a privy. You perceive that at Quinsigamond the proportion is very small, — about one-twentieth of one per cent., and it grows very small down there. I think the nose is as good a thing to detect a smell as any other instrument you can have, and if you have an educated nose, you can pick out smells very closely. Now, the smell which they refer to as a privy smell is the result of the decay of some of those lower orders of animals, which I have tried a number of times, and found always the same result. I put in the works at Long Island, which are supplied by a well, and the water never gets above fifty-two degrees; generally it is about fifty. In winter it is about forty-eight. It varies about four degrees. On that water, at one time, there formed a scum, like the frog-spittle which you see on those cold springs that you find under a hill. I wanted to find out what it was, and took it to a man to analyze. He analyzed it, and told me he could find nothing different from what he detected in the water itself. If you took it in your hand, it was perfectly smooth and impalpable. I bottled it and kept it a week, and when I opened it the smell of a privy, as most people would call it, was very perceptible. I had the same thing examined at Fort Richmond, and with the same result. This *spongilla fluviatalis* that you get in Boston is the same thing. When alive it has a cucumber smell, but when it is dead it has a very disagreeable smell. When the water settles down, these animals die, and when they do die they emit a very offensive smell. There is another thing also to be said, that I think is an acknowledged principle in sanitary engineering, or whatever you may call it. Von Pettenkoffer, of Vienna, who, I suppose, is an authority, says that there is more danger to a community from the lowering of the level of water than from any thing else. He says that in Vienna they can trace almost all their epidemics to lowering the strata of the water. I think you will find, almost always, that where a pond is lowered, whether there is any pollution in it or no, they have these low fevers, malaria. I think it is almost invariable.

Q. — What do you think would be the result of taking down those dams, and letting that stream flow? *A.* — I am not so certain about that; I have not made sufficient examination to be able to say that that is a remedy. I should want to make a careful examination before I should feel authorized in advising so great a change as that, and paying for all those mill-powers. I have only seen this water in colder weather. I should like to see the stream through the summer months, and see exactly how it was. The basis of my opinion I have given you. I want to say that I should infinitely prefer to try the mayor's remedy first. There is another thing which I think ought to be done, if you should take Mayor Stoddard's plan. There should be a deep sluice to every dam, so that in times of high water it could be raised, and the pond washed out, with any sediment that might be in it, and so on down.

Q. (By Mr. SMITH.) — Can you give us any reason why that scum to which you referred should rise upon the water; whether it was from pollution or otherwise? *A.* — No, sir: there is no pollution whatever. This water is cool spring water, and you will see it upon any other spring water you ever saw.

Q. — Have you any theory as to its cause? *A.* — Yes, sir. Perhaps light and air were mixed. I do not know how it operates, but I have never found cool water without it. You will find, in Newton and Waltham, which are both spring supplies, that in the coolest water there comes up a kind of jelly form, which, when it detaches itself, as it does sometimes, passes down with the current. They have to keep a rack to prevent it from getting into the pipes. As soon as it is laid upon the ground, it begins to create a smell. I do not know what it arises from, but the analyst told me that the analysis of this stuff was exactly the same as the analysis of the water; and he was the analyst of the Board of Health, a man of good standing.

Cross-Examination.

Q. (By Mr. MORSE.) — I want to call your attention for a moment to some figures that you gave. I understood you to state that three pounds of fæcal matter are discharged, on the average, from each person? *A.* — Yes, sir.

Q. — Now, assuming that there are forty thousand persons in Worcester who use the sewers, it would follow that one hundred and

twenty thousand pounds a day of fæcal matter are discharged?
A. — Yes, sir, including urine and all.

Q. — That would be sixty tons of sewage a day? *A.* — Yes, sir.

Q. — And three hundred and sixty-five days in the year would give as the total, twenty-one thousand nine hundred tons of fæcal matter? *A.* — If the multiplication is right, that is correct, sir. I cannot do that in my head.

Q. — Twenty-one thousand nine hundred tons of fæcal matter are discharged into the Blackstone River at the mouth of the sewer? *A.* — Yes, sir.

Q. — You would not consider it a surprising fact if a considerable part of that great mass should find its way into the ponds and above the dams below Worcester, would you? *A.* — Not a bit, sir.

Q. — And it would not surprise you that it should be found in the mills, and in the various places where the water is used? *A.* — It would surprise me very much; the percentage is small. You see, you are giving the mass, and do not give the percentage. It would surprise me if I should find one-twentieth of one per cent. of hard matter, or one-third of one per cent of urine. I said I doubted if chloride of sodium could be detected, which would show the presence of urine.

Q. — Suppose that reputable persons, whose word was to be taken, assured you that fæcal matter was found in the mills, or stuff that appeared to be composed in part of fæcal matter: you would readily believe it came from Worcester, would you not? *A.* — I should not. I should want to go myself and see it.

Q. — You would not take their word? *A.* — I would not.

Q. — Your faith in your theory is so strong, I suppose? *A.* — It is so strong. I do not know whether any of you are conversant with the workings of a water-closet; but if you would go down when some party is in the closet, and look at the outflow, you would be utterly astonished to see how little fæcal matter shows there. I have never been able to see it, where the water-closet system was in operation. Occasionally, where a privy is located right over a stream, the matter which drops preserves its form, and goes floating down some distance; but, so far as my experience goes with water-closets, I never saw that in my life. I never saw it when I have looked at the mouth of a sewer, where the ejections were merely water-closet discharges. The matter was all broken up. So far as drinking the water is concerned, you do not drink it, because you know what it is, but people who are dry do not notice it.

Q. — I want to see how far you do or do not agree with other authorities on this subject. You are familiar, I assume, with all the literature of the subject, as, indeed, you have stated? *A.* — Not with all. There is no branch of the profession that I have not investigated.

Q. — I assume that you are largely acquainted with it. I call your attention to the language of the State Board of Health, in their report of 1873. They say, on p. 96, —

"It is a wide-spread popular idea, that no matter how much impurity is discharged into a running stream, yet, by flowing a dozen miles or so, the stream will for all practical purposes free itself from the impurity, and become fit for use, even as a source of water-supply. It has been alleged that the organic matter is almost completely oxidized by the oxygen of the air, and by that dissolved in the water, and that this oxidizing action is very much increased if the water be agitated by passing over weirs or natural falls. This feeling has gained considerable currency, and has been held by some men who are looked to as authorities, such as Dr. Miller, Dr. Odling, and Dr. Letheby. It is, however, unsupported by direct proof; in fact, the experimental evidence leads us to the contrary opinion. The Rivers Pollution Commission made this question the subject of direct investigation, and showed very conclusively that the commonly-received opinion was erroneous. They chose localities on several streams where the rivers, in each instance, flowed for almost a dozen miles without receiving additional pollution, and determined the amount of organic matter destroyed. They also made mixtures of ordinary sewage with different quantities of water, and in these artificial mixtures, which were by various devices exposed to the free action of the oxygen of the air, they determined the rate at which the organic matter disappeared. This they did, by estimating from time to time the organic nitrogen and carbon contained in the solution; also by observing the rate at which the dissolved oxygen disappeared. As a result of these experiments, they affirm that, —

"'It is evident, that, so far from sewage mixed with twenty times its volume of water being oxidized during a flow of ten or twelve miles, scarcely two-thirds of it would be so destroyed in a flow of one hundred and sixty-eight miles, at the rate of a mile per hour, or after the lapse of a week. In fact, whether we examine the organic pollution of a river at different points of its flow, or the rate of disappearance of the organic matter of sewage when the latter is mixed with fresh water, and violently agitated in contact with air, or, finally, the rate at which dissolved oxygen disappears in water polluted with five per cent of sewage, we are led in each case to the inevitable conclusion that the oxidation of the organic matter in sewage proceeds with extreme slowness, even when the sewage is mixed with a large volume of unpolluted water, and that it is impossible to say how far such water must flow before the sewage-matter becomes thoroughly oxidized. *It will be safe to infer, however, from the above results, that there is no river in the United Kingdom long enough to effect the destruction of sewage by oxidation.*'

"These results confirm the opinion arrived at from theoretical considerations, and expressed by Sir Benjamin Brodie in his evidence, given before the former Rivers Pollution Commission (First Report, River Tha..es, vol. ii., Minutes of Evidence, p. 49). His evidence was to the following effect:—

"'I should say that it is simply impossible, that the oxidizing power acting on sewage, running in mixture with water over a distance of any length, is sufficient to remove its noxious quality. I presume that the sewage can only come in contact with oxygen from the oxygen contained in the water, and also from the oxygen on the surface of the water; and we are aware that oxygen does not exercise any rapidly oxidizing power on organic matter. I believe that an infinitesimally small quantity of decaying matter is able to produce an injurious effect upon health. Therefore, if a large proportion of organic matter was removed by the process of oxidation, the quantity left might be quite sufficient to be injurious to health. With regard to the oxidation, we know that to destroy organic matter the most powerful oxidizing agents are required; we must boil it with nitric acid and chloric acid, and the most perfect chemical agents. To think to get rid of organic matter by exposure to the air for a short time is absurd.'"

Now, to begin with, you are familiar with the Report of the Rivers Pollution Commission? *A.* — Yes, sir.

Q. — Do you agree with their conclusions? *A.* — Not at all.

Q. — Do you agree with the State Board of Health? *A.* — I agree that the stream purifies it within a certain distance.

Q. — The State Board of Health express the opinion that that idea is erroneous. *A.* — I do not agree with the Board of Health.

Q. — Then you agree with the popular opinion? *A.* — I agree with the popular opinion. Not *as* the popular opinion, because I have good authority which I can give you on the other side.

Q. — You agree with what the State Board of Health calls an erroneous popular opinion? *A.* — Yes, sir. I believe there is a time when every thing turns over. Let me give one illustration. The water in the aquarium at Croyden has been there some four years, without any change whatever. It is merely pumped up and aërated; and those fish are fat, they swim about there, are in perfect health, and in good order and condition. That is, mere aëration has made that water perfectly good. The fish there do better than they do in the aquarium at Brighton, where they pump water from the sea every day, and do not give sufficient aëration. At Croyden they get an excess of oxygen in the water, and the result is, the fish there are healthy; while at the Brighton aquarium, where they have a fresh supply of water every day, they are not healthy.

Q. — I understand your position to be, that sewage may with safety be drained into a running stream? *A.* — I think so, sir.

Q. — And that no system is necessary to prevent this stream from pollution? *A.* — If the stream is large enough, no, none.

Q. — According to what you understand of this case, you consider that the Blackstone River is large enough? *A.* — At the present time, yes.

Q. — You think, then, that the draining of the sewage of Worcester into the stream does not pollute that stream? *A.* — I refer, now, to the amount that comes from house use, and all that. I don't know about the other. Yes: I mean to say that the quantity that enters the Blackstone River to-day, with what I understand to be the flow of the river, is not sufficient to pollute it.

Q. — Will it, in your judgment, in the natural course of things, be a source of pollution? *A.* — It will.

Q. — Can you give any judgment as to the time? *A.* — That I am not prepared to answer, because I have not made that part a study. I have only considered its present condition.

Q. — Do you think it would be reasonable to take some precautions in advance to prevent pollution? *A.* — I think some experiments ought to be made, because I do not believe a system has yet been invented to purify sewage. That is my opinion from what I read. Here is a little periodical that brings up the question in a new form. It is "The Cosmos" of December, 1881, and January, 1882. That describes a system which the man has tried for twenty years; and, if it could be applied on a large scale, it would answer all your questions exactly, with very small expense, and without offence.

Q. (By Mr. GOULDING.) — Is it in French? *A.* — Yes, it is in French.

Q. (By Mr. MORSE.) — Will you state, in brief, what it is? *A.* — In brief, it is this: He applies it particularly to houses; for instance, the house sewerage, say, of twenty people. He states that the sewerage is to discharge into a tank, say of a cubic metre capacity, which is perfectly tight: it is closed with a water-seal at each end. As the water goes in, it displaces the other water, which flows out. There is a certain lapse of time, according to the size of the tank, in the course of which the water becomes purified. He says it then flows out with hardly any color, and no deposit. He says another thing: that the more water you put in with it, the better it works. He says, in this article, that it has been in use now some twenty years, privately, and has been successful all the way through; and he makes a little experiment in the laboratory to show how it works, in part.

Q. (By Mr. MORSE.) — What do you think of that plan, sir? *A.* — I think I would try it, if I had a chance.

Q. — To go back for one moment to the same question which I asked you before, about the amount of pollution. The present population of Worcester I understand to be, in round numbers, sixty thousand, of which, in round numbers, forty thousand use the sewers. I understand, further, that the increase of population, according to the State Board of Health report, has been twenty per cent. in five years. Taking those figures as a basis, can you express any opinion as to the time when, in your judgment, the sewage from the city of Worcester would be a source of injury to the water of the stream?

A. — That is to say, you mean, what is the injury?

Q. — You have given an account of the condition of things to-day, with forty thousand, say, draining there. How many, do you think, could safely be added? *A.* — If twenty thousand more were added, I should want something done. I should do something now.

Q. — Then, if the whole population of Worcester to-day were to discharge into the sewer, you would consider it a proper case for some action? *A.* — For *some* action, yes, sir: I think I should do that.

Q. — Let me ask whether your attention has been called specially to what plan you would recommend the city to adopt? *A.* — Not any: I have not been shown any plan. They have had no plan, so far as I have heard, as the mayor states. I have been thinking over the matter myself; and what I should try first, I think, would be the simple plan of keeping the ponds up, and, at the time of storms, clean them out by the flow of water.

Q. — Have you any doubt, that, with a reasonable opportunity to do it, you could carry out some practicable plan which would improve the condition of things very much there? *A.* — I should like to make one or two experiments first. I think I should want to try this de Moura's plan first. If you would give me a chance to experiment, I think I could. But there are a number of things come in there: for instance, the State Board of Health, Lunacy, and Charity recommend pumping; Col. Waring does away with pumping. Certainly, as far as that is concerned, if we could do away with pumping, it would be a gain. But I am not able to say myself, from any levels, or any thing taken there, whether that could be done. It would be well to avoid that, if it could be avoided.

Q. — You would agree with us in this, that it would be wise for the city of Worcester to get an opportunity to try such a plan by appropriate legislation? *A.* — Certainly.

Q. — And if the city were allowed a reasonable time to experiment and adopt a plan, have you any doubt that some plan could be adopted which would very much improve the present condition of things? *A.* — No, sir: if you gave them the right to take property, and any thing of that sort. They could not do any thing there without authority to take property for the purpose: it may be mill-power or land. I should not be able to state any thing about that.

Q. (By Mr. SMITH.) — What would you consider a reasonable time for making experiments, such as you would contemplate? *A.* — Well, if you would give them the privilege of having that dam kept up, that would start one thing. That would not be very expensive, anyway; and it ought not to add much to the expense if they had the right to put in a sluice there, with which they could wash out that pond. That thing would last quite a little time. If this stuff that comes down, partly from the dirt of the streets, and partly from the manufactories, could be washed out at the time of freshets, there would not be so much offence in that pond. That would purify the river very considerably, and at very little expense. How long that would last, I am not prepared to state.

Q. (By Dr. HODGKINS.) — You spoke about the purification of this sewage on land making a good deal of smell? *A.* — Yes, sir.

Q. — Have you examined any other places where that has been done, and found that to be true? *A.* — Yes, sir: at Barking, the smell comes from a pretty large surface of country. The dilution is very large there. The smell is the same that you get at the mouth of a sewer, — just about that sweet smell that comes from a sink-spout, a sort of sickish, sweet odor, the same as you get in the sewers of Paris. You could not mistake it for any thing else. You know the water runs through ditches there. It is first put on one side into one ditch, and then into another; and, as it comes along through that district, you smell that same smell. You see, they do not concentrate the sewage at all: every thing goes in. The London sewage, you know, is all pumped up, and goes down below. That smell goes over quite a little district.

Q. — That is over territory where vegetation is grown? *A.* — Yes, sir.

Q. (By the CHAIRMAN.) — Do you think it possible to detect the same smell at Millbury that you get down at the mouth of the sewer? *A.* — I do. You get that same smell which I call sewer-smell. It is different from any other smell. You get that smell down below the woollen-mill. You cannot mistake that smell after you have got used to it. You know what the sewer-smell is: it is a distinct smell. It does not belong to fæcal matter or animal matter, or any thing of that sort. It is an entirely distinct smell.

Q. (By Dr. HODGKINS.) — I would like to ask what effect the large amount of filth that runs in from the street has upon the purity of the water? *A.* — I suppose, absolutely, there is a great deal more manure, horse-dung, goes into the sewers from the streets than there is of sewage from houses. I should think so, and it is said by some English writers that it is as much or more. There must be a great deal of that get into the sewers, and every thing that is thrown into the street gets in in that way.

Q. — What is your opinion in regard to allowing the street washings to empty into sewers? *A.* — I should not allow it. The city of Worcester should keep its streets clean, and let as little of any thing of that sort get into the sewers as possible. The storm-water should go in, but not the droppings of horses. That would keep out one cause of pollution. The mere sand and grit that come in fill up the stream down below somewhat, but do not pollute it.

Q. (By the CHAIRMAN.) — How do you account for the prevalence of the same sewer-smell down as far as Millbury that you get at the mouth of the sewer, in view of the vast amount of water that comes in? *A.* — It is not quite so strong. You will find that all that kind of smell is a very pervading smell. It comes from a very small amount. All that goes in from the streets helps to make that smell.

Q. (By Mr. HODGKINS.) — What is the effect of that odor upon the health of people? *A.* — I have been particular in my inquiries about that. In all my experience in connection with sewers I have never found a man who was in ill-health from working in sewers.

Q. (By the CHAIRMAN.) — How long were you in the sewers of Paris? *A.* — Perhaps two hours.

Q. — How long do the workmen stay in them? *A.* — There they work any time. In New York, if a plumber makes a sewer-connection with the street, he has to go into the sewer to mend the pipe; and some of our old sewers are very bad. When I was in the Board of Health, I knew all that were bad. The old sewers in Canal Street and Amity Street are extremely bad sewers, but I have never known of a person who worked in sewers who contracted any disease. I do not know why it is so, but it is so. I have found another thing: that, however abnormal it may be, all scavengers are healthy. I never saw a scavenger connected with our board — they are all licensed — who was not a large, hearty man. They seem to run in that particular line.

Mr. MORSE. That is the kind of men who are ordinarily selected for the business.

WITNESS. That is a mistake, — a very great mistake. I think it is very well established. This is out of my line, and I guess I will not go on. I was going to the doctors' part of it.

Q. (By Mr. MORSE.) — Your remark suggests a question I forgot. As I understand you, the sewers of New York discharge into the river? *A.* — Yes, sir, the late sewers are carried out into the river. The old sewers ran into the slips.

Q. — Is it not a fact, that, owing to the enormous current that comes down that river, and the wash of the tide, there is no trouble in the city from the sewage? but has not very great trouble been found with reference to the sewage and filth of all kinds being washed up on the beaches, — so much so that complaint of the insalubrity of the sewers has been made there? *A.* — No, sir. Our trouble there has been from this: all our garbage is carried down by boats, and is supposed to be deposited pretty far out; but, if a fellow gets a chance, he dumps it anywhere; and then it comes up on the beaches. I do not know that they have had any trouble from the sewage.

Q. — Has not a plan of intercepting sewers been proposed in New York? *A.* — Yes, sir: I proposed one myself.

Q. — What was the occasion for it? *A.* — At that time all the sewers debouched into the slips; and my idea was, to carry them down below, — to have a reservoir, as it were, and let the sewage go out with the outgoing tide; but it was never carried out. I go on the slips very often now, but not so much as I used to; and I find there is very little trouble now.

Q. — You say there is no plan in contemplation now of intercepting sewers? *A.* — Not that I know of. All the late sewers have been built on the plan of throwing the mouth of the sewer clear out to the end of the pier. The old sewers run into the slips. Canal-street sewer runs into a slip, and Fulton-street sewer runs into a slip.

Q. (By Dr. HODGKINS.) — They all open under water, I suppose? *A.* — It is not necessary, but they do.

Q. (By Dr. WILSON.) — Did you see Singletary Brook? *A.* — No, sir: I only went down as far as Mr. Morse's. I went down to see Mr. Morse, but he was not at home.

Q. (By Mr. GOULDING.) — You were treated, however, with great courtesy there? *A.* — Oh, yes, sir.

TESTIMONY OF CHARLES A. ALLEN.

Q. (By Mr. GOULDING.) — You are city engineer of Worcester? *A.* — Yes, sir.

Q. — How long have you been so? *A.* — This is my fifth year.

Q. — What experience have you had as a civil engineer, in connection with sewers and other similar structures? *A.* — Well, an experience that has extended over about fifteen years.

Q. — What works have you had charge of in engineering? *A.* — I have had charge of a good many. I built the railroad viaduct in Worcester, and the one at the lunatic hospital: I had charge of a good deal of work for the city of Worcester before I was city engineer.

Q. — Built large sewers extensively, as well as smaller ones? *A.* — Yes, sir, I built the largest sewers that the city has.

Q. — I want you to state to the Committee what part, in your judgment, of the outlay that has been made upon the sewerage-works in Worcester would have to be practically abandoned if the plan recommended by the State Board of Health, Lunacy, and Charity were adopted. *A.* — You refer, I suppose, to the sewers that would have to be given up in the side streets?

Q. — All those sewers that would have to be given up, or practically given up, such as would not have been constructed if this plan had been adopted to begin with. *A.* — Probably a hundred and fifty thousand dollars.

Q. — I want to ask you how long, in your judgment, it would take, if you were to go to work with reasonable attention to economy, to construct the works necessary to put into operation the plan of downward intermittent filtration, as stated in the Report of the State Board of Health, Lunacy, and Charity? *A.* — It would take at least three seasons. We have no details at all of the work, nothing except a general idea of what is wanted. It would take at least one season to work those up carefully, and two seasons for the work.

Q. — Have you any suggestions to make with regard to either of the plans proposed, as to any of the difficulties to be encountered in carrying them out? *A.* — Well, I don't know exactly what you mean.

Q. — Whether you have any criticisms to make upon the plans, or any suggestions? *A.* — One suggestion that I have to make is, that, in estimating the cost, the fact that quite a large portion of our sewerage-works would have to be abandoned, ought, of course, to be taken into consideration. That I have already mentioned. I would like to say, in connection with the matter of the dilution of the sewage, that the sewage of the city of Worcester (I take this from the Report of the Commission) "is twice as dilute as fifty-four English cities and towns where irrigation has been resorted to." After it enters Mill Brook, it is diluted with a daily flow of water of from 2,000,000 (which is a very low estimate as the minimum) to 40,000,000 gallons. When it reaches the Blackstone, it is still further diluted by a daily flow of from 7,000,000 to 300,000,000 gallons of water. I state that to give some idea of the amount of water flowing in the stream. I see, by the Report of the State Board of Health, that they estimate the minimum flow of the stream at 750,000 gallons. I think that is a mistake. That undoubtedly refers to the Piedmont-district sewer.

Q. — Were these flows given by you or your assistant to those parties? *A.* — Yes, sir: the figures were all obtained from us; and we sent them to New York in a mass, and undoubtedly it was an error in putting them in the report in that way.

Q. — Can it be possible that the minimum flow of Mill Brook is only 750,000 gallons a day? *A.* — No, sir: that refers to the Piedmont district.

Q. — Have you spoken to Dr. Walcott about it? *A.* — Yes, sir, I spoke to Dr. Walcott; and he said it was undoubtedly a mistake.

Q. — The lowest flow, you think, is about 2,000,000 gallons? *A.* — Yes, sir: that is a very low estimate indeed. We have, by our gaugings at Lincoln Square, which is above the point where the sewers enter Mill Brook, nothing that shows less than 4,000,000 gallons; but then, these gaugings were probably taken at a time when the water was not at its very lowest. Of course, that 2,000,000 gallons is full stream; not the amount of sewage that is turned into it, but all the stream together.

Q. — Mr. Taylor, your assistant, has taken these gaugings? *A.* — Yes, sir: he is here, and can show them to you. There is another point to which I did not call attention, in relation to this, which is also mentioned in the report; and that is, that there are twenty-six woollen and cotton mills, besides wire-works, iron-manufactories, and two or three shambles, on the stream. What I want to call attention to, in relation to this, is the fact that quite a large portion of them are out of this territory.

Q. — Read your list, and state where they are. *A.* — There is a wire-mill at South Worcester (I do not refer to the Quinsigamond wire-mill: there is a small one at South Worcester); the Crompton Carpet Company; Hopeville Woollen Mill; Curtis & Marble Iron Works; and the Trowbridgeville Shoddy Mill, on the stream in Worcester.

Q. — On the Blackstone above the sewage? *A.* — Yes, sir. And then we have Stoneville. There are two mills there, — one a small tape-mill and the other a cotton-mill.

Q. — What town are they in? *A.* — Those are in Auburn. Then we have, on a branch of the stream, known as Ram's-horn Brook, two woollen-mills that are not down upon this list. They are in Auburn.

Q. — Do you know their names? *A.* — I don't know their names. The mill at Jamesville; John A. Hunt's mill; the Darling mill; the Ashworth & Jones mill, in Worcester. The stream bends around, and comes into Worcester again. Then we have Smith's mill, Olney's mill, Pierce's mill, what is called the Chappel mill, E. D. Thayer's Bottomly Mill, Kent's mill, Mann & Marshall's mill, — all in Leicester.

Q. — Those are on Kettle Brook, and other tributaries of the Blackstone River? *A.* — Yes, sir, those are on tributaries of the river.

Q. — Do you know where Blackstone River begins to be called by that name? *A.* — Well, I always supposed that it was called "Blackstone River" after Mill Brook entered it. It is called "Middle River" from the point where Kettle Brook and Tatnuck Brook come together.

Q. — Now go on with the other streams. *A.* — There are on Tatnuck Brook —

Q. — That comes into Kettle Brook or Middle River at New Worcester? *A.* — At New Worcester. I commence with A. G. Coe's wrench-shop, Loring Coe's wrench-shop, Loring Coe's forge-shop, the woollen-mill at Charles Ballard's privilege, and the woollen-mill at the A. L. Whiting privilege, I think it is. I don't know but those are the old privileges. Those are all in Worcester.

Q. — Any above Worcester on that stream? *A.* — No, sir.

Q. — Then take Mill Brook. *A.* — The first establishment on Mill Brook proper is S. Warren & Sons: it isn't a mill, it is a tannery. With the exception of Washburn & Moen's, I am speaking now of mills outside of the city proper, — S. Warren & Sons' tannery, A. C. Butterick's mill. There are three mills above the city; and then there are innumerable iron-works and dye-houses, and other industries that get into the stream in the city.

Q. — Is there any other point? I don't desire to occupy the Committee's time by going over things that have already been gone over. *A.* — I don't know that there is.

Q. — You have taken some specimens of river-water at various points, have you? *A.* — Yes, sir.

Q. — They are in those bottles, are they? *A.* — Yes, sir.

Q. — Were they taken at the times and places indicated on the bottles? *A.* — They were.

Q. — Have you got any washings from filters of the Cochituate water? *A.* — Yes, sir: I obtained some yesterday.

Q. — Will you produce them? *A.* — There is a sample of Cochituate, after running through a filter one hour.

Q. — Where was that obtained? *A.* — That was obtained on Tremont Street here. There is one after the filter had run about fifteen minutes.

Mr. MORSE. We don't object to the purification of Cochituate water.

WITNESS. It was simply to show that other water as well as Mill Brook will show impurities when filtered.

Q. — That model of the city of Worcester was made under your supervision? *A.* — Yes, sir.

Q. — What is the vertical and horizontal scale? *A.* — The horizontal scale is three hundred feet to the inch, and the vertical, fifty.

Q. — It is an exact model, with that correction? *A.* — Yes, sir.

Q. — If you have any special specimens that you desire to show the Committee, you can show them, and tell where they were taken. *A.* — Mayor Stoddard, in his testimony, referred to specimens taken from Mill Brook below Washburn & Moen's establishment, the day after the Committee were there. These were all taken before the storm. This was taken above the point where any sewage comes into it. Of course it has settled now. There is no sewage whatever, except, possibly, something that Washburn & Moen may have put in themselves. What I mean is, no public sewage.

Q. — Give us one taken down by Cambridge Street, after it has got all the sewage, and before it has been polluted by the Blackstone River. *A.* — Here are two. This one was taken out of Mill

Brook at Cambridge Street, where Quinsigamond-avenue sewer comes in, Feb. 23 of this year, at 3.30 P.M.

Q. — That was before the storm, and before the Committee were there? A. — Yes, sir; and it was taken just above the point where the brook comes into the sewer. That [showing another bottle] is a specimen from the gas-house. Then, here is a specimen that was taken at the outlet of the big sewer, Feb. 23, at 3.45 P.M.

Q. — If you have them, show us some specimens from Burling Mills. A. — There is a specimen of water taken at Mr. Morse's dam, Feb. 24, 3 P.M., when the water was running over the rollway eight inches deep. He showed a specimen taken when the water was running over ten inches. That was taken, of course, before the storm.

Q. — Have you got any specimen from the Burling Mills? A. — This specimen was taken at the Burling-mill Pond, which is above Mr. Morse's, Feb. 23, 4.15 P.M., when the water was running about two inches deep over the dam.

Q. — Do you know about what quantity of water was going by Morse's Mill the day the Committee were down there? A. — Well, probably in the vicinity of between five and six hundred million gallons in twenty-four hours. There is a specimen of the river-water below the Cordis Mills, just below Millbury, taken Feb. 24, at 4 P.M. This is a specimen of Singletary-brook water, taken Feb. 24, at 3.15 P.M. This is another specimen of Singletary-brook water, taken March 3, at 4.30 P.M. That was after the storm. Here is a specimen of water taken from the brook at the Worcester & Nashua freight-yard, which is just below the Washburn & Moen works, before any of the sewage enters the brook, Feb. 23, at 2.45 P.M. Here is a specimen of water from Kettle Brook, taken below Hunt's Mill, Feb. 25, at 3.30 P.M. And also one taken below the Washburn & Moen wire-works, at Quinsigamond Village, where the stream crosses the highway, at what is called the iron bridge, Feb. 23, at 4 P.M. That is about half a mile below the point where all our sewage comes into the river. This is a specimen of Salisbury-pond water, taken Feb. 23, 1882, at 2.30 P.M., above the Washburn & Moen wire-works. That is where they cut ice.

Q. — Where is Hunt's Mill? A. — Hunt's Mill is on Kettle Brook. It is the last mill before you get to the Leicester line.

Cross-Examination.

Q. (By Mr. FLAGG.) — Do I understand that you took all these samples yourself? A. — Yes, sir.

Q. — Did you take that Singletary-brook sample March 3, yourself? A. — I beg pardon: I did not take that one. My assistant took that one. I can't say as to that. My assistant is here. I had forgotten I didn't take that. I took the others myself.

Q. — I understand that samples taken at different times will show a different character? *A.* — Yes, sir.

Q. — These show only the state of the river at that particular time? *A.* — Certainly.

Q. — Taken at other times, they might appear different? *A.* — Certainly: of course.

Q. — I do not understand you to say that you prefer Blackstone-river water to Cochituate? *A.* — I have never thought that I did: no, sir.

Q. — And these mills that you speak of — some of them are very small? *A.* — Yes, sir.

Q. — Some of them two-set mills? *A.* — I think very likely.

Q. (By the CHAIRMAN.) — How many mills are there on the Blackstone above the sewerage system of Worcester? *A.* — We had twenty-six reported in the report. Then there were two on the Ram's-horn stream that I didn't get in my report.

Q. — Those are factories, not saw-mills, or any thing of that kind? *A.* — No, sir. They are all manufacturing establishments.

Q. (By Mr. MORSE.) — I want to know whether you have given any attention to the preparation of any plan for the purification of the sewage? *A.* — Well, not to any extent: no, sir.

Q. — You haven't yourself considered what plan you would recommend? *A.* — No, sir, I have not. I made all these surveys, or rather had them made, superintended them, for the State Board; but so far as developing any plan is concerned, that I have not done.

Q. — Are you prepared to say that the plan recommended by the State Board is not the best plan, under the circumstances? *A.* — I can say this, Mr. Morse, that I consider any plan, that with the others, as being entirely experimental.

Q. — I did not ask that question; but my point was, whether you were prepared, from such examination as you have made, to say that the plan recommended by the State Board was not a wise plan? *A.* — I have not considered it a wise plan for the city to adopt. I can say that conscientiously.

Q. — Are you prepared to say that the plan recommended by Col. Waring is not a wise plan? *A.* — I am, most decidedly.

Q. — But you are not prepared to say what would be a wise plan? *A.* — No, sir, I am not.

Q. — You have not considered that question? *A.* — No, sir.

Q. (By Mr. GOULDING.) — Will you finish the list of manufactories which you began to read? *A.* — These are not all on Mill Brook, but some of them are on tributaries that run into Mill Brook.

The Washburn & Moen Manufacturing Co., wire-works; Richardson Manufacturing Co., iron-works; Ames Plow Co., iron-works; Wetherbee, Rugg, & Richardson, iron-works; N. A. Lombard, iron-works; Wheeler Foundry, iron-works; Johnson, Bye, & Co., iron-works; Pond's Iron-works; H. C. Fish & Co., iron-works; Merrifield's buildings, devoted principally to the manufacture of iron in its different branches; Rice, Barton, & Fales, iron-works and foundry; Worcester Felting Co.; Wheeler Foundry Co., Mechanics Street, iron; Knowles's Iron Works; Crompton Loom Works; two smaller foundries, that I do not know the names of; Colvin's Foundry at junction; Colvin's Foundry on Gold Street; Earl & Jones Foundry; Adriatic Mills, woollen; Fox Mills, woollen; Worcester Copperas Works, two mills; Worcester Gas Works; besides three or four large iron-works at what we call the Worcester Junction; and then, there are other smaller concerns that I have not got.

Q. (By Mr. FLAGG.) — Those all empty their sewage into Millbrook sewer, or into sewers that empty into Mill Brook? Yes, sir.

Q. — And, in addition to that pollution, there are the sewers emptying into it in the city? A. — Yes, sir.

TESTIMONY OF LUCIEN A. TAYLOR.

Q. (By Mr. GOULDING.) — You are assistant engineer of Worcester? A. — Yes, sir.

Q. — How long have you been in the engineer's office in Worcester? A. — Fifteen years.

Q. — Were all these gaugings of the flow of Mill Brook and the sewers which are mentioned in the report of the State Board of Health, Lunacy, and Charity made by you, and furnished to them? A. — I believe they were.

Q. — Now, I want to call your attention to this statement of theirs: that the average daily flow, dry-weather flow, at Cambridge Street, exclusive of sewage, for four months of the year, may be stated at about 3,500,000 gallons, and its minimum daily flow at 750,000 gallons. Can that be true? A. — I don't think that can be true; at least, there have been no gaugings to show any such thing. That is a self-assumption. The lowest gaugings I have taken were in 1871, in June, July, and August, at Lincoln Square, possibly in September. They cover a period of about four months. Where the drainage district is stated in the report as eight square miles, it is 7¾. Those gaugings show something over 4,000,000 gallons. I don't remember the exact figures. That is the average. In 1875 I took some gaugings, and I think the average was about 8,000,000.

Q. — I don't care to go into the average flow, or to question this report, except to find what the minimum daily flow of Mill Brook is. *A.* — The lowest flow that I have ever recorded I have not at hand now.

Q. — Do you know how much it was? *A.* — I can't say; that is, it would be governed, perhaps largely, by Washburn & Moen. If they were not running their mills, and the gates were shut down, it might be a small quantity for a short period of time, twenty-four or twelve hours, or some limited period of time; but the average flow for a month could not be any such quantity as stated there, I am positive. In 1875 I took gaugings in August and September, and the lowest amount was 8,700,000 gallons.

Q. — Is there any other point to which you have given special attention? *A.* — I don't know of any thing very special.

Q. — I suppose you agree with Mr. Allen in what he has said in regard to the time it would take to construct the works necessary to put the downward filtration scheme into operation? *A.* — Yes, sir: I think it would take at least three working seasons to do it in a proper manner.

Cross-Examination.

Q. (By Mr. FLAGG.) — Leaving out of account the dry-weather street-flow into Mill Brook, it might be, might it not, that the dry-weather flow of Mill Brook for any one day would be 750,000 gallons? *A.* — It might perhaps be arranged so. Of course I don't know but what it might be so. That is not the flow, but what might be stopped.

Q. — As a matter of fact it is sometimes stopped, isn't it? *A.* — I don't think so.

Q. — Do you know upon what surveys the State Board of Health make their statement? *A.* — I am very certain from information they received from me: I am very positive about that.

Mr. GOULDING. Dr. Walcott stated to Mr. Allen this morning that he presumed that 750,000 gallons was a misunderstanding.

Mr. FLAGG. We were content to let it remain so; but now, if you are going to attack the State Board of Health —

Mr. GOULDING. Not at all. This 750,000 gallons we supposed was a misstatement; the rest we supposed to be correct. Mr. Taylor furnished the gaugings.

Q. (By Mr. FLAGG.) — The statement that the average flow in four months was 3,500,000 gallons is right? *A.* — As I said, the lowest recorded flow is over 4,000,000 gallons at Lincoln Square.

Q. — The lowest you have recorded, or *they* have recorded? *A.* — That anybody has recorded.

Q. — Do you know every thing that has been done by the State Board of Health? *A.* — No, sir, I do not; but I think that all the gaugings that ever have been taken of Mill Brook, or the sewers of Worcester, I have taken. That I am quite positive of.

Q. — You suppose the State Board of Health, Lunacy, and Charity deliberately allege what is untrue? *A.* — No, sir: I do not mean to be understood as saying any such thing.

Q. (By Mr. GOULDING.) — Is there any doubt that all the figures of the gaugings of those flows that the State Board of Health, Lunacy, and Charity obtained this last year were obtained from you? *A.* — I don't think there is.

Q. — Could they get into the sewers without your knowing it? *A.* — I don't think they could. I think you will find in a former report the statement of my gauging of 4,000,000.

Q. (By Mr. FLAGG.) — Speaking generally, the rest of the report is somewhere near accurate, isn't it? *A.* — I have no reason to suspect that it is not.

Q. — Do you know any thing about the flow of water in the sewers, whether it looks clear, or not? *A.* — In the lateral sewers, it is generally quite clear. Mondays you will notice a difference. You will notice a soapy appearance, and you will often notice that in Mill Brook.

Q. (By Mr. CHAMBERLAIN.) — How do you account for that water being clear? Does running the sewage into it make it clear? *A.* — No, sir.

Q. — How do you account for it? *A.* — The only reason is, it is so much diluted by pure water, partially water from the water-pipes and partially water from the under-drainage. That is, the sewers do take more or less of the under-drainage of the city.

CHARLES A. ALLEN. *Recalled.*

Q. (By Mr. GOULDING.) — Have you noticed any thing about the water running clear in the sewers? and, if so, is there any explanation that you can suggest? *A.* — I don't know that I ever examined it on Monday; but I should be inclined to think, that of course it would present a different appearance on that day from others: but ordinarily, in our lateral sewers, where there are no manufacturing establishments, the water is very clear. Of course, there is sediment at the bottom of the sewers, and the quantity of it depends something upon the grade; but I think it is undoubtedly due partly to the fact that the sewers that were built previous to three or four years ago were not built tight at the bottom, and we get a tremendous quantity of under-drainage, sub-drainage; and probably the sewage matter is diluted to that extent that you would not notice it particularly.

Cross-Examination.

Q. (By Mr. FLAGG.) — The previous witness stated that the lowest flow was 4,000,000 gallons of water in Mill Brook; and you say that with an average daily flow of 3,000,000 gallons of sewage it is so diluted that it looks all right? *A.* — No: I wasn't speaking of Mill Brook. I said the lateral sewers, where there were no manufacturing establishments.

Q. — Now, how about Mill Brook? How does that look? *A.* — It looks black and filthy, just as you say, of course.

TESTIMONY OF ROBERT H. CHAMBERLAIN.

Q. (By Mr. GOULDING.) — You are the superintendent of sewers in Worcester? *A.* — Yes, sir.

Q. — And have been for a good many years? *A.* — Yes, sir.

Q. — I want to call your attention to one or two points. First, how many catch-basins are there in the city? *A.* — About nine hundred.

Q. — How much material is taken out of those catch-basins in the course of a year? Have you any means of estimating? *A.* — My only means of estimating is by knowing about how much can be got out in a day, etc., and the number of times we go around in a year. In all probability, 4,000 to 5,000 loads.

Q. — Where is that carried? *A.* — That is used for filling, and disposed of in any way we can.

Q. — Is it dumped into the river or brook at all? *A.* — No, sir.

Q. — Have you noticed the flow of water in the sewers, how it appeared? *A.* — Yes, sir.

Q. — What do you say about it in the lateral sewers? *A.* — Usually clear.

Cross-Examination.

Q. (By Mr. FLAGG.) — This material that you take out is such that you can use it for filling? *A.* — Yes, sir.

Q. — And it is not fæcal matter? *A.* — Not at all: only street-wash. No sewers enter into the catch-basins.

Q. — Have you ever observed any ill effect of the sewers on the health of the men who work in them? *A.* — No, sir. I have had men work in them for ten years in succession. They have their hands in the sewage, and are wet with it every day; and they are in as good health as ordinary laboring men.

Q. (By Mr. GOULDING.) — Are you about the sewers yourself a good deal? *A.* — Every day, more or less.

Q. (By Mr. FLAGG.) — Ever observe any ill effects upon your health? *A.* — No, sir.

Q. — Well, you would rather see this sewage flow down the river than to have it remain in Worcester? *A.* — We don't desire to keep it.

Q. (By Mr. CHAMBERLAIN.) — Is there any city ordinance regarding cesspools, or do they run them directly into the sewers? *A.* — They run directly into the sewers, sir.

TESTIMONY OF BENJAMIN WALKER.

Q. (By Mr. GOULDING.) — You are one of the aldermen of Worcester? *A.* — I am.

Q. — Your business is that of an ice-dealer, I believe? *A.* — Yes, sir.

Q. — Pretty extensively engaged in that business in Worcester? *A.* — Considerably.

Q. — Have you had any different specimens of ice from different places analyzed this winter? and, if so, with what results? *A.* — Yes, sir.

Q. — Tell the story. *A.* — I didn't know of being called here at all, and I haven't brought any statistics with me. I had five samples carried to Professor Thompson about three weeks ago, that were taken from three separate ponds.

Q. — What ponds? *A.* — One from what is called Crescent-street Pond; one from Salisbury's Pond on Grove Street, just above the wire-mill; and one from Coes's reservoir, in New Worcester.

Q. — Coes's reservoir is on Tatnuck Brook? *A.* — Yes, sir, on Tatnuck Brook, — a brook which the city is talking of taking for pure water for the city.

Q. — No sewage from the city of Worcester goes into that stream above that pond? *A.* — I think there is no sewage enters the stream at all anywhere. There is a little factory in Tatnuck. Possibly the sewage from that may work into it.

Q. — Where is the Crescent-street Pond? *A.* — Crescent-street Pond is near Lincoln Street, right north of where the Boston, Barre, & Gardner Railroad Depot is.

Q. — Now, what were the comparative results of that analysis? *A.* — That is all I can give you, because I haven't got the statistics with me. I have the report of Professor Thompson, but I can't give you the exact statements contained in it. His general statement was, that the No. 1 (he had it by numbers: he didn't know any thing about where the ice came from) shows indications of sewage.

Q. — What pond did that come from? *A.* — That came from Coes's reservoir. The other ponds were all reported as slightly polluted. There was no sewage in it. That is my recollection of the report; and, so far, it is correct. He gave it in numbers, and the analysis he gave was $\frac{85}{100000}$, taking the city-water of Worcester for a standard. That was the most impure water that he found,—the water coming from that ice.

Cross-Examination.

Q. (By Mr. FLAGG.)—Mr. Walker, you don't furnish your customers with ice from Blackstone River below the mouth of the sewer, do you? A.—It would not be fit for domestic purposes.

Q.—Why wouldn't it? A.—That depends more on scientific men. I never tried it to see.

Q.—You wouldn't want to buy it? A.—No: I have no occasion for it. I believe you have some down in Millbury that has been offered me.

Q. (By Mr. MORSE.)—Did you personally take these specimens of ice that you had analyzed? A.—My agent took them to Professor Thompson.

Q.—Perhaps you misunderstood my question. I didn't mean whether you personally took them to him, but whether or not you personally cut them from these different places. A.—I didn't do the work myself; but I saw that they were cut from those places, saw them cut, and saw them taken from the pond. After the five pieces were carried, which represented three different ponds, I received a note from Professor Thompson, wishing me to carry him other samples from the same places, saying that he was not satisfied with the analysis. Then I went and took the ice myself, as my agent was away, from these various ponds, and carried them to him; and he did not alter his report at all.

TESTIMONY OF LORING COES.

Q. (By Mr. GOULDING.)—You live in New Worcester? Yes, sir.

Q.—You are the owner of Coes's reservoir, so-called? A.—Yes, sir.

Q.—And the wrench-factory, etc.? A.—Yes, sir.

Q.—Do you know any thing about the use of those sprinklers? A.—I never have had any in my mill.

Q.—Have you known about their use? A.—Yes, sir: I have seen them put up and seen them up; never have seen them used.

Q.—Were you a practical mechanic in your earlier days? A.—Well, I have been always a mechanic.

Q.—What do you say about the utility of them, if you know any thing about them? A.—I don't know very much about them. I suppose where they have been put up, if they are always kept dry, they will work; but, where they have been wet, a small hole through a piece of iron, if I understand it, will fill up.

Cross-Examination.

Q. (By Mr. FLAGG.) — Do you know that those sprinklers are put up extensively over the country? *A.* — Yes, sir.

Q. — By recommendation of the insurance companies? *A.* — Yes, sir : I believe they are.

Q. — And that they remain put up until a fire occurs? *A.* — Yes, sir.

Q. — And that they have been used with good effect? *A.* — Yes, sir, when they are in good order, they have been used with good effect, I believe.

Q. — Can you get manufacturing property insured in a first-class mutual company unless you have these very sprinklers? *A.* — I don't know about that. I haven't been insured in a mutual company.

TESTIMONY OF JOSEPH M. DYSON.

Q. (By Mr. GOULDING.) — What is your business, Mr. Dyson? *A.* — State inspector of factories and public buildings.

Q. — Do you know this rendering establishment down there on the Millbury Road near the Blackstone River? *A.* — I do.

Q. — Won't you describe what it is, and what they do there? *A.* — It is a place, I have always understood, that cleaned tripe, and rendered grease and bones and such like. I have noticed a smell from it a great many times within the last year in driving by there evenings, and even daytimes.

Q. — How far away can you smell it? *A.* — I have smelled it just after leaving Millbury. I have also smelled it up above Quinsigamond, in going down that way. The strongest was opposite the watering-trough, directly opposite the building there. I have found it worse in the evening than I have in the daytime. Some two months ago it was so bad that I drove over there to see what they were doing. I found they were emptying their vats. It was somewhere about ten o'clock at night when I came by there.

Cross-Examination.

Q. (By Mr. FLAGG.) — When you speak of "leaving Millbury," you mean, leaving the town of Millbury? *A.* — Leaving the town of Millbury.

Q. — That is three miles away? *A.* — That is three miles away, at the turn of the road. It is about half way between the village and Burling Mills.

Q. — That would be three miles from this establishment? *A.* — No : I shouldn't say it was over two.

Q. — At the same time, the river would be within a few feet of the road? *A.* — It is off a little piece from the road at that point.

Q. — How far? *A.* — I should say it was two or three hundred yards, or more.

Q. — And you ascribed the odor to an establishment two miles away? *A.* — Yes, sir.

Q. (By Mr. CHAMBERLAIN.) — Are those steam-tanks, or not? *A.* — I should say they were not.

TESTIMONY OF ALBERT A. LOVELL.

Q. (By Mr. GOULDING.) — Do you know any thing about that rendering establishment? *A.* — I visited the rendering establishment of Jeffard & Darling yesterday, in company with the city marshal of Worcester. We were informed by Mr. Jeffard that their business was the collecting of refuse from the markets in Worcester. This refuse is collected every day, and taken to the establishment, and submitted to a process of boiling, I think, for extracting the grease. After the grease is extracted, this material is put into vitriol, and then it is placed upon steam-coils, for the purpose of drying it. In that process of drying, it throws off a thick, heavy vapor, which has a very disagreeable smell. In order to get rid of that smell, as far as possible, they have constructed some earth-closets; and, with a revolving fan, they force what they can of this vapor into those earth-closets. Those earth-closets will not, of course, take all the vapor. A great deal escapes into the air; and Mr. Jeffard pointed to some houses some three-quarters of a mile off, and said, when the wind was in that direction, the people who occupied those houses could distinctly discern the odor there, and for a distance of half a mile or a mile in that direction they could clearly discern the odor from that factory.

Q. — What is the state of the atmosphere around the establishment there as you come to it? *A.* — It is very disagreeable.

Q. — Where does the drainage of that establishment go to? *A.* — They have constructed a pipe-drain from the factory to the Blackstone River; and I went to the mouth of that pipe-drain, and it was throwing into the river a bad-looking mixture of blood and little scraps of meat and every thing that was disagreeable.

Q. — Has that any thing to do with the sewage of Worcester? *A.* — It has no connection with the sewage of Worcester.

Q. — You are on the Board of Health of the city, and have been for a number of years? *A.* — I am, sir.

Cross-Examination.

Q. (By Mr. FLAGG.) — This place, I take it, must be a nuisance? *A.* — To some extent.

Q. — But the Board of Health of Worcester allows it to exist within its limits? *A.* — We do not consider that the odor from that establishment is a sufficient nuisance for the Board of Health to interfere in the matter at all. They have taken all the pains that we can reasonably expect from them. If we suppressed every industrial enterprise in Worcester on account of some offensive odor, we would drive half the population out of the city. There is a strong pressure brought to bear upon the board all the time to suppress this establishment and that establishment all over the city; and, if we acted upon that pressure, we should drive half of the population out of the city.

Q. — This tripe establishment was formerly at Quinsigamond? *A.* — I don't know where it was. I have heard it said that there was an establishment there.

Q. — Which the Board of Health drove out of Quinsigamond? *A.* — I wasn't on the Board of Health at that time. I don't know anything about that.

Q. — It cannot be on account of the fact that this is so near the Millbury line, that it is allowed to exist? *A.* — I don't think it is. I think a person driving on the Millbury road, when he got into that neighborhood, would discern a strong odor; and nine persons out of ten would say that it came from the sewer, whereas it came from that establishment. My attention was never called to it until yesterday.

Q. (By Mr. GOULDING.) — How far is it from any house? *A.* — I can't say whether there is any house close by. I think one of the firm lives near there, and Millbury line is half a mile from there.

Q. (By the CHAIRMAN.) — Did anybody around there ever complain of that establishment as a nuisance? *A.* — I have never heard of any complaint.

Q. (By Mr. CHAMBERLAIN.) — What kind of pipe is that? *A.* — It is cement pipe laid on the surface.

Q. — How large is the pipe? *A.* — Eight inches, I should judge, by the looks.

Q. (By Dr. HARRIS.) — Was it full all the time? *A.* — It was full yesterday, the only time I was there, and discharging very offensive matter into the stream.

TESTIMONY OF GEORGE A. BARNARD.

Q. (By Mr. GOULDING.) — What is your business? *A.* — Slate-roofing and asphalt.

Q. — Ever work on the Burling Mills? *A.* — Yes, sir.

Q. — When did you work on the Burling Mills? *A.* — I can't give you the exact date.

Q. — What year? *A.* — I work there almost every year, not particularly on the Burling Mills; but I am slating and asphalting the roof at Fishersville, and my men are down there every day, and I am down over the road.

Q. — Ever notice any offensive odors from the river? *A.* — No, sir. I have in the cars, when the wind was west and the river on the other side of me, noticed an offensive odor from this tripe-shop, as they call it.

Q. — You usually drive down? *A.* — I go both ways, — in the cars, and drive.

Q. — Where have you noticed it? *A.* — Soon after leaving Quinsigamond, and from there until I got beyond the watering-trough.

Q. — Where did you make up your mind that it came from? *A.* — I always supposed that it came from the rendering establishment. I supposed it was simply a tripe-shop.

Q. — You have smelled that same smell when you have been in the cars, with the river on the other side? *A.* — I have smelled it when the windows were open. You don't notice it when the windows are closed.

Q. — When the wind was west? *A.* — Yes, sir.

Q. (By Mr. FLAGG.) — Both the railroad and the highway run along by the river, and you never noticed any smell? *A.* — I didn't notice it until I got below Quinsigamond.

Adjourned to Monday, March 20, at 10 o'clock.

281

SEVENTH HEARING.

BOSTON, Monday, March 20, 1882.

THE hearing was resumed at 10.15 A.M.

MR. FLAGG. I have a letter from a witness of ours who was unable to be present. I understand my brother Goulding does not object to its coming in as part of our evidence now. It is from Dr. Joseph N. Bates of Worcester, a physician of more than local reputation.

WORCESTER, March 16, 1882.

GEORGE A. FLAGG, Esq.

Dear Sir, — I regret that sickness has prevented my attendance before the Legislative Committee now in session, the question before the board being the disposal of the sewage of the city of Worcester by the Blackstone River and its tributaries.

For quite ten years last past, the unpleasant consequences of contamination of deleterious gases from sewage, and various contaminations from impurities from manufacturing materials, have caused impurities unsuitable for purposes of cleanliness, or for use for live-stock, or for use with machinery, or the ordinary uses of a living stream of water. To refer to diseases generated by the use of impure water need not here be dwelt upon, as all understand the dangers to which man and the brute creation are subjected by such exposures. The diseases incident to such exposures have been well considered by gentlemen who have preceded me. I am pleased to indorse the declarations uttered by gentlemen of my profession, scientists, and others, and most sincerely trust that judicious and timely measures may be speedily adopted for the radical relief of this important measure.

Respectfully yours,

DR. J. N. BATES.

CLOSING ARGUMENT FOR THE REMONSTRANTS BY FRANK P. GOULDING, ESQ.

MR. CHAIRMAN AND GENTLEMEN OF THE COMMITTEE, — Upon whatever other matter the parties here differ, they agree that this is a very important question. The petitioners believe that it is important, because they profess to believe that their business enterprises, and the salubrity of their homes, are in some measure involved in the disposal of this question by the Legislature. We believe that it is important, because it is a proposition to embark Worcester in a course of expensive experiment, which, so far as we are able to see, gives but little promise of certain success. But it is further, in our view, important, because it is a step in a path of new legislation, involving important consequences, involving important questions of law and of science, involving, in short, a new departure with the Legislature.

I shall best attest my sense of its importance, I think, by proceeding at once, without any preliminary observations, to present the views that I have to present on behalf of the city of Worcester. But I desire to premise what I have to say by the statement of a few facts, and the furnishing of a few statistics, — facts which are not in dispute substantially, statistics that cannot be controverted, and which must be the subject of constant allusion in the course of my argument, and, I think, must be a subject of constant reference in order to arrive at any just conclusion upon this question.

(1) The sewage of Worcester is about twice as dilute as the average of fifty English towns, before it mingles with Mill Brook, which is by law a sewer.

(2) It then becomes diluted by an addition, on the average, of more than four times its own volume of the natural flow of that stream, and, on no very extraordinary occasions, with fourteen times its own volume, and, in times of freshets, with forty times its own volume.

(3) Mill Brook, thus polluted, empties into Blackstone River, and adds to its pollution.

(4) The pollution of the river at Morse's sash-factory (which includes all the pollution of Kettle Brook, Ram's-horn Brook, and all the other streams which empty into Kettle Brook), as compared with the reservoir, is as 9.0899 to 4.4120, or more than 2 to 1.

(5) As showing the effect of the flow of the stream in purifying the water, the pollution of Mill-brook sewer, as compared with the river at Morse's sash-factory, in 1875, is as 28.3109 to 9.0899, or more than 3 to 1; or, taking the analysis of 1881, is 41.8790 to 12.6410, or $3\frac{1}{3}$ to 1, — a comparison which again charges to the sewage all the pollution of the mills and other pollutions of Kettle Brook and other streams.

(6) The results generally stated by the Board of Health, Lunacy, and Charity are as follows: —

(a) By means of a dam midway between Worcester and Millbury, the stream is ponded; and there the solids held in suspension are deposited, and a nuisance is created. And this is the only nuisance which the board report in terms, and that is caused by a dam.

You will search this report throughout in vain for the word "nuisance," whatever they mean by nuisance there, except, I think, at this point; and that is caused by a dam.

(b) At a number of dams in Millbury some further deposition occurs, presumably with no nuisance resulting; at least, there is no nuisance stated.

(c) Along the whole course of the stream, for some miles below Worcester, putrefaction of the organic constituents of the sewage takes place (most rapidly in the summer months); and, as a consequence, offensive gases are liberated, which are largely the cause of complaint.

(d) The deposits stimulate the growth of aquatic plants, and thus, and by their own bulk, are filling up the ponds; and this the millowners complain of.

(e) The stream is very offensive at times; and this, with the filling up of the ponds, will soon depreciate the property in the vicinity. You will observe that both this statement and this prophecy are of an indefinite nature.

(f) It is the belief of the people dwelling on and near the banks of the stream, that a perceptibly injurious effect upon the general health has been produced; and this belief is shared in, to some extent, by the resident physicians.

This belief is not, at least, so far as appears, shared in to any extent by the board.

(g) The stream, four miles below the sewer outlet, is unmistakably polluted; twenty-five miles below, the impurity is all but lost to chemical tests.

(h) It appears that an increase of the pollution of the stream since 1872 has taken place, and a much less marked increase since 1875, showing a diminished ratio of increase.

(i) The deposits, except at the sash-factory, have not increased so considerably as to be the nuisance, of themselves, which might have been expected.

This statement, gentlemen, which is taken from the Report of the State Board of Health, Lunacy, and Charity, is, I submit, the strongest statement of the case that can be made against Worcester. I shall have occasion later on to discuss this evidence; but I submit now that I must carry with me every member of the Committee, when I say that the State Board of Health, Lunacy, and Charity, on the facts, have stated this case out to the extreme limit of possibility consistent with the truth.

Now, in addition to this statement, I wish also to premise some other facts, which must be equally beyond dispute, geographical, topographical, and historical.

A. Mill Brook, with its tributaries, Pine Meadow and Piedmont Brooks, is the natural and only channel of drainage of the city of Worcester. The territory on which that city is situated will continue to discharge its surface drainage into Mill Brook while the laws of gravitation continue, unless some system of drainage of an artificial nature shall be devised on a much more stupendous plan than any thing yet suggested.

B. The city of Worcester was settled much earlier than Millbury, and the towns lower down on the Blackstone. Worcester was settled in 1684 to 1718, Millbury in 1743, Grafton in 1730 to 1735, Sutton in 1716 to 1718, Northbridge in 1772, Uxbridge in 1727; and therefore, whatever superior strength of title and right arises from priority in time belongs to Worcester.

C. The growth of Worcester has been a natural, steady growth, obedient to the ordinary laws that govern the accumulation of large masses of people in communities; and that growth covers a period of nearly two hundred years. It is not a case of Alexander, or any other conqueror, taking a body of colonists out and planting them on a stream up above an older town or city. The city has grown by those natural processes, more rapidly at times, less rapidly at other times, but obedient to the laws that govern the growth of communities in this country.

D. During all this time the pollution of Mill Brook and Blackstone River has gone on as the natural and necessary result of the existence of the town. *Necessary,* I say, because any project of otherwise disposing of sewage than by emptying into the nearest river, or into the sea, was never heard of in any land (if we except Edinburgh, which is a notorious failure) until within a comparatively few years. We have heard from Dr. Folsom, since this hearing, that it began about fifteen or seventeen years ago, on the Croydon farm in England; and not until within a very short time could the most enthusiastic friend of precipitation, downward intermittent filtration, or broad irrigation, or "willow walla winding through the meadow," according to Col. Waring's plan, claim that any thing like success was attained by any form of disposing of sewage except the natural form of diluting it with as much water as possible, as quickly as possible.

E. Long years ago, by the natural results of the growth of Worcester and Millbury, and the other communities on its banks, the Blackstone River had become a foul stream.

F. The Blackstone River never was and never will be a source of water-supply for domestic and other similar uses.

Along with these statements of facts, I desire to read some more facts. And now I come to the point of statistics, for the purpose of comparing the Blackstone River with other streams in this Commonwealth, and for the purpose of determining what is the actual amount of pollution of this river, and what is intended by the general statements contained in State reports, and what exact facts exist which justify, or fail to justify, the vague imputations that are laid upon the Blackstone River.

(1) Analyses of different streams and basins reporte the Board of Health in 1876 (the same report that most of the analyses in this

report of 1882 are taken from) show the following facts, which I desire to set over against the condition of the Blackstone at Morse's sash-factory, which is the foulest point, I think, of the Blackstone, chargeable to Worcester sewage.

And let me say here, that I take the analysis of 1875 of the Blackstone River for this purpose, for three reasons: —

First, Because it is the analysis, or earlier analyses, to which I am obliged to resort in order to get the analyses of most of the other streams in the State, with which I propose to compare it.

Second, This report of 1882 says that the increase of pollution since 1875 has been much less marked; and it is fair to presume, therefore, that the other streams have increased in pollution as much as this.

Third, Because this report of 1882 is really largely taken, so far as its facts are concerned, from this Report of the State Board of Health of 1876.

Let us disabuse ourselves, if we suppose the State Board of Health, Lunacy, and Charity, in obedience to your resolve of last year, went into any thorough and extensive original investigation. They did not do it. They have made a very able and thoroughly impartial report; but they derive and draw most of their facts from that mine of information, to wit, the Report of the State Board of Health in 1876.

It is, therefore, fair that I should, for the purpose of showing what is meant by the proposition that the Blackstone River is probably more polluted than other streams, take that analysis found in that report of 1876.

(*a*) In Neponset River, at Milton Lower Mills, above Neponset Village, the impurities are 6.7363, nearly three-fourths (or exactly 74 per cent.) the pollution of the Blackstone (p. 96).

(*b*) Charles River, below Bellingham, 7.5758, more than three-fourths (or exactly 83 per cent.) the pollution of the Blackstone (p. 107).

(*c*) Connecticut River, immediately above Springfield, 6.6254, or more than two-thirds (exactly 72.8 per cent.) of the pollution of the Blackstone (p. 122).

(*d*) Winixetuxet River, above Taunton, 8.1320, more than six-sevenths (or 89 per cent.) the pollution of the Blackstone.

(*e*) Taunton River, near North Dighton, 17.5902 (or very nearly 1.93 per cent.), almost twice the pollution of the Blackstone (p. 140).

This report of 1876 says that the Blackstone River is probably more polluted than any other stream in Massachusetts, and that is quoted by the report this year, showing that they have not gone beyond that, and that they founded their opinions upon that. The

State Board of Health, Lunacy, and Charity, in their report of 1882, make the following statement of analyses in two other basins: —

(*f*) Deerfield basin, at seven points. I will not stop to read them all; but the average of the seven points is 11.4914 (1.26 per cent), or more than one-fourth in excess of the pollution of the Blackstone, on an average. Several of these Deerfield-basin points, selected at random, have much more; and in one place there is twice the pollution of the Blackstone River.

(*g*) In Miller's-river basin eleven points were taken, which show an average of 8.8350; that is, 97 per cent. of the average pollution of the Blackstone; and in several places largely in excess, and in some more than twice the Blackstone standard.

In the report of the State Board of Health for 1874, they show the condition of several other streams: —

(1) The Merrimack, below Lowell, 8.3891, or more than nine-tenths the Blackstone standard at sash-factory (exactly 92 per cent.).

At another point, 7.007, or more than three-fourths the Blackstone standard (exactly 77 per cent.).

Consider, also, the vastly greater volume of water in the Merrimack.

(2) The Sudbury River: —

(*a*) Above Ashland, 7.5041, or more than six-sevenths the pollution of the Blackstone at sash-factory (exactly 85 per cent.).

(*b*) At Framingham, 7.6711. This is one of the sources of Boston's water-supply, and its pollution 85 per cent of that of the Blackstone at Morse's factory.

(3) The Concord River: —

(*a*) At Concord, 6.3731, nearly three-fourths the pollution of the Blackstone (exactly 70 per cent.).

(*b*) At Lowell, 8.5754, or more than nine-tenths the Blackstone standard (exactly 94 per cent.).

(4) Cochituate Lake and its sources of supply: —

Look at some comparisons in connection with that, without taking Pegan Brook, which is really a nasty place. We will leave that out, as exceedingly filthy, and take other places with which it is fair to compare the Blackstone, if we want to see where the Blackstone stands comparatively with the rest of the rivers of the State, and what the State Board of Health mean when they say it is *probably* the most polluted stream in the State, and what the Board of Health, Lunacy, and Charity mean when they quote the words of the State Board of Health of 1876 to the same point.

(*a*) In the channel through the bar which separates the main lake from the basin into which Pegan Brook empties, average [filtered]

8.5621, more than nine-tenths the Blackstone standard (94 per cent.).

(b) Outlet of Farm Pond, 6.8027, nearly three-fourths the Blackstone standard (74.8 per cent.).

(c) Beaver-dam Brook, 7.8867, nearly seven-eighths the standard of Blackstone impurity (86.7 per cent.) (pp. 116, 117).

(5) Mystic Lake and its sources: —

(a) Bacon's Bridge, average, 12.5178, or more than one-third greater pollution than the Blackstone (1.37 per cent.).

(b) Outlet of Horn Pond, 7.6657, more than five-sixths the Blackstone (84 per cent.).

(c) Abajonna River, 11.6973, or one-fourth more than the Blackstone (1.28 per cent.) (p. 130).

(6) Water as delivered in Lowell from the Merrimack, after filtration, 7.8271, or six-sevenths the pollution of the Blackstone at Morse's factory (86 per cent.) (p. 135).

These analyses, you understand, are made by taking a hundred thousand parts of water, and then they state the ammonia, the albuminoid-ammonia, the chlorine, and the volatile and fixed solid residue, and sometimes they put in other elements; but those are the only elements that are in these analyses that I have referred to. That is the basis on which they proceed.

The average of nineteen different places in different basins in the State, taken at random, excluding such places as Pegan Brook, and including several sources of water-supply, and some filtered water, shows a pollution expressed by 8.8662, or about .99 as much pollution as the much-offending Blackstone River at Millbury (exactly 98.6 on an average).

Another fact that I desire to put in, which appears in the analysis of 1881, reported in the Report of the State Board of Health, Lunacy, and Charity of 1882 (pp. 122, 123), is, that the pollution of the Blackstone River, above the Worcester sewage, before it has received the contamination of the city, is 8.7579. It is, therefore, twenty-nine-thirtieths as much polluted before it receives the Worcester sewage, as it is when it reaches Morse's sash-factory. Add, from the analysis of 1881, Singletary Brook in Millbury, and this pollution is 8.9560, or more than ninety-eight per cent. of the pollution of the Blackstone at the sash-factory. And this board themselves say that "it must be remembered, that a very large portion of the pollution below Millbury comes from Singletary Brook, which is a very foul stream." It is .985 as polluted as the Blackstone at Morse's factory.

I have read these statistics, that cannot be controverted. They are taken from the State Board of Health Report; and they exhibit the condition of Blackstone River, with reference to other rivers, as far as chemical analysis reveals any thing.

Now, in view of this state of things, the petitioners come here, and they ask for legislation which shall prohibit the citizens of Worcester from emptying its sewage into the Blackstone River, so as to pollute it; and that involves, of course, its purification; for the city must empty into the Blackstone River the waste of all the water it uses, and it is not in the power of this Legislature to prevent it, because it cannot repeal the laws of nature. They ask for legislation which will compel the city of Worcester to purify it. Now, seeing that this is enormously expensive, to begin with, according to anybody's view, I suppose they would receive no considerable attention from this Legislature, unless there is some ground of morals, or of law, that would impose this duty upon Worcester as a community. And I want, in the first place, to discuss this question very briefly, of the relations of Worcester to this thing, in a moral point of view.

My friends have assumed here, with great confidence, and I may say with great nonchalance, that Worcester is to blame in this matter, that she is in fault. The motto of Millbury, and of our friends below, has been this: " Let Worcester cease to pollute the Blackstone River, or let Worcester cease to exist! You are committing this injury, remedy it!" They have not stopped to discuss any rule of morals which imposes an obligation. They have reiterated the injury; they have dwelt upon the inconvenience; they have not discussed the rule of morals, if any exists, that imposes any obligation upon Worcester, in connection with this matter. They have reminded me a little, in this respect, of the proceedings of Dogberry, in the case of the Commonwealth of Messina against Borachio and Conrade, when he said, " Masters, it is *proved* already that you are little better than false knaves; and it will go near to be *thought* so, shortly." They come here, and they seem to think or assume that it is proved already that we are guilty of this thing: it is only a question how we are to absolve ourselves. My friends on the other side regard Worcester, perhaps, as having, out of executive clemency, the privilege to choose the mode of her execution, but not to plead to the indictment, or to be heard in defence; and perhaps there is a little temerity in my position in undertaking to question this position, inasmuch as they have brought here a gentleman who has discussed in writing, before you, the law and the ethics, as well as the science, of this question, and apparently disposed of the whole matter. Now, for Col. Waring as a sanitary engineer, I should have no feeling and no words but of the utmost respect; and if he came here, and expressed an opinion on this case, upon the scientific view of it, or upon the law, or the morals of it (although I do not know why he should undertake to instruct this Committee of educated gentlemen, either in law or in morals), I should feel that he was entitled to my simple respect; but when the exigencies of his case, and the inducement of his retainer, require him to ingeniously construct a studied insult to the city of Worcester, and come here and read it when it has grown cold, and then allow it to be published in the public prints, to be circulated, I feel that I am absolved from the obligation to treat him with any more respect than his virtues and his intelligence and his apparent knowledge of the subject appear from his statement to entitle him to. I shall accord him that, and no more. What I mean is this. He went on to describe the Paris method, and read this : —

"In Paris, until very recently, the water-closet matter, and even the chamber-slops of houses, was by law delivered into tight cesspools, to be emptied from time to time. Even now there is only an insignificant exception in the case of persons who adopt a certain prescribed straining apparatus, which allows the liquid portions thus produced to pass into the sewers. In many towns in England, generally as the result of judicial or legislative restrictions, the devices above indicated, or their equivalent, have been adopted, and are systematically carried out with the direct purpose of preventing the pollution of rivers. In nearly every city on the continent of Europe, sometimes with this object, but more often with the view of preserving a valuable manure, there is and always has been an entire withholding of such wastes from the sewers, which are constructed to remove storm-water only. In fact, more precedent by far can be found for the above-prescribed course than for any other method of treating domestic and industrial wastes.

"Please understand" (and this is the point) "that I do not make this suggestion as a recommendation. I realize very fully, that, for this restriction to be placed upon a community like that of Worcester, would be nothing less than an economical and sanitary calamity. It would inevitably lay a cumbersome tax on all its people, and would lead to serious injury to the public health. I suggest it only as a possible means by which that community may, without sacrificing its existence, and without destroying its property, concentrate upon itself the disadvantages which it seems not averse to inflicting upon others."

When he comes here, and goes out of his way to impute motives to the city of Worcester, and to say that they are not averse to inflicting a "sanitary calamity" upon others, I say he has relieved me from any obligation of dealing with his paper in any other way than as its merits deserve.

He starts out with a display of learning which I desire to look into : —

"Under all ancient practices a sewer is only a drain, a channel for the removal of waters which the proper enjoyment of territory requires to be removed. Until well into the present century this was probably the only meaning of the term; and up to that time the office of a sewer was simply to furnish a safe outlet for rain-water, for soil-water, for the overflow or backing-up of streams, etc. The use of these sewers for the removal of excrementitious and other refuse matters is very recent. The use of common sewers for foul drainage is an assumption of recent date, which has grown up largely through neglect, and with no well-determined conception of the ultimate effect to be produced."

Now, that is set in the fore-front of his statement, and it is important, if true; but I suppose that it would be impossible to condense into the same space more charlatanry and ignorance than that displays. It has just enough of truth to show that his whole investigation of that matter is on the surface, and specious. It is undoubtedly true, that, in the beginning, the term "sewer" included any clean drain as well as an unclean one, and that there was an ancient statute of sewers which had particular reference to those drains which drained such fens as exist in Lincolnshire and other parts of England. But that the term "sewer" has meant any thing else, for the last four hundred years, than a drain to carry off filth, is a proposition that no man in his senses, whether he was a sanitary engineer, or whatever else he was, would put in writing, and read to an intelligent committee. There was one Shakspeare, who lived in England, and wrote quite a number of plays; and Artemus Ward says that he was "the pride of his native village." He died as early as 1616, and he wrote a large number of his plays in the latter part of the preceding century. In his play of "Troilus and Cressida" he gets the Trojan and Greek chiefs together — somewhat improbably, perhaps — for an interview; and, when they separate, Hector, the Trojan, turns to Menelaus, who was the husband of Helen, as you remember (and that adds to the improbability), and says, "Good-night, sweet Menelaus." Thersites, who is a Greek cynic or buffoon, and hates them all worse than he hates poison, as he goes out, says, "Sweet draught;" and "draught" in that connection means precisely what it does in the seventeenth verse of the fifteenth chapter of Matthew, where the text is, "Do not ye yet understand, that whatsoever entereth in at the mouth goeth into the belly, and is cast out into the *draught?*" "Sweet draught: sweet, quoth 'a! sweet sink, sweet *sewer.*" Shakspeare thus places in the mouth of Thersites *sewer*, sink, and privy, all coupled together, as the very opposite of sweet. "In Pericles, Prince of Tyre," Marina, the beautiful daughter of Pericles, is kidnapped, and taken to a brothel for purposes of prostitution, where, by her purity and her wit and wisdom, she protects herself against the influences of that vile place: and at last she makes an impression upon Boult, one of the panders of the house; and he says, —

"What would you have me? Go to the wars, would you? where a man may serve seven years for the loss of a leg, and have not money enough in the end to buy him a wooden one?"

She replies, —

"Do any thing but this thou doest. Empty
Old receptacles, common sewers, of filth;
Serve by indenture to the common hangman:
Any of these ways are better yet than this;

> For that which thou professest, a baboon,
> Could he but speak,
> Would own a name too dear."

What a pity Col. Waring did not live in those days! Mark Twain said, when he went to the grave of Adam, "What a pity he could not have been spared to know me!" What a pity Shakspeare could not have been spared to know Col. Waring! Then he would not have put into Marina's mouth such a malapropos speech as that, in which a sewer, which, according to Col. Waring, was only used for the purpose of carrying off clean water, is represented as a receptacle of filth. Then, there was old John Stow, who wrote a work called "A Survey of London," which is the source of nearly all the information we have about the ancient history of that city. He says, that, as early as 1307, the Earl of Lincoln applied to Parliament to have Fleet River cleaned out, because it had got choked up: and a commission was appointed, consisting of the constable of the Tower and the sheriff of the city; and they investigated, and found it was blocked up by mills and other obstructions. The mill-owners, as far back as that ancient time 1307, had squatted down upon that river, and polluted it, and tanneries had filled it up with their filth; and those mills were cleared out. Now, my purpose was to quote from Stow, whose work was published in 1598, — two hundred and eighty-four years ago. It is current English, and it is presumed that he might have known something of the meaning of the word "*sewer*." He says, speaking of Fleet River, "but still, as if by nature intended for a common *sewer*, it was soon choked up with filth again." That was away back in 1598. How far ahead of the times he was! He didn't know, that, according to Col. Waring, a sewer was not used for filth until this century, or until very recently; but away back in 1598 we find him using those words as applicable to the condition of the river nearly three hundred years before that time.

There was Milton, a very good writer in his day, who wrote poetry and other things, and had a very competent knowledge of English. He was not a sanitary engineer, that I know of; but he had a tolerably good knowledge of the English language. He wrote a poem called "Paradise Lost," still extant in choice English; and when he describes the joy of the serpent as he discovered Eve in the morning alone among her flowers, dressed in but little beside her own loveliness, he says,—

> "As one who long in populous city pent,
> Where houses thick and *sewers* annoy the air,
> Forth issuing on a summer's morn, to breathe
> Among the pleasant villages and farms
> Adjoined, from each thing met conceives delight; . . .
> Such pleasure took the serpent to behold
> This flowery plat, the sweet recess of Eve,
> Thus early, thus alone."

Dr. Johnson — not a sanitary engineer, but a man of great learning, and who laid the foundation of the lexicography of the language — made a dictionary, and he did *not* dedicate it to Lord Chesterfield. He published it in 1755. That was a time when Col. Waring did not know a great deal about sewers; and in 1755 the only definition that Dr. Johnson gives of the word " sewer " is " a passage for foul drainage to run through."

Now, when there is so much wisdom embodied in the foundation proposition of this report as all that comes to, I think it is not necessary for me to pursue it further; and I will venture, notwithstanding the opinion of Col. Waring, to say a brief word upon the moral aspects of this case. I suppose that the most general proposition is, that no large community can exist without inconvenience to its neighbors, and some degree of injury. There is the liability to the outbreak of disease, or to the propagation of disease, which is attendant upon a great community. There are noxious exhalations and gases that to some degree permeate the air. To live around such a city as Pittsburg must be very much more of an annoyance than it was in 1755, when Braddock took his army over there to capture Fort Du Quesne, and didn't take it back. There is the accumulation of the criminal classes, and the liability to predatory excursions, and many other things which I will not stop to enumerate. And there is this very common pollution of water-courses and the sea in the vicinity.

Now, what is the rule of obligation? I take it that the plain rule of obligation is, that the city should take such measures to prevent these injuries, or to reduce them to their minimum, as ordinary diligence requires; and I suppose that ordinary diligence, as applied to a city, is like ordinary diligence applied to an individual. Ordinary diligence, applied to an individual, is, that he shall do what other men of prudence do about the conduct of their own affairs. It is not a question whether we in Worcester are not bound to live cleanly lives, and not get " between the wind and the nobility " of our friends in Millbury, in an unclean condition; but the question is, what measure of obligation have we? what shall determine our obligation? If we do what cities from the beginning of time have always done, and what they do at the present time, with the exception of a few which you can count upon your fingers, we have discharged the obligation of ordinary diligence; we are not guilty of negligence because we have omitted to do what some scientist reports is possible, and another scientist says is impossible; what one set of experts, of great ability, I admit, says may be done, and another expert, of equal ability and more experience, says cannot be done, or would not be likely to succeed. In other words, we are not bound to put out into the realm of the unusual and the untried, to penetrate and explore the regions of the extraordinary. We have done with the sewage of Worcester what every other city has done since Rome emptied its sewage into the Cloaca Maxima; and I know not what other ancient examples I might cite, down through all the ages to the present time, and what all the cities on the face of the earth are doing at the present time, except a few that you can count, as I have said, on your fingers, and those under circumstances where it is a matter of debate what success they have attained, — and I shall show you before I get through, from the reports of the State Board of Health, that they have attained no success whatever that would have any bearing upon this case. I shall show you that the best of them failed to purify their waters, and they leave them more impure than the Blackstone is when we have done with it. It will appear, before I finish my argument, that the State Board of Health have apparently overlooked one important consideration; and that is, the probability of a result being attained that will be at all commensurate with the expense, or that will reduce the impurity of the Blackstone to any degree, so that it shall be substantially better than it is now.

I say, the rule is in exact analogy with the rule of law that applies to individuals: adopt the precautions that other people of common and average prudence adopt under similar circumstances, and it is grounded — if there were any need of a more general proposition as underlying it — it is grounded upon the proposition that everybody has reason to expect his neighbor will do as other prudent men do, and will guard himself upon the supposition that he will do that, and will not expect him to fail to do it. It is according to reasonable expectation; and the rule that applies to individuals must apply to communities, and no other reasonable rule can apply. You will find Worcester doing exactly what other cities have done, and nothing further. There are many moral and many physical evils in this world, and they are not all preventable. "Use your own so as not to injure others," quotes my friend from Mayor Verry. Nobody doubts that maxim. But how does my friend apply it in his own practice? How do the courts apply it? That you shall not use your own at all? No: use your own as other men of common prudence use their own, and then you have made the application of that rule. You have a horse that is spirited, and, uncontrolled, is a dangerous animal. If he gets loose, he may do more injury in five minutes than all your accumulated means can settle, or the means of ten men like you. There is the pent-up power capable of producing such effects if it once escapes from your control. Have you not a right to use that horse? Leather may break, harness-makers are not infallible, the breeching may not hold; but, if you use the prudence of ordinary men, you have discharged your liability. You have used your own so as not to injure another, within the meaning of the law.

I want to say a few words upon the legal aspect of this case. My learned friend quoted several cases. I do not know yet whether they claim that we are violating any law or not. I am not able to say. There is one case before the case of Merrifield *v.* Worcester, which I desire to cite, as leading up to the position taken in Merrifield *v.* Worcester; and that is Wheeler *v.* Worcester. In that case, the plaintiff complained that the city had filled up Mill Brook in such a way as to set the water back upon his premises and injure them. It was referred to a very able commission for the ascertainment of the facts to be reported to the court, consisting of John Wells, Charles S. Storrow, and Asaph Wood. The case is reported in the 10th Allen, p. 591. That commission made an elaborate report, and found that the drainage from the sewers had done something, but not much; that the wash from the streets had done a great deal towards filling up; that the building of the railroad bridges had done more; and that the building of the bridge on Front Street had also added to the evil. I only desire to call attention to the opinion of the court, which was given by the late Justice Colt. It says, —

"The plaintiff is a riparian proprietor upon Mill Brook, a natural watercourse flowing through the city of Worcester, and has the right to have it flow through and from his premises in a free and unobstructed channel. He may maintain this action against those parties who interfere with that right, or against any one of them who by his unlawful act contributes substantially to the injury which he suffers, unless the party or parties charged with creating the obstruction can claim the protection of the statutes known as the Mill Acts, or those other statutes which provide compensation in a particular mode for injuries done by public authority in the exercise of the right of eminent domain. If the injury is produced by the joint action of several parties," etc.

And then it goes on at considerable length, which I will not read, until we get to p. 602, where the court says, —

"Of these co-operating causes, thus briefly indicated, the case requires us to consider only those which it is alleged the city is responsible for.

"1. The surface-wash from the streets. This is stated to be incidental to the growth of the city and the construction of the streets. It finds its way naturally into Mill Brook, which furnishes the only channel for the accumulated surface water of the vicinity. No new water-course has been diverted into it. It receives no more water than would be collected by the natural surface of the land; but, by the changed uses to which a dense population have appropriated it, the soil of the numerous streets has been more rapidly carried into the stream. To hold the defendants liable to an action from such course would be to say that the owner of land must be restricted to such uses of it as will not, by the ordinary action of the elements, cause the soil to wash in and fill to any increased extent the adjacent brooks and streams. The injury which results to the plaintiff from this cause must be regarded as *damnum absque injuria*. There is another answer to this claim of the plaintiff. The city, by their proper authorities and agents, are charged with the public duty of constructing and maintaining the public streets. They must construct and maintain them in such places and in such manner as the public convenience and necessity require. They must provide for and dispose of the surface water which falls upon them; and, in the discharge of this duty, neither the city nor their agents can be proceeded against in an action of tort for damage sustained by a private citizen. In the construction of streets, highways, and bridges, it is the right of the public to take all private property necessary, and do all other necessary incidental damage to the individual. The laws of the Commonwealth provide compensation for such injury; but the remedy must be sought in the manner pointed out by the statutes, and not by action of tort against the city or their agents. If the public work is built so as to cause unnecessary damage by want of reasonable care and skill in its construction, then the right of eminent domain will not protect the parties by whom the work is done, but they may be liable in tort for such unnecessary injury. The case does not find that the surface-wash from the streets was not the necessary and inevitable consequence of their construction, or that the streets were laid out and built without reasonable care and skill."

That point I shall elaborate later on. The point of that case is, that the city is not liable for the surface-wash of its streets washing into Mill Brook. The question of sewers they did not pass upon, because it did not become necessary. I now refer to the case of Merrifield v. Worcester, 110 Mass., p. 216, which my friend says, on p. 16 of the report, "will very likely be referred to as much upon the other side as this." My friend referred to it for the purpose of showing, apparently, that it decided, in the first place, that, for diverting Mill Brook, under the statute of 1867, his proper action was by petition under that statute. Nobody disputed that. The court adverted to that incidentally, and neither the counsel for the plaintiff nor the counsel for the defendant made any question about that. He also referred to it upon this other point, that if the sewers were not properly laid the city might be liable. Nobody disputed that. But that case decided an important question, which was not alluded to by either of the learned counsel on the other side; and the principal purpose of it was to decide that question. Mr. Merrifield brought this action against the city of Worcester, and alleged "that the defendants, on April 5, 1861, and on divers days and times since, ' wrongfully and unjustly cast, carried, and deposited, and caused to be cast, carried, and deposited, into said Mill Brook and the waters thereof, at points in the channel thereof above and higher than the works of the plaintiff, great quantities of filth, dirt, gravel, refuse material, matter discharged from sewers, privies, water-closets, stables, sinks, and streets, and divers other noxious materials and ingredients,' by reason of which the water became greatly corrupted and unfit for use in the plaintiff's business; 'said water so corrupted, among other things, corroding the plaintiff's boilers and engines and fixtures, causing an adhesion of sediment and other materials to said boilers, and greatly increasing the expense of making the necessary amount of steam for said works, and greatly increasing the danger of explosion in said boilers, and causing thereby frequent breakages in the engine fixtures and works, and deterioration thereof, and causing great expense in the repair thereof and in the interruption to the running of the works, thereby causing great injury to all of the plaintiff's establishment;' and that 'the waters of the brook so corrupted are thereby rendered so offensive that it is difficult and expensive to procure competent engineers and workmen to operate said works.'"

In other words, he said, that, by reason of the sewers, they had committed a nuisance in that water; that is, an offence describing a nuisance. Now, the court says, —

"From the report, we infer that the ground of liability is that the dirt, filth, and other materials were carried into the stream by means of certain drains or sewers constructed under authority therefor conferred upon the city council by the charter."

He then cites several statutes : —

"The statute of 1867, chap. 100, authorized the taking of Mill Brook and the entire diversion of its waters from the channel by which it passes the plaintiff's works. So far as he has suffered damage from any proper exercise of the power and rights conferred by that Act, he must seek his remedy by a different proceeding from this, under the special provisions of the Act itself. But the stream had not been so diverted at the time when this action was brought, and it does not appear that the injuries complained of were the result of any proceedings under that Act."

Now the point of fact appears : —

"It appears that in 1850, more than twenty years before the date of the writ in this case, a drain or sewer was constructed, by order of the city council, discharging from Thomas Street into Mill Brook, a short distance above the works of the plaintiff. [You remember that Thomas Street runs down from Main Street to this Mill Brook. This sewer of 1850 extended up to Main Street.] This drain extended back to, and ran a short distance along, Main Street. In 1857, and at various times subsequently, this drain has been extended farther along Main Street; and drains running along several other streets have been connected with it. The plaintiff contends that the injurious effects of the drainage into the brook have thus been constantly increasing, down to the time of action brought. This question, so far as material, it is agreed shall be submitted to assessors."

In other words, Worcester then had a system of sewers which included Thomas Street, Main Street, and various other streets, emptying into Mill Brook. Mr. Merrifield was a riparian proprietor; he had just the same rights as any Millbury riparian proprietor had, — no greater, no less; he had just as much right as any or all of them. The fact that he lived in Worcester did not alter the case. The court then proceed to discuss the question of his rights against the city of Worcester by reason of their emptying sewage into Mill Brook; and the proposition of Mr. Justice Wells — whose early death took from the bench a judge who had already established his reputation as a very eminent jurist, who gave as much promise of a career of unsurpassed brilliancy as any judge who ever sat there — was this: —

"The right of which the plaintiff alleges a violation, is not that of acquired property in possession. It is not an absolute right, but a natural one, qualified and limited, like all natural rights, by the existence of like rights in others. It is incident merely to his ownership of land through which the stream has its course. As such owner he has the right to enjoy the continued flow of the stream, to use its force, and to make limited and temporary appropriation of its waters. These rights are held in common with all others having lands bordering upon the same stream; but his enjoyment must necessarily be according to his opportunity, prior to those below him, subsequent to those above. It follows that all such rights are liable to be modified and abridged in the enjoyment, by the exercise by others of their own rights; and, so far as they are thus abridged, the loss is *damnum absque injuria*. . . .

"So the natural right of the plaintiff to have the water descend to him in its pure state, fit to be used for the various purposes to which he may have occasion to apply it, must yield to the equal right in those who happen to be above him. Their use of the stream for mill-purposes, for irrigation, watering cattle, and the manifold purposes for which they may lawfully use it, will tend to render the water more or less impure. Cultivating and fertilizing the lands bordering on the stream, and in which are its sources, their occupation by farmhouses and other erections, will unavoidably cause impurities to be carried into the stream. As the lands are subdivided, and their occupation and use become multifarious, these causes will be rendered more operative, and their effects more perceptible. The water may thus be rendered unfit for many uses for which it had before been suitable; but, so far as that condition results only from reasonable use of the stream in accordance with the common right, the lower riparian proprietor has no remedy."

Now we come to the point that seems to me to settle this whole question of the legal right of Worcester to do what it is doing, independently of the statute of 1867: —

"When the population becomes dense, and towns or villages gather along its banks, the stream naturally and necessarily suffers still greater deterioration. Roads and streets crossing it, or running by its side, with their gutters and sluices discharging into it their surface water collected from over large spaces, and carrying with it in suspension the loose and light material that is thus swept off, are abundant sources of impurity, against which the law affords no redress. . . .

"It may readily be supposed that a small stream like Mill Brook, with a considerable city like Worcester upon either bank, and the adjacent lands descending rapidly towards its bed, would cease to preserve its waters from impurity, and become valueless for any purpose, except that of drainage, and the creation of power by its head and fall."

In other words, the court recognize that a city built upon a small stream will condemn that stream to the purposes of sewerage.

"All this may result, even though no unjustifiable act be done to effect it. To enable a riparian owner to maintain an action for damages, he must show, not only that the defendant has done some act which tends to injure the stream, and which he has no legal right to do, or which is in excess of his legal right so as to be an unreasonable use thereof, but also that the detriment of which the plaintiff complains is the result of that cause."

Then he goes on to cite Child *v.* Boston to the point, that, if the sewer is laid out and constructed with reasonable skill and diligence, the defendants are not liable; and so the court decided, in that case, that the city of Worcester was not liable to Mr. Merrifield for polluting that water by reason of a system of sewers that extended back into Main Street, through Main Street, and received lateral sewers from several other streets. That case covers, it seems to me, the whole proposition as to our common-law right.

I now desire to refer, for a moment, to the case of Washburn & Moen Manufacturing Company *v.* City of Worcester, 116 Mass. 458. This action was brought after the construction of the sewerage system under the statute of 1867. The plaintiffs owned this mill-pond down there at Quinsigamond Village; and they said, by reason of the construction of this sewerage system, that pond had become filled up with filth, and "great quantities of sewage matter and filth, both solid and liquid;" and so they alleged that a public nuisance and a private nuisance were created in their pond. It was a bill in equity; and Senator Hoar and Judge Nelson argued for the plaintiffs, and the late Judge Thomas and Mr. Williams for the defendant. Chief Justice Gray says, —

"Where a city, or a board of municipal officers, is authorized by the Legislature to lay out and construct common sewers and drains, and provision is made by statute for the assessment, under special proceedings, of damages to parties whose estates are thereby injured, the city is not liable to an action at law or bill in equity for injuries which are the necessary result of the exercise of the powers conferred by the Legislature. But if by an excess of the powers granted, or negligence in the mode of carrying out the system legally adopted, or in omitting to take due precautions to guard against consequences of its operation, a nuisance is created, the city may be liable to indictment in behalf of the public, or to suit by individuals suffering special damage. . . .

"The case at bar, as now presented, does not require the court to define the limits of the application of either of these rules to the discharge of the Millbrook sewer into the Blackstone River. The only acts charged against the city of Worcester in the bill before us are the converting of the channel of Mill Brook into a sewer, and the opening of other sewers and drains into the same. These acts were expressly authorized by the stat. of 1867, chap. 106. Butler *v.* Worcester, 112 Mass. The only further allegations in the bill consist of a conclusion of fact, that a nuisance to the plaintiff was thereby created; and a conclusion of law, that the acts of the city were unauthorized and in violation of the plaintiff's rights. The bill does not allege any negligence of the city, either in the manner in which the sewage was discharged from the mouth of the sewer, or in omitting to take proper precautions to purify it. The necessary result is, that the *demurrer must be sustained.*"

That case decided that whatever was the natural and necessary consequence of doing the things provided for by the statute of 1867, laying out Mill Brook as a sewer, and discharging other sewers into it, there was no remedy for it unless it was under the Act itself. That settles that question. The court say that the plaintiffs did not allege that the defendants were guilty of any negligence either in the construction of the sewers or in omitting to take proper precautions to purify the sewage. Now I want to inquire what that means, for I apprehend that some stress will be laid upon that. The principle is universal, and my friends and I do not disagree about it, — it is not possible that we should disagree, — that when the power is conferred upon a corporation or otherwise, to construct works, to take land and construct works, or to construct works without taking land, they are bound in the construction of those works to exercise ordinary care; they must not be guilty of negligence. If such a work is constructed or maintained negligently, it is a nuisance: you have a right to abate it. If they fail, in managing or maintaining it, to exercise ordinary and reasonable care, then so far forth it is a nuisance. There is no use in disputing about it: it is well settled. Now, they did not allege in that bill that we had been guilty of any negligence in constructing the sewers. They could not; because we had employed the best engineers to construct them, according to the best modes of engineering. Nobody disputes it at all. Then, we had a right to construct and maintain them. We were not guilty of negligence in omitting to take proper precautions in purifying the sewage. Now, that raises the question, what is the meaning of negligence and ordinary diligence? I have discussed it in another connection perhaps sufficiently, but I will touch upon it here.

We were bound to do what cities similarly situated have ordinarily done in that behalf; and, if we did that, we exercised ordinary care. They did not allege, because they could not allege and prove, that we had not taken proper precautions to purify the sewage. There were no methods known at that time; and I believe you will think, after the testimony of Mr. Worthen, and after the contradictory reports of Mr. Waring and the State Board of Health, that to-day the method is not invented to purify sewage. There were no methods known by which it could be purified; or, if there were, that would not impose upon us the obligation to adopt them. The obligation would not begin to bear and press upon us to adopt such a measure until the adoption of such a measure had become the ordinary, the natural, the general course adopted by cities of ordinary prudence in the management of their sewage. Can there be any dispute about this proposition among lawyers or among intelligent men? It is not to be decided whether we have taken proper precautions, by calling a few experts, more or less eminent, to say they guess it could be done this way or that way. Has the method of purifying sewage now become so general in this climate and in this country, so pronounced a success — is it adopted everywhere, in such a way that the omission to adopt it is a failure to exercise that ordinary and common prudence and skill which is the measure of the legal obligation? To state the proposition is enough; to state the contrary proposition is to refute it, — to say that you fail in ordinary diligence if you do not resort to extraordinary measures, is a contradiction in terms. How would it be if you were driving your horse, and your horse — which is a tractable animal, and can be driven by a good driver — gets frightened and runs away, and you show that you did what everybody does under such circumstances when a horse is frightened by an unusual thing? Suppose, now, somebody should say, "Oh, but you didn't adopt Mr. Jones's patent method of throwing a horse down, or of picking him up and putting him into the carriage, when he runs away! If you had adopted that patent method and appliance, a model of which he has on file in the patent office, which shows how it might have been done, you would have used due care. We can call half a dozen experts here to say that that thing might be done; and how do you pretend that you are using ordinary diligence and skill in driving your horse, when you didn't adopt that patent? You ought to put that on your carriage, and then you would not be liable." The answer is the answer of common sense: "I did what other men do who drive horses. I was a good driver; I learned how to drive when I was a boy; I had a perfectly good harness; and I laid out my best strength to hold that horse, and used my best skill to manage him. He was a well-broken horse; but this unusual thing alarmed him, and set in motion those powers which were beyond my control." Now, do you say, that in not adopting this patent process of purifying sewage, which one expert said would work, and the other said would not work but would be a failure, we did not use ordinary skill and diligence? The same rule applies to cities which applies to individuals. We have adopted the same measures that have been adopted by every city that was ever organized, from the beginning down to the present time, with the exception of a few which you can name and count by the units without going into the tens, and they within a very few years. Not a city on this continent, without exception, has ever adopted any thing of the kind, unless you take this little sewer over here at the Mystic, which was adopted in obedience to an Act of the last Legislature, and which you are modifying by this Legislature. Nobody knows or pretends that it is a success, and it is on an exceedingly small scale.

Now, the petitioners come in here and say, "Oh! we have got a patent process; and we call Mr. Waring: and Mr. Waring not only knows all the law, and all the ethics, and all the every thing, but he knows what the English language is; he knows that Shakspeare, he knows that old Stow, he knows that Milton, he knows that Dr. Johnson, didn't know any thing about the language, but it was all reserved to George E. Waring, colonel, and he knows all about it; and if you don't adopt his plan to purify your sewage, why, then you are guilty of negligence." Now, would anybody allege under such circumstances, and undertake to prove before a jury, that we were guilty of negligence? From time immemorial the city of Worcester has emptied its sewage into the Blackstone River through Mill Brook. At first, the impurities were emptied out upon the surface, and carried along by the street-gutters and other channels to the brook. Later, sewers were constructed, beginning with a single sewer, which was gradually extended, and received other sewers, until, in 1867, this system was adopted; and it has become the elaborate system it is. But the same rules of law will apply.

But my friend, Mr. Morse will say undoubtedly, "You have no right to commit a nuisance, public or private; and I will cite Badger v. Boston." I see he has put into his argument several cases that he did not cite; and I was surprised that he did not, because there were some things in the opinions that might seem to make in his favor. He refers to Haskell v. New Bedford, Boston Rolling-mills v. Cambridge, and Brayton v. Fall River. He will say that we have no right to commit a public or a private nuisance. Gentlemen, I do not propose to discuss a question that I do not understand to be settled exactly. I do not know what the court are going to say when the question comes before them. Suppose a right is granted to a corporation or municipality or otherwise, to do a certain thing, and the necessary and natural consequence of doing that thing is to create a public nuisance; what then? Are you to read into the statute the proviso that you are not to do it, although you have express authority by the statute to do it? That question I do not understand to have been decided. The court, however, has decided, that, if you can do the thing authorized without committing a public nuisance, then, if you commit a public nuisance, you are violating the law. Undoubtedly that is so.

Now, I shall not discuss this question, because I shall argue that no public nuisance has been proved here affecting the public health. I want to cite the statute of 1878; but before I come to that, however, I desire to say a word in regard to these cases. Take the case of Badger v. Boston; that is as good for illustration as any of them. That was a case in which a urinal was authorized by statute to be constructed, with a provision for assessing damages to parties injured by such construction ; and a man brought his petition to recover for damages, alleging, and offering to prove, that it was a nuisance. That proposition was broad enough to include two propositions : first, that it was a nuisance by reason of the necessary result of constructing a urinal, and, if that was so, he would be entitled to damages ; but it also included the other proposition, that it was a nuisance by reason of negligently constructing or maintaining it. And if it was a nuisance, because of negligent construction or maintenance, the remedy was not by a petition for damages, but by an action of tort. The court say, in effect, that it was not, as a matter of law, a nuisance to construct the urinal, and it might be that the nuisance resulted from the negligent maintenance of it, or by an improper construction of it ; and, in this case, his remedy was by proceedings other than petition for damages.

In regard to the cases of Haskell *v.* New Bedford, and Brayton *v.* Fall River, those were cases where the cities of New Bedford and Fall River respectively had the right to construct sewers along streets. They debouched them into the private docks of individuals, and created a nuisance there in such docks ; and the court said they had no right to do it. Those cases are distinguishable from Merrifield *v.* Worcester ; they were not on a small running stream on which a city was built ; the facts are entirely different, and it cannot be possible that they overrule Merrifield *v.* Worcester. Take the case of Haskell *v.* New Bedford. That case (108 Mass. 208) was decided in 1871 : Merrifield *v.* Worcester was decided in 1872 by the same court, and by the same judges. It is not to be supposed that the judges did not know of the prior decision. If they had intended to overrule it, they would have said so. It is, therefore, undoubtedly true that they did not intend to overrule it at all. But if Haskell *v.* New Bedford is at all in controversy with the case I have cited, Merrifield *v.* Worcester, then, upon familiar principles, the later case overrules the earlier one. Brayton *v.* Fall River was decided in 1873, Washburn & Moen Manufacturing Company *v.* Worcester in 1875 ; now, if there is any controversy between those cases, the later case overrules the earlier. There is none stated ; the court consisted largely of the same judges ; it is not likely they were ignorant of the previous decisions. There is undoubtedly a distinction between the two ; and one obvious distinction is, that one related to the sea, the other related to a running stream, and the rights of riparian proprietors below.

This case of the Boston Rolling-mills *v.* Cambridge, 117 Mass. 396, has no relation to this case in any way. There was a private channel owned by private individuals, a canal which they had dug, and of which they owned the fee ; and I know not what right Cambridge had to empty sewage into it at all. My friend Mr. Morse alludes to this case out at Mystic Brook, and says that it has been held to be constitutional by a single justice. I don't know whether it has been held to be constitutional by a single justice, or not: I am not aware that it has been argued and decided. It appears that the parties agreed upon a decree, so that there was no occasion to decide upon the constitutionality of the statute: it is not now necessary to discuss whether the Act of 1881 is constitutional or unconstitutional. My friend says that is a stronger case than that against the city of Worcester, because it was an artificial channel. It seems to me that is a strange *non sequitur*. This is a natural channel into which, the court say in Merrifield v. Worcester, we have a right to empty our sewage. They have decided we have a right to, as against any riparian proprie'ors; but that was a channel, as I understand, by which Boston undertook to take, by an artificial conduit or sewer, a stream that polluted their water-supply, around into lower Mystic Pond; and then, after the Legislature had granted that right, it provided that they should purify it. The cases are entirely distinct and different. One was the right of a city upon a natural stream, on which it has grown up from infancy; the other was the right of a city to corrupt, by a sewer that the Legislature had authorized it to build, a pond to which it had no other relation than that created by the sewer.

My friends say the burden does not rest upon Millbury to show how this evil should be avoided. No: if Worcester is doing you any damage by acts which it has no legal right to do, if it is violating any law, why, then, it does not rest upon you to show how we shall stop it, I admit. But when you say we are guilty of some negligence, that we are doing something that we have no right to do, then the burden *does* rest upon you to show that we are; and, if we are doing what everybody else does, it is upon you to show how that result can be avoided. We claim, therefore, that the city is in the exercise of its legal rights, and the proposal is to deprive it of its legal rights. We say that the statute of 1867 is in the nature of a grant of a franchise; or, rather, it is the identical thing. On that statute we have laid out a million and a half of money. You cannot take away the rights you grant to a railroad corporation, in which it has invested money: it would be a violation of the Constitution of the United States, against the impairment of the obligation of contracts, but for the fact, that long ago, before any of the railroads received their charters, there was a general law which applies as a condition, giving the Legislature the right. There was no condition annexed to the statute of 1867. Can you now proceed to take away the grant of that franchise? Is it a constitutional position?

I do not propose to stop to discuss this constitutional question, or any question of constitutional law : I commend it to your careful consideration. It may be that my friends can find the authority for this measure in that general power of police regulation, which is vague, I admit, in its limits ; but, it seems to me, this action exceeds any limits that have ever been reached by any valid legislative enactment, as yet. It may be clear that this authority can be derived from that power of the Legislature to impose local taxation for local benefits ; but such taxation is usually imposed upon the persons that are benefited, and not on other persons. I submit that it is difficult to find the legal and constitutional ground on which you can found this statute.

That is all I care to say upon that branch of the subject. We submit, that at common law, by the decision of Merrifield v. Worcester, we have the right to empty our sewage into this stream, even if it causes pollution in the Blackstone River, because to do it necessarily causes that pollution to some extent; and, to the extent that it is necessarily caused, our right is extended and proceeds.

But, gentlemen, there is another proposition ; and I must hurry on. Independently of the constitutional questions, there are questions of public good faith. Nobody will doubt that in 1867 this whole question of sewage purification was in its infancy so far as any practical solution of the problem was concerned, whatever they may think now as to its present condition. Worcester has put a million and a half of money into her sewerage system, and it will take another million and a half if you adopt this scheme and we adopt the State Board scheme, including the value of such parts of the sewerage system as will have practically to be abandoned. This large sum of money they have put into it, and there are considerations of good faith involved. That money was expended, that outlay was made, upon the faith of our people in the permanency of that policy on which the statute of 1867 was adopted. And that is an appeal that I think will not be made to any committee of this Legislature or to this Legislature in vain.

Now, if there exists a nuisance here which the public health requires the abatement of, then I say it should be undertaken by the State itself. One gentleman of the Committee asks whether it would not be a bad precedent for the State to undertake to dispose of the sewage of the city. If a public nuisance is created there, the precedent is already established in the case of the Boston Back Bay here, the purification of the Church-street district, I think. That expense was indeed imposed upon the city of Boston, but it was only a single district of the city of Boston which was directly benefited. The Legislature imposed the expense of it upon the city of Boston, but for the direct benefit of a district of the city of Boston. A portion of the public which received a peculiar but indirect benefit was selected to bear the burden, but no reason in the nature of things can be suggested why any larger portion of the public might not have been selected to bear the burden.

In Talbot v. Hudson, in 16 Gray, 417, the case of the Concord and Sudbury meadows, the improvement of those meadows at one time became an object of public interest to such an extent that it was thought proper to pass a statute, and to take the rights of the mill-owners for the benefit of the public, at the expense of the public; and it was done, and a provision made for compensation of the mill-owners by the State. If an improvement, whether by draining and bringing into cultivation tracts of marshy land, or by purifying water-courses, or otherwise providing for the increased salubrity of a region, is of such a nature as to be a benefit to a large number of people, such improvement is so far a public improvement as to be properly chargeable to public expenditures. So that the precedent, so far as that goes, is established. If we are doing what the law gives us a right to do, and thereby a nuisance is created, why, then, there exists an exigency for the public to abate the nuisance at the public expense, and not at the expense of Worcester.

But, gentlemen, the evidence in this case falls far short of showing any public nuisance substantially affecting the public health. Do not misunderstand me. I have been quoted by the local papers, which you undoubtedly do not read, as saying that I proposed to defend this case on the ground that there was no offence there; that the river was no more impure than it was thirty or forty years ago. That is not a proposition that I ever maintained anywhere, and do not propose to here or anywhere else. It is a fact that the river has been growing more polluted year by year for a great many years; and, in proportion to the rapid growth of the community there in Worcester and below, the increase in this impurity has gone on more rapidly.

But this is a question whether there is any evidence here to establish to your minds that we are committing, by that sewage, a nuisance substantially affecting the public health. This is the Committee on Public Health. Now, there may be any degree of nuisance. It is a nuisance to one man to hear a Lancashire bag-pipe, and it is a nuisance to another man to hear a hand-organ, while still another man dotes on a hand-organ. It is a nuisance to come into the presence of some men on account of their offensive breath; and a person of sensitive nerves finds nuisances in his path all along through life. A stream may be offensive: it may even rise to the character of a nuisance in one sense, in that it is offensive, that it emits smells, that it is offensive to a great many people, so that the causes of that nuisance may be indictable, and still it may not be a nuisance affecting the public health. The question here is what this evidence tends to show.

Well, they called several witnesses; but I am not going to dwell upon their evidence. I will relieve you as soon as I can; but, as I did not make any opening, I must say all I have to say now. They called Nathan H. Greenwood: he lives near Burling Mills; his mother and his aunt are not well; his mother is seventy-six years old, and has been suffering from debility a good many years. His aunt is seventy, and she was very sick last fall. His cows, when they are very thirsty, will drink the water of the river, not otherwise. George D. Chase, he works for Morse: the hands in the mill cuss the river like any thing. Now, that is very curious. We find that in Morse's mill the hands curse and swear like the army in Flanders, which swore terribly. They swear about this river. It is a curious thing, but Mr. Morse says it was a subject of prayer in the churches! Strange what opposite effects the impurity of the Blackstone River produces! At Morse's mill it stimulates profanity; in the churches it promotes piety and prayer. I suppose that it will remain forever unsettled how the balance stands,—whether the impurity of the Blackstone River has on the whole been conducive to profanity, or on the whole has advanced the interests of piety and prayer. Then this man tells about Harrington, who is eighty-one years old, and his aunt died there at ninety-two. She was cut off at the early age of ninety-two; and she lived within a few feet of Morse's pond, right in the midst of this pollution. It is perfectly clear what killed her.

Levi L. Whitney, he is one of the selectmen of Millbury, a good specimen of vigor and health. He tells the story about the dead fish: he is brought here to originate a fish story. Well, we happened to have Mayor Pratt down there from Worcester on that day; and, if there is anybody that can beat Mayor Pratt at a fish story, he has got to get up exceedingly early in the morning. You know that James Russell Lowell said in one of his poems, "You have got to get up airly if you mean to take in God," and so you have got to get up exceedingly early if you are going to tell a fish story that will beat Mayor Pratt. But they bring this man to tell a fish story. It appears that when the river was very low, on a hot day in August, down below Morse's mill, there was an abundance of dead fish floating around. The quantity we do not know about, but they make it very large. But you will remember that the sewage had been going down there before and afterwards, and yet that is the only instance of such an occurrence, and that is the only place on the river where the thing ever occurred. If we had had at that time our friend's scheme of intermittent-modified-patent-what-do-you-call-it downward filtration, and this thing had happened, it might have been attributable to the sewage, because it might have happened on a day when there was some intermission in the filtration; but the thing occurred on a solitary occasion, although the sewage, like the river itself, flowed on forever. Dr. Walcott's scheme hadn't been adopted at all. But why didn't this mortality of fishes occur at some other time and place? The cause was operative all the time. Why was the effect so singular? But fish died in the river on that Sunday in August, and on no other day. And that is solemnly paraded here as a fact of great significance! There were also twenty musquash died up there one afternoon, a week or two ago. They died right there in the Blackstone River; but it appeared in the papers that a man shot them. That fish story was lugged in as a make-weight; but the failure to connect it with the sewage is complete.

Then they called Samuel E. Hull: he was sick last summer, — I knew a good many people who were sick last summer: I was sick myself, but I didn't go near that river. Thomas Wheelock said that two years ago this winter, and a year ago, the schools were largely affected with canker-rash, so much so that there was talk of closing the schools. We have proved by the documentary evidence that there was no falling-off in school attendance. I take it that the statistics of school attendance are evidence of school attendance. There was no way to dodge that, there was no way to hedge against that. The official returns of school attendance do show the attendance of the schools with some degree of trustworthiness; although, of course, the death-rate doesn't show any thing about the health-rate.

Then they call Rev. Philip Y. Smith. He tells of several cases of sickness down there (his evidence is on pp. 59-65) in Saundersville, after the river had passed Millbury and received the water of Singletary Brook, and after all the settling in the various ponds, when I submit the sewage must be all lost to any perception. He tells about one school that was reduced to six members, and one up in Grafton which was reduced as low as fifteen; and it appears there was a great reduction in the attendance at the schools in other parts of the town. I will not go over this, for I do not wish to detain you; but I submit that there is not a community in this Commonwealth, that extends over as large a space as that to which this evidence applies, where at least as striking a story could not be framed. I submit to the candor of this Committee, that you could, upon short notice, if you could get the people enlisted with any common motive to assist you, without any dishonesty, without any making of evidence, make out as strong a case against any community in this Commonwealth.

Then they called Dr. Wilmot; and Dr. Wilmot's evidence is on p. 65. Now, Dr. Wilmot is a very bright man; and if anybody can paint any thing, he can paint it. He knows how to paint it; and this is one of those cases where it all depends upon the painting power of the witness. He has got a good deal of this power of word painting, and he can make any river look as black as Phlegethon or the river Styx. It is all a matter of imagination. When you hear about the "exhalations of the stream," it is the man's power of metaphor; it is in his command of language; it is not in the stream. But where are your facts? What effects have been produced? He says that there was a mild typhoid fever prevailing there; and, when you come to particular cases, there were two sisters in Rockdale, where a pond had been drawn off, and the bottom exposed: they were right over that pond which had been drawn off. Then, he says that there were several other cases, where he would say, "I think it was living by that nasty, stinking river." He tells about his wife's attack of cholera-morbus, which began by being Asiatic cholera, which shows the man's lively imagination. "Pure Asiatic cholera," says he; and when Dr. Wilson asked what he meant by "pure Asiatic cholera," it simmered down to a case of cholera-morbus! Well, didn't anybody else have cholera-morbus? "I don't know but what it was that stinking pond." But what evidence is there that the Worcester sewage had any thing to do with it? Whoever saw the bottom of a large pond exposed without smelling odors that would make anybody sick? And then he says that the diphtheria that he has seen there was not caused by the river: he excludes this as a cause. That is Dr. Wilmot, and he is one of their best witnesses; and all there is to his testimony is, that two people were made sick by a pond which had been drawn off. Then he says, when speaking of the greater prevalence of malaria, or that kind of mild, low fever, there than he had seen before, "But that, gentlemen, ought not to weigh very much; because where I have lived I could sit in my office and throw the stump of a cigar into the sea." That is Dr. Wilmot.

Next they called Nehemiah Chase of Wilkinsonville. His mother also has lived there sixty-seven years, and she is ninety-one years old. She is still weathering it; but, if this sewage is not stopped, she may die before she is a hundred and twenty. He had an Irishman that was sick: he had come over from the other country, and had been there two or three months. He had come into a new country. Well, who goes into a new country without exposing himself to typhoid fever or some other disease that may be incident to being acclimated? He came here to a new climate; he didn't know how to protect himself; he got sick with typhoid fever, and got well again. Does that make out any case against us?

Rev. John L. Ewell of Millbury tells about Whitworth's rowing on the river and getting sick; but his own family are all right, and he ascribes that to his wife, — not because the river wouldn't be willing to make them sick, and probably, if it could have its own sweet will, it would make them all sick; but his wife is a very good nurse, and apparently it doesn't think it is any use. His family is well because his wife is a good nurse, and not because it is a healthful region.

Then he calls Mr. Esek Saunders, a man for whom we all have the greatest possible respect; but Esek Saunders has the idea that he has a mission on earth, and that is to compel Worcester to stop emptying its sewage into that river. Well, here is Mr. Esek Saunders : he is now in his eighty-second year. He was born in 1800 ; has attended to his business up to last May; and there is one of the lamentable cases. That old man doesn't feel so well as he used to, and therefore it is probable that a case affecting the public health has been made out! What does he know, when asked for statistics, about the health of his mill-hands? Has he looked into it at all? He has got that "impression." You will see, gentlemen, that I am running over this thing. I could dwell upon it for about a week, showing the fallacy of most of it; but I will just touch it as I go along.

Joel Smith is superintendent of the Sutton Manufacturing Company, and has been so for three years. He says there is more sickness in the mills. He has made no figures, nor any examination of any books: he has got that impression. Does that make out a case of a nuisance causing substantial injury to the public health? Is it possible that it does, — such a vague impression as this superintendent has?

George W. Fisher — he is a native of that region, or has lived there for some time; and I guess he is a native there. He is the agent of the Fisher Manufacturing Company, and he tells about the pollution of the stream. "I have not noticed about the public health." He says there is a common talk, but he has not noticed it, he has not discovered it, — this intelligent man has not discovered it.

Then they called in Dr. Lincoln, and Dr. Lincoln is one of the most intelligent witnesses of whom they called a great many ; but his evidence, so far as it has any tendency to prove any thing affecting the public health, is this: that when he went to Millbury sixteen years ago, there were two physicians in town ; and they thought there were no more there than they could well attend to, and that it was no place for a new man. That was a very singular thing, for those physicians to entertain that idea! I think the members of this board who are physicians will say that the idea that they could probably wrestle with the Millbury sickness and attend to it considerably well, was a very singular idea for them to have! Well, now there are six; and he thinks any one of them has more than either of the two had before. I don't know about that. It is possible that when they didn't have but two there, neither of them were good for any thing, and the people came up to Worcester for their doctoring; and it is possible that now they have got such good physicians, that they rather like their company, and call them often. I don't know how it is that these things grow, but that is all the doctor has got to say about this matter. He says that the common sicknesses have been mostly of a zymotic type. "Have you in mind any particular cases which you can call to the attention of the Committee? *A.* — No: I don't know of a case that I should be warranted in saying was the result of the sewage or any thing of that kind. The general health-rate is not as good among our people." That is his answer. He does not know of any prevailing sickness nor a single case that he can ascribe to this influence, and he lives right in the midst of it.

Then they call Dr. Webber; and Dr. Webber came here cocked and primed, — a very able man, a nice man, an entirely honest man, — no question about that. But he came here all prepared to make a statement, and he has made the strongest statement that could be made. When you come to examine his statement, it is that he had four cases of typhoid fever that he could not ascribe to any thing other than the river. How many doctors on this board have had cases that they couldn't account for? He couldn't ascribe them to any thing else but this river. He had four cases of dysentery also; and those are the cases, if we except Benjamin Flagg who was cut off at seventy-five or seventy-seven years of age with an organic trouble of the heart. The old gentleman was working in his mill, and got tired, and died by reason of diarrhœa. All Dr. Webber says about this thing is, that it is liable in the future to produce an epidemic. But is it a thing now affecting the public health? That doctor, right down in the midst of it, has only these eight cases to speak of: and he simply cannot account for them. He knows, and we proved by him, that there were typhoid cases all over the town; and you know and I know that there are cases on hills and in valleys and in every possible situation where the resources of science cannot furnish any means of discovering what causes them.

William H. Harrington is next called; and the point of his testimony is, that there was a time when he drank the water of the river at Atlanta Mills in the winter. He used to drink it, but it has grown more impure. But he doesn't testify any thing about the public health.

Then Capt. Peter Simpson tells about having compromised seventy thousand yards of goods at five cents a yard discount, making twenty-three hundred dollars' loss; and he thinks it affects public health. But, when you come to facts, he says that Mr. Wilmarth died, and that Dr. Gage said he died of this river. Now, you see what kind of evidence we are exposed to here. They put in the hearsay of this one, and the understanding of that one; and Capt. Simpson, one of the most intelligent and honest men down there, — a man who would not state any thing that he did not suppose to be true, — comes here, and lugs in that rumor, that Dr. Gage had condemned that river as the cause of Mr. Wilmarth's death! We put in the letter of Dr. Gage, in which he says that he died of pneumonia caused by exposure after a warm bath, and that he had no reason to suppose the river had any thing to do with it, and never said so. That disposed of Simpson, so far as the public-health question is concerned. Simpson has got some extraordinary cows which present a case almost as singular as the opposite effects produced by the river upon the morals of Millbury. The effect of the river upon cows is very singular indeed. Some of them won't drink it; you can't get them near it; but we called Mr. Harrington, who says that they drink it up there at his brother's place at Morse's pond constantly, — they don't drink any thing else; always drink it; never heard any trouble about it. That is one case. Then we called Deacon McClellan from Grafton, who has lived there twenty-seven years (he was not our witness properly). He believes that the Millbury people are suffering a good deal; for he tells the truth anyway, wherever it cuts. He says his cattle drink that water, and his neighbors' cattle drink it; and, although he has springs and brooks in his lots, they will go to that river and drink that water, instead of drinking from the springs or from the aqueduct. Now, Capt. Peter Simpson of Millbury, who, along with Mr. Morse, must be regarded as the Castor or the Pollux of this fight — they are the twin champions of it — and he tells you that not only his cattle drink it (he has got a fine herd of Ayrshire cattle), but you can't keep them away from it with clubs! You have got to fence them away; and he had a fence clear down the river there, to keep the cattle from it. And why? Why, because they drank it, and it affected the milk; some of the Blackstone River got into the milk! How, I don't know; but it is Emerson or somebody who says, that when you find a trout in milk, that is circumstantial evidence to which there is no answer whatever. Now, when you find Blackstone-river water in Capt. Simpson's milk, you can account for it (and that is circumstantial evidence) by saying that it went through the cows, and retained all its peculiar odor as it was deposited in the milk!

Then they call Thomas Heap, who testifies nothing about health. John Gegenheimer is called, the superintendent of the Cordis Mills. He speaks about "more spare hands;" and, when you come to inquire of him what he knows about this thing, he says he does not know as there are any more spare hands than any mill where they run all the machinery. Herbert A. Pratt, civil engineer, testifies nothing about health.

Charles Whitworth had some hard rowing in a boat, had a headache, felt unwell, and stopped the rowing, and felt better. His companion rowed in the boat, and felt sick. The doctor told him he better stop rowing, so he ceased rowing; yet he lived right on the bank of the river, and has become robust in health, and hearty. This is to show that the Worcester sewage is producing an epidemic that is affecting substantially the public health!

Then they call Charles D. Morse; and he is, as I said before, either Castor or Pollux; he is one of the champions; let me say, a man of perfect honesty, but he has got this bee in his bonnet, and his testimony is to be taken with a great deal of allowance, — I say it with the utmost possible respect; and he has come here to turn the specimens over, and keep them roiled up well, and perform other little offices in aid of this case before the Committee. The most that he testifies to is this: he did not feel very well himself, and then he tells the story of its effect upon the churches. I am not going to stop to discuss the matter of his standing-pipes. They were there in the mill, and probably got rusty. I believe it is a pretty clear proposition, that they are a very unreliable kind of security to rest upon in case of a fire; and they did not work very well: but I am not discussing the question whether or not there is more or less of impurity in the Blackstone River, but another question, to wit, whether there is any evidence that warrants a finding that we are committing a nuisance which substantially affects the public health; and that is all, as I understand, this Committee has any jurisdiction over. Therefore, gentlemen, I do not care about any other part of his testimony.

Then they call Dr. Robert Booth, who had cases of typhoid fever, diphtheria, and intermittent fever, and stated that matters were worse in Blackstone (when the report of the State Board say the impurities are all but lost to chemical tests) than in Millbury. So I do not think that his testimony advances their cause any.

They now call Dr. Charles F. Folsom, or the Committee called him; and of course whatever he says he believes. He says the public health may possibly be affected; that there is no great quantity of sewage deposit at the dams. He says it would be his opinion that to a certain extent the public health may be affected. Well, that I suppose may be, — "to a certain extent;" but nobody knows to what extent. You remember in "Pinafore," after it is found out that Buttercup had changed the children in their boyhood's happy hours, and that Ralph was the captain, and the captain was Ralph, and the admiral, — I forget his name, — who was going to marry the captain's daughter, repudiates the engagement when he finds she is nobody but a poor peasant's daughter, it was suggested to him that "Love levels all ranks." — "Yes," he said, " it does, to a certain extent, but not to this extent;" and so "to a certain extent," according to Dr. Folsom, the public health may be affected; but we certainly claim that there is no evidence that it is to any such extent as to constitute a substantial impairment.

Mr. FLAGG. Brother Goulding, that admiral's name was not Folsom.

Mr. GOULDING. That is the evidence of Dr. Folsom, that to a certain extent the public health is affected; and you are asked to base a finding upon that, that the public health is substantially and seriously affected there, so that you should now embark upon this new class of legislation, and should impose upon the city of Worcester this burden.

I pass over Mr. Waring, and come to Dr. Henry P. Walcott, another gentleman whose evidence is to be taken as exactly the truth so far as he states any fact as he understands it. He says it " affects the public health" — you remember how he waited, when you asked him the question, to give it full consideration — "to a slight extent at present, not to any great extent." That is the whole of Dr. Walcott's testimony on that point, perfectly cautious and fair. What does it amount to? I mean, as to establishing the proposition we are arguing; not whether it would not be desirable if you could have this cause wholly removed, but whether there has been proved the present existence of any thing that substantially affects the public life.

Then they put in letters from Rev. B. J. Johnston and Elijah Thomson. The only thing of importance in Thomson's letter is the story he tells about the explosion of bubbles on the pond there. They would have you believe that statement (I do not know what effect it may have upon this proposition), that you can go along there at any time, and if you want a little 4th of July entertainment, all you have got to do is to apply a match to the bubbles, and they will pop off like fire-crackers all over the pond there anywhere. That is the story. Sulphuretted hydrogen, Mr. Elijah Thomson says is the cause of it. I suppose the doctors upon this Committee know something of chemistry; and, with such a volume of water as is contained in that pond, I submit to you, if all the fish in that part of Blackstone River should suddenly die, and be buried under that dam, and all the musquash were to follow them, that there would not be enough sulphuretted hydrogen rise up through that water to produce any such effect at all. It would be impossible. I submit to the intelligence of the doctors upon this Committee, and the other gentlemen who have paid any attention whatever to chemistry, that the effect of the water is to absorb it and destroy it. You may bury one hundred dead horses under there, and it might possibly produce such an effect.

Now, we have an answer to this, in some vital statistics that have been presented here. I will dwell upon them for a moment. They show that the death-rate of Millbury has not increased; that it has rather improved, on the whole, compared with other places similarly situated; but how carefully they have hedged and guarded against this thing! They seemed to scent it in the air, they seemed to anticipate it. Without knowing what the death-rate was, they were carefully guarding against it, and hence they would parry it before it came. But it has come. Now, what do we claim with regard to this? They put in a book of Dr. Simon of England who says that it does not bear a constant ratio to the health-rate. Perhaps not; nobody claims it does. It is not common sense to suppose it would, a constant ratio. But is it true we have in this nation gone into the question of vital statistics as affecting the health-rates for the purpose of seeing if we can do something for the public health, and have a National Board of Health to investigate, and that it is no criterion whatever? If these statistics had been inculpatory of Worcester, I guess you would have seen them in here produced by the other side.

We called Dr. Martin and Dr. Rice. They know of no epidemic disease there; and we called Mr. Charles B. Pratt, and he disposed of the story about the fishes. We called Deacon John McLellan, whose health is all right, and who says the smell there when the water is low is like the smell which comes from the bottom of any pond when the water is drawn off. We called Mr. Harrington. I have had already occasion to refer to his testimony. That is all there is about public health so far as the evidence goes.

Then we come to the report of the State Board of Health, and I want to go over that hastily. It does not report any thing about the existence of any condition of things which is producing any effect upon the public health, except to say that the population along the banks of the river believe it has a perceptible effect, the resident physician sharing to some extent the belief; but they do not add any thing to this point. Now, what does that board recommend? Do they recommend in that report that this scheme of theirs should be adopted? I do not understand that they so recommend. I do not understand that that was the question submitted to them; and the board had no function, as I understand, to recommend whether any scheme should be adopted, nor at whose expense; but that a specific question was put to them, and they answered it.

Now, let me, before I proceed to discuss their report, as I shall do as hastily as possible, refer to a case in the western part of the State concerning the Lenox Pond, not for the purpose of reading any considerable part of it, but for the purpose of calling your attention to one or two sentences. The Smith Paper Company have been subjected to an indictment which is now pending in court in Berkshire County. It came on for a hearing last May or June, and was postponed; but the prosecution propose to prove, as establishing a public nuisance affecting the public health, the following: They propose to show that " malaria began to develop so as to be noticeable about 1877, and a little the year before. It began on the north end of the east side near the Pittsfield line, the following year spreading down the east side, then across to the west side, till now in nearly every house there have been cases, sometimes whole families being afflicted. About a hundred and eighty persons have been sick in one year." Now, I simply allude to that as an instance where it is proposed seriously to prove a public nuisance. Is there any such thing existing here, or any thing that at all resembles it?

Now I come to the Report of the State Board of Health, to see what they have reported. The resolution of 1881, chap. 67, directed the board to consider and report " with recommendations as to a definite plan for the prevention of such pollution."

In other words, as I construe that resolution, and as I suppose the board construed it, they were required to examine and report what definite plan would best prevent such pollution. The resolution did not require them to report whether any definite plan ought on the whole to be tried, nor at whose expense. The question is, What is the most practical plan to prevent it? Assuming that the thing exists, assuming that it is desirable to prevent it, what is the most desirable method of preventing the cause? Now, in answering, the board say, —

First, That " all methods proposed are, to a certain extent, experimental. . . . The board accepts with great confidence the conclusions stated in the report of its experts." How much time do you suppose that board gave to this subject? They went up there to Worcester to settle a controversy, and turned it over to these experts; and these conclusions of the experts are the conclusions of the board, and of course they would accept them with great confidence; but they do not add any thing to the report. " Being convinced that ' the system of intermittent, downward filtration,' supplemented, if necessary, by broad irrigation, is best adapted to the condition of things," they therefore recommend the system " submitted in the report of the experts " as . . . " in the judgment of the board the best method of disposing of the sewage of the city of Worcester."

Now, the conclusions of the experts, what are they? These experts are Drs. Folsom and Walcott, and Mr. Davis. They refer to five possible methods, of which they reject two, leaving three, to wit, chemical treatment or precipitation in tanks; second, intermittent filtration through natural soil; third, broad irrigation. Of those it is to be noticed that Col. Waring, whose eminence as an authority the board recognized in the most flattering terms, says, The first, that is, chemical treatment or precipitation in tanks, is not worth considering. You will find that in his report. Thus in their report they expressly recommend a method which this very expert says is not worth considering.

Second, The experts say, that, considering the climate, dilution of sewage, the difficulties in the way, etc., are far beyond those of any town where the question has already been met. And when men of the caution, coolness, and judgment of Dr. Walcott, Dr. Folsom, and Mr. Davis say that, it means the whole length of that proposition: it does not mean any thing less, — that they believe the difficulties are far beyond any that have been met, and the thing is experimental.

Third, Any scheme that may be proposed, they say, may be said to be experimental to a certain extent.

Fourth, To be successful, and not create a greater nuisance than it abolishes, the outlay must be costly, etc.; then there is danger that you may create a larger nuisance if you abolish this. It may be that you had better bear the ills you have than fly to others you know not of. They carefully guard themselves; and you will see all through this report, which is drawn with the greatest possible judgment, that they have carefully guarded every bridge over which they can escape, if this thing fails. They have not left an avenue which they have not guarded. There is not one of them that would invest five hundred dollars in the scheme, if he could get five thousand dollars if it succeeded. They carefully hedge and guard themselves against all possibilities, and make good the avenues of retreat.

Fifth, After some description of downward filtration and broad irrigation, the latter is dismissed as not available for Worcester. Then the experts say this, and say it with deliberation, — I regard it as the most extraordinary sentence in their whole statement. You will not regard me as reflecting in any way upon these gentlemen. I am commenting upon their report. They were asked by the Legislature, if I understand it, "Gentlemen, suppose you have got to dispose of the Worcester sewage without polluting Blackstone River, what is the best scheme by which that can be done?" and they say this is the best scheme they know of. They do not say any scheme will work, but, "This is the best scheme we know of." They say this, and I want to call your particular attention to it. The experts then say, "We know of no scheme so practicable as being able to provide for all the ordinary sewage by modified, intermittent, downward filtration, and procuring several hundred acres upon which surface irrigation may be attempted and extended from year to year." I consider that a most remarkable sentence, because it was drawn with great care. As Dr. Walcott says, "They did not leave any sentence to any one of their number, but it had the personal attention of all." They do not say, "You will be able to do it," but, "We know of no scheme so practicable as being able to do it." Well, I should think not. Now, if Capt. Bunsby ever gave an opinion that was more guarded, and could be defended at all times, and could be protected, whatever the final result might be, if he ever gave an opinion that was more calculated to stand upon experience, and go out and survive all storms, and ride safely into harbor, then I am not aware of it. Now, I do not mean to compare this board with Capt. Bunsby in any other sense than that they have guarded their opinion; so that it will mean whatever you want to have it mean, according to circumstances.

Sixth, "Probably there is hardly another place in the State" — You see how they are guarding themselves in this thing. It is like the reports we get from Old Probabilities. Cautionary signals on the coast are set up; and they set up all along the coast of this report their cautionary signals, "Beware of danger at this point," "Lookout for man-traps." "Probably there is hardly another place in the State where the conditions of the problem can be as readily met as in Worcester to remedy an evil which is fast becoming a general one." And they say, — in view of the statistics I have put in here, showing you that in twenty places in this State the pollution is $98\frac{1}{2}$ per cent. as great as in the Blackstone, and in many places a great deal more, — they say, "This is fast becoming a general one;" and I submit a general law would be the right sort of measure to remedy it, instead of singling out Worcester, and laying this burden upon her.

Seventh, "We therefore recommend, as the most practicable and least expensive method of disposing of the sewage of Worcester, intermittent downward filtration." Well, there is their report, and there is the whole of it.

Now, concurrently with this report, having sufficient dignity to be printed in their report, is Col. Waring's report. I shall not discuss the scheme by which he proposes to take this water through a winding drainage, backwards and forwards, see-saw, fifteen miles, and plant it with osiers dank and precious-juiced flowers, to suck the contagion out of the sewage. They are going to remove all contagion out of this thing, and make it delightful and sweet. I will not go through the report. It is not my purpose. I do not allude to it for any such purpose. But his report is dated Dec. 15, and the report of the board is dated Nov. 17. He knew all about their report, and had it before him, of course. It came out a month before.

Mr. FLAGG. Do you state that as a matter of fact? It is not so.

Mr. RAND. Dec. 5 was the date of Col. Waring's report, not the 15th.

Mr. GOULDING. Very well, Dec. 5. Now, he either knew of their report, or else he is more conceited than I suppose he is, to go on and recommend a plan without knowing what their report was. Their report was out Nov. 17, it is dated then; and if he did not look at it, — if he regarded it as not worth looking at, — he is a great deal more conceited than I gave him credit for; and, when I heard the statement he made here, I gave him credit for a great deal: but I do not think he would have made his report upon this matter without consulting the Report of the State Board of Health. He is commended as high authority by the State Board of Health, and he was retained by the town of Millbury to find out a plan for them by which this pollution could be removed. Now, what does he do? He totally ignores intermittent downward filtration. He does not return the compliment of the board, though his report is made a month later. He does not pay the scheme of that board, which they have been setting forth, the poor compliment of a passing notice.

Mr. FLAGG. As a matter of fact, he did not know of their report.

Mr. GOULDING. Then, he is more conceited than I supposed he was. I do not believe it. If he went to work to report, when he knew this board of experts of Massachusetts were seeking to solve this question, without examining what their report was, and gave his report independent of theirs, he paid them very little respect, and exhibits a sublimer self-confidence than I supposed he had. He testified in his report right here to you, gentlemen; and he must have heard of it by this time. And whether Col. Waring knew, or did not know, the substance of the Report of the State Board, when he made his report, the argument here advanced is not affected. For if modified downward intermittent filtration is, demonstrably by science, the method best adapted to Worcester's case, then this great scientist ought to have reached that conclusion by independent scientific processes. He does not pay it the poor compliment of any notice except to say, "The State have reported one method." He does not tell you how this other plan will work; but he has got a scheme which, he says, will easily settle this question. This expert, I say, does not notice them.

On the other hand, they proceed to demolish his report, as you will see on pp. ccix., ccx.; and, without reading it over, I will say that they devote a page to considering a few objections which they consider his plan open to. They go right to the heart of the matter, and, by a few elementary objections, demolish it entirely, — that is what they do with Col. Waring. Now, if such an illustration could be used consistently with the utmost respect for the State Board, I should say I am indifferent which of these Kilkenny cats kills the other first; and I will say, that, when scientists who are so eminent, and who are honestly seeking for a plan, differ so diametrically, it does seem to cast ominous conjecture on the whole success of the scheme.

Now, we say that this scheme is so experimental, so uncertain, that it is unjust to the people of Worcester to enter upon it, or upon any scheme that would be an expense to them for the benefit of the public. We called Mr. Worthen. Mr. Worthen is an engineer who knows about this whole subject. He was an experienced engineer, was an eminent civil engineer, before my friends Dr. Walcott and Dr. Folsom were out of college, and before they were in it, and, I think, before they had got very far along in the common schools. Mr. Worthen was an eminent engineer; and it is no disparagement to any of them to say that he is the peer in intelligence of any of them, and vastly their superior in experience. He says that he does not agree with this plan. He does not believe in it. He has told you about its effect.

Now, I will not dwell upon Mr. Worthen's evidence except to say he is an expert of high ability, and he totally differs with them. Rudolph Hering says on the seventeenth page of supplement 16 of the National Board of Health Bulletin, dated Dec. 24, 1881, "Whereas comparatively little has been written on the previous subjects, there exists a considerable amount of literature on this one, owing mainly to the controversies that have arisen regarding it. I shall therefore merely state the conclusions to which the opinions of those engaged with the subject" — I want to call the attention of this Committee, who are practical men, and who propose to deal with this question as practical men, to this sentence, as embodying the whole matter in a nutshell: "I shall therefore merely state the conclusions to which the opinions of those engaged with the subject" — concur? A conclusion in which they concur? A conclusion in which any two of them agree? Not at all, but towards which they "converge." That is, they are converging towards these conclusions, "and which a personal inspection of the various methods and works warrant, adding a few elucidating remarks." He goes on and says the best way of disposing of sewage is into a large river or the sea. He goes on to show the expectations formerly entertained about utilizing sewage, but says they have not been realized, etc. Then he goes on and gives the names of a number of places where the best examples of irrigation were found; and among others he enumerates Bedford, Doncaster, Croydon, Wrexham, and Leamington, in England. It is important reading, in connection with the state of this science. We say there is no condition of science to warrant anybody to come to any other conclusion than that come to by Mr. Worthen; to wit, the method of purifying sewage has not yet been invented.

Now, these experts come here full of enthusiasm, full of learning, full of theories: they are able men; but when you come to the question of whether they agree that there is practically any method, and what, they totally fail to agree, or to come anywhere within hailing distance of any agreement. Now, we say it might be proper to impose this burden, if it promised certain success; but it does not promise certain or assured success; it promises possible failure, probable failure. Failure at Worcester we are not ready to pay for. If this thing was going to be a perfect success in every particular, it would be an entirely different question whether you would impose it upon Worcester; but, when you remember that it may fail, you will also remember that failure is just as valuable to the rest of the State as success. They would gain as much by way of the warning derived from our example as they would by success. It is not a question whether we ought to pay for success, but whether we ought to pay for the failure for the benefit of the State. This board was repeating the language of the board of 1876; and I assure you they did not do any great amount of original investigation, for they did not need to: they had the facts in that report. The report says, "This is a fast-growing evil." We have shown you by the evidence that the evil exists as much in other places as in Worcester; and shall we be the pioneers? shall we be made the paschal lamb to be sacrificed for the benefit of the rest of the community, and, if we fail in the attempt, yet furnish them with just as much instruction by our failure as we should if we had succeeded? Failure would be no answer to Millbury. They will come here five years hence, supposing we were to adopt this scheme, and supposing the effluent is not satisfactory, and they will say, "You are putting your sewage into our stream. We don't believe in downward intermittent filtration." They will then say, "We have got Col. Waring here with new views. Don't talk to us about downward filtration! Don't you know that in other towns they have got a new process? In other towns they are adopting more recent plans of purification, which cost only two, three, or four hundred thousand dollars. Modified intermittent downward filtration is obsolete."

Now, we have shown you that there are fifty or sixty manufacturing places that pollute this stream; and the proposition is that we shall purify this stream of the pollution caused by the manufacturing establishments, as well as that caused by the city. The proposition is, in other words, that the people of Worcester, which is doing only what she has a right to do, shall purify this stream, not only from the impurities which she puts into it by her sewage, but also purify it of the impurities put in by the mill-owners who have no right whatever to do so. And my friends on the other side know that if Mr. Morse, or anybody else in Millbury, is injured by the pollutions put into that stream by any manufacturing concern at Millbury, or Worcester, or elsewhere, they can get the parties doing the injury enjoined, if they can prove it, or bring an action against them for damages. They can join them all, or as many as they have a mind to, in a single action; and if they can prove that either alone, or in conjunction with others, the mill-owners have injured their property, they can recover damages.

Merrifield v. Lombard, Bryant v. Bigelow Carpet Company, Wheeler v. Worcester.

These cases, the last of which I had the honor to defend, settle this proposition; that is, where several parties do independent acts which go to constitute an injury, although one of them might not alone constitute the injury, if they are independent yet combined acts, and produce an injury, you may join them all, or any of them, or sue one of them alone, and recover your damages. There is not a mill-owner putting pollution into that stream that has a right to do it; and shall the city of Worcester be compelled to purify the stream from the impurities that we put in there as a matter of right, and also of the impurities which are put in by mill-owners as a matter of wrong?

Mr. FLAGG. We only ask that the sewage be purified. We will take care of the manufacturers that are polluting it.

Mr. GOULDING. Perhaps that is worth while. That illustrates the paralytic state of mind they are in upon this subject, to suppose that it is possible to pick out of this stream the sewage, and purify it of that, and not purify it of the impurities of the mills that flow into it. Is such a proposition possible to be maintained? And yet they say that! They have nothing else to say. I say the proposition is, to compel Worcester to make it pure, not only of the impurities of the sewage put into the stream under the law, but also of the impurities which mill-owners put in there against the law.

Another thing: Millbury, from the beginning of its history, has done just exactly what we have done to this stream, to the extent of her ability. Singletary Brook, which empties into the Blackstone, is almost as impure as the Blackstone. They propose that we shall wash this stream pure — for what purpose? In order that they may empty their filth into it. We are to wash this stream pure in order that Millbury may have it clean for the purpose of emptying her filth into it. The nymph of the Blackstone, which has become impure by the embraces of Worcester, is to become purified in order that she may submit to the foul embraces of Millbury. This is the object of their application.

Now, I will call your attention to the Report of the State Board of Health of 1878, p. 66, where they say this: —

"There are some places where harm is already done by pollution of watercourses, although not to that extent which may commonly be seen in England, for instance. As a whole, throughout the State, the evil from the pollution of streams is small as compared with that arising from the accumulation of filth in privies, cesspools, etc., near dwelling-houses."

Who doubts that there is vastly more danger to public health of Millbury from the contamination of air and water by means of accumulation of filth than from the pollution of the Blackstone?

There is another proposition that I desire to submit to you. Either there is a nuisance affecting the public health, or there is none. If there is a nuisance, the statute-book is full of remedies for all the parties in the courts. There is no deficiency in the statutes. They can proceed by action, by a bill in equity, by indictment, and bring us into court, and convict us of maintaining a nuisance, if there is any such thing. If there is a nuisance, there is a plenty of remedies. If there is no nuisance, then I submit it would be the height of injustice to select us out, and subject us to legislation which, if it should be enforced, would be a calamity to the people of the city which I cannot undertake to measure. But now they anticipate that argument; and my friend said in his opening, "Mr. Goulding said that last year, and he will probably say it this year." Well, I think I should be likely to say it. If there is a nuisance, indict us, bring us into court. If there is no nuisance, then do not select us out for special legislation, which is worse than criminal law.

But they say they are actuated in this by a feeling of neighborly friendship. That is, they do not propose to bring us before the courts, where we would have the protection of judges learned in the law, and be tried by the rules of evidence; where evidence that was competent would be put in, and not rumors, like Capt. Simpson's story of Dr. Gage's diagnosis of Wilmarth's case, and where evidence would not be put in at random. We may there get our case postponed for a time, long enough to construct some of these works. They are doing this now altogether out of friendly feelings; and, as a manifestation of their kindly regard, they haul us here before the Legislature, and propose a bill which provides that the city shall have four months to do this work in. I say, we have shown you that it would take three years to construct the works; yet in four months we are expected to leave the Blackstone River free from the defilement of the city, either from the city alone, or in conjunction with other sources. That is the kind of tender mercy that the people in Millbury show to us. They go to their prayer-meetings, and pray over this matter; and then their Christian charity leads them to do such a thing as that! Why did they do it? I won't stop to read from my brother's opening, as I intended, because I have already consumed a great deal of time; but they did it because two or three mayors have said something in their inaugurals, and Senator Hoar has said something. I will read that passage, because I do not want to misrepresent my brother Morse. On p. 20: "Mr. Chairman, while there are such men in Worcester, and such a spirit shown there, it would certainly be in the highest degree unfair and unneighborly for the people injured by this evil to take any legal action; and that is why they have forborne to do it."

They won't take any legal action where we can be protected by regular laws and by counsel, where we can take such steps as may be necessary to protect our rights; but they haul us before the Legislature, and ask the enactment of this cruel law, which would put a stop, if it could be enforced, to the life of the city. They haul us here, and in doing so they are actuated by that tender mercy because one or two mayors have in their inaugurals said something which they think is favorable to them. I do not know how many of you have ever been elected as mayor; but, if you were elected mayor in Worcester, you would be expected to deliver an inaugural on the 1st of January, and in it you would be expected to discuss all questions relating to the treasury department, the investment of money, etc.; you would be expected to go over the school question, and offer suggestions as to the educational interests of the city; you would be expected to have something to say upon the sewer question, and all such questions; and, if you were a man of any prominence and capacity, as you would be, of course, if you were elected mayor, you would want to say something that would be new and sensational, and not present a dull statement unenlivened with any thing striking and original. Now, at the time those inaugurals were written, much was being said about the utilization of sewage; and these mayors thought if they could say something about the waste matter which was running down the sewers, and suggest that it might be utilized, they would have touched a topic that would make the councilmen prick up their ears, and the board of aldermen would say, "We have got somebody here now that is going to do something." Then they read Mr. Hoar's speech, which I shall not read again. No person has any higher respect than I for Senator Hoar, and I do not think there is any thing in this case that would induce me to say any thing disrespectful of him. Senator Hoar, as far as I am aware, has never investigated the facts involved in this question, nor the law of it, only as he knows the law as one of the most prominent lawyers in the State and the easy leader of our bar. So far as I am aware, he has expressed no opinion on the subject whatever.

Now, gentlemen, suppose that the town of Millbury, and the towns constituting that representative district, had suddenly discovered the merits of a gentleman who belonged to the minority party, whose merits before they never had conceived of, who had been quietly left out of office all his life, and then the people, sinking all political differences, had united upon that man, an able lawyer, for the sole reason that he was supposed to be able to deal with this question. Suppose they had selected Mr. Hopkins for that purpose, a man who did not represent their political opinions, but who was an able lawyer, and a man who, they supposed, could aid them in the Legislature. Now, suppose in Worcester the minority party, not representing the political sentiment of the majority of the people of Worcester, put up another able lawyer as their candidate, as able as Mr. Hopkins, as adroit, acute, and far-seeing, able to deal with anybody on any question. Now, supposing some members of the party in the majority were using the argument through the street, "Your Republican candidate isn't able to deal with this question. Don't you see they have put up Hopkins down there, and you want Mr. Verry down there to meet him in the Senate?" It was in that situation that Mr. Hoar made the speech read here. In that state of the question he said in substance, "We do not want to retain a lawyer to defend us as though we were doing a wrong, but we want to send a representative to represent us." Worcester does not want to send down to the Legislature a senator "to get her off," as if she were inflicting a wrong. We want our representatives to represent Berkshire as well as Worcester. That was the question Senator Hoar was discussing; and to wrest his remarks into an indorsement of their views upon this subject seems to me to show the condition into which they are driven by the exigencies of their case. There may be other citizens of Worcester who have the opinion, — the present mayor of Worcester, for aught I know, may have the opinion, — that if we could get rid of this claim of Millbury by paying them money, or if we could settle this question once for all, so this should be all and the end of all here, or if there were somebody qualified and able to represent everybody concerned, and to sign a receipt, a discharge in full, my friend Mr. Stoddard, the mayor, might be willing, and go with all his heart for paying them a sum of money that they would receipt for in full, and let that end it; and in that respect he may represent many of the citizens of Worcester. And this he and others might be willing to do wholly apart from, and independently of, legal rights and obligations. But so far as this proposition goes, a proposition that proposes to invade our rights, and take away from us that which we have a right to do, the city government is unanimous in their opposition to it. I am authorized by the vote of the Committee having the matter in charge to oppose all these schemes as unjust.

The CHAIRMAN. Do I understand you that this Committee ought not to report a bill authorizing the city of Worcester to do certain things, but leave it entirely to them?

Mr. GOULDING. I am not aware that any members of the city government have any objection to this Legislature conferring upon the city all or any authority it sees fit upon any subject. I will answer that question directly: I am not arguing any such thing. I simply make the remark that the mayor and other gentlemen may be of the opinion that they would be willing to pay something if anybody was present authorized to give them a receipt in full, and settle this matter; and I understand the mayor is not opposed to the Legislature giving specific power in the premises, and I do not know that anybody is opposed to that. But I should suggest that the legislation should be general, and should apply to all cities and towns which may have now or hereafter sewerage systems.

Now, I want to speak of one thing more, and I will close. It seems to me there is one thing these experts have not taken any notice of; at least, they have not developed it in their report: and that is, the results of this sewage purification. The Croydon farm in England is always cited as one of the best examples of sewage purification works in England. It is on the Wandle, and there is a sewage farm there. You will find on the 17th page of the report of Hering, that he says what they all say: that broad irrigation is better than downward filtration. Croydon has broad irrigation. Now, I want to ask you what results have been obtained in Croydon, and I will turn to the 376th page of the State Board of Health for 1876. On that page we find that the effluent of the Croydon farm, where, you understand, they had provided broad irrigation, is 47.0821. That is, it is a good deal worse than that of Mill-brook sewer at the worst point, after they got through treating it, as shown by this analysis.

Mr. FLAGG. You mean to say it was in 1875, or whenever those statistics were made up.

Mr. GOULDING. I mean what I say. It was succeeding as well as ever it was, before or since. There is no evidence one way or the other except what you get from that report; and I mean to say when that analysis was taken, after it had been treated at Croydon, the best result is 47.0821, about four times as polluted as the Blackstone at Millbury. Our sewage at the worst point, according to the analysis last year, was 41.8790, while the effluent of the Croydon farm after irrigation was 47.0821; or, in other words, the effluent of that farm is twelve per cent. in excess of the impurity of Mill-brook sewer at the very worst point. Now, in Merthyrtydvil, in Wales, they use a downward filtration scheme, I believe. The result there is, after treatment with lime, 65.854, or about one-half in excess of our sewage at the worst place, and five or six times as corrupt as at Millbury. At Merthyrtydvil these figures show that, after treatment with lime before the intermittent filtration, it is 65.854. That leaves out nitrogen, and some other elements which are not mentioned in the analyses of the Blackstone, to which I have referred. This report shows the result to be, after intermittent filtration, 37.675, nearly as bad as our sewage at Cambridge Street, and nearly four times as bad as at Millbury.

Now it is proposed we shall enter upon this experiment of downward intermittent filtration, to purify for the benefit of Millbury, and prepare it for their sewage, a stream which is four times as pure as the effluent of the best sewage farms in England, which effluent is "no way objectionable," "is quite pure."

The truth is about this, gentlemen: this investigation into the impurity of streams was started originally for the purpose of purifying such streams as furnished sources of water-supply for towns or cities. When we compare the Blackstone and other streams in the State with the English streams, we do not find that there is at present any such degree of pollution as demands immediate relief. The words of the State Board are carefully guarded, and they speak rather of a condition which may become a serious evil rather than a present existing evil. There is no occasion for hasty and precipitate legislation. It is something relating to the future you are dealing with. The attempt made here to establish the proposition that a great danger exists in the present has, I submit, failed. There may be a remedy for this, by taking down the dams, and making the Blackstone River accomplish what nature intended it for, — that by its currents and motion it may carry away the off-scourings and impurities of the land.

Gentlemen, I have to thank you for the attention you have given me during this long speech. I could not see how long it was going to be. I have gone over the whole subject. What Worcester wants is fair treatment. Pass a law that shall apply to us as to others. Worcester is too influential to be treated with injustice, and this great Commonwealth cannot afford to treat her with injustice; and it could little afford to treat her with injustice if she were the weakest community in the land. The blood in our veins is red. We have the usual supply of the corpuscles that make it red. If the time has come when this growing evil that my friends speak of has reached a point where there ought to be a general investigation by a River Pollution Commission for the purpose of inquiring into this matter (not taking an old report and transferring it into another), a commission having this subject specially in charge, and having ample time to make a thorough investigation, and then report some general legislation, there is not any community in this land that will more cheerfully join in aid of any scheme that the public necessities require than Worcester. She has always borne her share of the public burdens cheerfully. She is not greater than the law. If she is violating the law, there are remedies enough; and, if new legislation is required, let it be general legislation, not legislation levelled at a particular community, which is likely to retard her progress; and not merely retard her progress, but, if it could be enforced, prevent her very existence.

Adjourned to Tuesday, March 21, at 10.30 A.M.

EIGHTH HEARING.

STATE HOUSE, BOSTON, March 21, 1882.

CLOSING ARGUMENT OF HON. R. M. MORSE, JUN., FOR THE PETITIONERS.

MR. CHAIRMAN AND GENTLEMEN, — The establishment of this Committee marked a new era in the legislation of this State. It is only within a few years that a Committee on Public Health has existed in this Legislature. Its first name was "Committee on Water-Supply and Drainage;" and under that name it continued for some time, attending specially to the matter of water-supply for cities and towns: but the reach of its inquiries and investigations was constantly extending farther and farther, until finally the name was given to it under which it now serves, — the "Committee on Public Health,"— the most comprehensive name, perhaps, which is held by any committee of the Legislature. Experience has shown that the class of subjects committed to its care, and the investigation which it has been necessary to give them, have been of the largest character; and to-day there is no organization under the control of either branch of the General Court which has to do so closely with the public welfare as this Committee. Upon the breadth of its recommendations, upon the wisdom of its policy, very much of the future prosperity of the State must depend. As you, gentlemen, are laying the foundations of wise legislation, or unwise legislation, in the future, I ask you at the outset to look at this question, as upon all questions, not with any narrow or selfish view, but with a very broad regard to the principles that are involved, and to the effects of their solution, or your solution of them, upon the legislatures and the people that are to come after you.

There is certainly nothing that of recent years has attracted the attention of public men and of intelligent writers to a larger extent than sanitary questions. The whole tendency of the medical profession, as you, gentlemen, are fully aware, has been, in these latter days, to prevent disease. Essays have been written on the subject, speeches have been delivered, legislation has been invoked and has been passed, the special point of which has been *to prevent disease*. In old times, the only duty of a physician, the only function which he had to discharge, was to cure disease, as well as he could, after it had come. The modern physician looks to the sources of disease, and undertakes to prevent it. The profession in this country have been fully up to the profession abroad. Everywhere the tendency has been to examine into the causes of disease, and to endeavor to eradicate them, to limit and control them, and so to prevent disease. There are certain fundamental principles upon which all this preventive policy has rested. Pure air and pure water have been regarded everywhere as essential; and both in the management of private establishments, and in the conduct of large municipal corporations, the principle has been invoked by all medical men, and, through their instrumentality mainly, has been acted upon by governing bodies, that every cause of impurity, so far as practicable, should be removed from the air we breathe, the water we drink, and the places we frequent.

I cannot do better than to quote the words of two distinguished English statesmen on the importance of this class of questions. In 1877, on the occasion of the opening of one of the Victoria Dwelling Association Buildings in London, Earl Beaconsfield said, —

"I have touched upon the health of the people; and I know there are many who look upon that as an amiable, but mere philanthropic, expectation to dwell upon. But the truth is, the matter is much deeper than it appears upon the surface. The health of a people is really the foundation upon which all their happiness, and all their power as a State, depends. It is quite possible for a kingdom to be inhabited by an able and active population; you may have successful manufactures, and you may have productive agriculture, the arts may flourish, architecture may cover your land with temples and palaces, you may even have material power to defend and support all these acquisitions; but if the population of the country is stationary, or yearly diminishing in number, it diminishes also in stature and in strength, — that country is doomed. Speaking to those who I hope are not ashamed to say they are proud of the empire to which they belong, and which their ancestors created, I recommend them by all the means in their power to assist the movement which is now prevalent in this country to ameliorate the condition of the people by improving the dwellings in which they live. *The health of the people is, in my opinion, the first duty of a statesman.*"

And Mr. Gladstone, addressing his neighbors and friends at Hawarden, in August, 1877, says, —

"I have lived in the West End of London for forty-six years; but, although there is a greater number of people there, and the town has spread in all directions, yet when you open a window now the air is purer and fresher, and fewer 'blacks' come in, than forty years ago. The reason is, that Acts of Parliament have been passed to prevent people from wantonly and wilfully making smoke, and compelling them to consume it. This is now done to a great extent, — *not quite so much as it ought to be;* but still a great improvement has been effected. God made this world to be pleasant to dwell in. I don't mean to say he made it to be without trial or affliction, but he made our natural and physical condition to be pleasant. The air, the sun, the skies, the trees, the grass, and the streams, these are all pleasant things; but we go about spoiling, defacing, and deforming them. We cannot, it is true, make the town as pleasant as God has made the country; but most of you can do something to prevent the pleasant things which have been vouchsafed to us from being deformed and defaced by the hand of man in the future."

I assume, Mr. Chairman, that the great prerogative of this Committee is, — and more than any committee of the Legislature does it have it in its power by the recommendations which it shall make, — to keep things pure where they are now pure, and to redress evils which exist, so far as those evils affect the health and comfort of the community.

We do not any of us expect that in an age of development, when we are gradually increasing the population of our cities upon and along the banks of all our running streams, — when industries are growing up, farms are being cultivated, civilization, in fact, is creeping in, — we do not expect those streams will run precisely as they ran in earlier times. There will be more or less impurity. But the impurity that comes from the natural and ordinary growth of population, the drainage from the banks, is one thing; the impurity that comes from the deliberate, systematic act of a municipal corporation is another thing. The one is not, in the ordinary nature of things, either injurious on the one hand, or preventable on the other. The other causes serious injury, and at the same time it is easily prevented.

The case, gentlemen, that comes before you here is the most serious one that has ever been presented to a Legislature in this Commonwealth, as regards both the present evil and the necessary future consequences. It is the case of a great, rich, powerful, and prosperous city, deliberately — not through malice, of course, but still *deliberately* — relieving itself of all its filth and nastiness, which would be a source of disease if kept within its own limits, and putting it where it must inevitably, in a greater or less period of time, become a source of disease to others. That is the proposition which I propose to establish; and I must pause here at the outset, although I shall have occasion again to refer to it, to ask the Committee to consider how the parties stand with reference to this controversy. On the one hand are these petitioners of Millbury, the inhabitants of a small town, and their neighbors in the little towns below them upon the river, who confessedly are without influence or power in the Legislature, as compared with the power which a great city can bring to bear, coming up to the Legislature, for a second time, to state their grievances, pointing to the long record of the State Board of Health, which has been for a great many years calling the attention of the public to the serious pollution of the Blackstone River by the city of Worcester, — coming up here and saying that the health of their community has been affected; that their children have sickened and died; that their schools have been dismissed; that the workmen in their mills have been unable to continue in the regular discharge of their duties; that their physicians have advised them that there is great danger of more extended disease, — bringing the trusted representative of the State upon the State Board of Health, and the experts whom the State Board of Health have appointed, to confirm their statement, and saying that, in their belief, *in their honest belief*, this evil has increased, and is likely to increase. They have also said — and I took particular pleasure in saying it at the outset of this hearing — that they did not consider, and intended to make no charge, that Worcester had intentionally done them any wrong; that they had refrained from all legal proceedings against the city because they knew that there were representative and leading men there who admitted the wrong, and were looking about for some proper way to remedy it. They came to the Legislature, therefore, in the full expectation that, upon the statement of their case and the production of their evidence, the city of Worcester would join with them in asking the Legislature for something practical that would lead to the reform of this abuse.

I stated, in opening for the petitioners, that we did not suppose, in view of the Report of the State Board of Health, and of the evidence that was taken by this Committee last year, that the question of the seriousness of this nuisance would be contested at all. I was very much surprised when I was told, by the learned counsel for the city, that they denied that a nuisance existed; that they proposed to contest that fact. And now, in his closing argument before you, he has undertaken to assert that no nuisance exists; that nothing calling for the attention of the Legislature has been proved; and every suggestion of the evils that have been caused, and that are apprehended, is treated with the utmost levity. From the beginning to the end, the position of the learned city solicitor — and I desire to speak of him, not as an individual, but as the representative of the city authorities — from the beginning to the end of this hearing, the object of the learned counsel has been to turn our case into ridicule. When our witnesses testified that there was illness in their families, that the little children were sick, that this horrible stuff that was being put into our river at the rate of twenty-two thousand tons a year was the cause of disease, and was affecting our people, the answer has been to ask our witnesses, "Didn't you have a grandmother who lived to an old age?" "How old was this or that resident?" And the attempt has been made to turn off with a laugh the suggestion that this terrible pollution caused any real trouble.

I said, in the beginning of this case, that some of the most prominent citizens of Worcester fully appreciated the character of this injury that was being done; and I quoted from Senator Hoar, to the effect that "he would rather Worcester should pay one million dollars than do a wrong to one of these towns. It is a great and serious thing to poison the air, to pollute the streams, or destroy the health of the homes of a town like Millbury or Sutton or Northbridge or Uxbridge or Blackstone. Worcester must call to her aid all the resources of science, all the experience of other cities and countries, all the ingenuity of mechanic art, to avoid such a result, whatever may be the cost. For one, he desired his representatives in the Legislature to meet the question in this spirit."

That was the position which Senator Hoar took. I am told that exactly the same position was taken, at the same meeting, by the Hon. Mr. Rice; although his remarks are not reported in the paper from which I cut this extract. But the point to which I am calling your attention, gentlemen, is this: that although prominent individuals in Worcester have taken this position, although to-day there are a great many men of humane instincts, and high sense of honor, and proper regard for the rights of people below them, who are in accord with Senator Hoar, and who see in what he said the utterances of a just and high-minded citizen, who despised the selfish view of the question, — yet, that when we come to deal with the city of Worcester, we must deal with it as a corporation, as with any other corporation; we must take the utterances of its agents who are presented before us. My point is, that the city, so far as its authorized utterances are concerned, is undertaking to ridicule the position of these petitioners, and to make light of the evils that they fear. Why, gentlemen, there has been placed before you a document of such an extraordinary nature, that, if it had not come in under the express approval of the city of Worcester, I should not have deemed myself justified in referring to it. I should have been told, I think with propriety, that such a screed as this was not worthy the attention of an intelligent committee. But, early in this hearing, the city solicitor took particular pains to lay before you a report of the Commission of Public Grounds of the city of Worcester for the year ending Nov. 30, 1881, in which, apparently by special request, a discussion of this question of sewage is made by that board. I beg the Committee to notice what is said, because it is in exact line with, and appears to have been the text for, the argument that has been addressed to you here. I ask you, gentlemen, whether or not it is in accord with the rules of morality or of law to which you have been accustomed. And, speaking of the rules of morality, I may say that they appear to have been changed in the Worcester code. As I understand the learned solicitor, the old law of morals, the Golden Rule, "Do unto others as ye would that they should do to you," is obsolete; and a new maxim has come into play, — "Do as the others do." But here is the document; and I read, from the 22d page of this extraordinary communication: —

"These petitioners from Millbury — owners of obsolete 'privileges' — assert that they have been sick at times; and they elect to attribute their ill-health to Worcester sewage. But all think it worse — i.e., the sewage — the nigher to Worcester. If so, the chief occupation of Worcester itself, instead of a demand for sash and blinds, should be the interment of its population. And, considering every thing, our last state does not appear to be much worse than our first.

"Is Worcester to be held answerable because Benjamin Flagg did not feel as vigorous or well at seventy odd as when a young man?"

That gentleman, the father of my associate in this case, died this last year from causes attributable, in the opinion of his physician, to this Worcester sewage.

"Shall Worcester respond in damages because medical men fancy that their town is not in quite as good sanitary condition as when it was one half or third its present size? Although the tables of mortality in Millbury show but 74 deaths in 1881, whereas there were 93 in 1880!

"Is Worcester to be subjected to the untold cost of repeating experiments that have nowhere proved successful, because mill-ponds fill up, and streams become sluggish and shallow, where dams are almost as frequent as the feet of fall?"

On p. 24, the writer of this report, the Chairman of the Commission, says, —

"When and where do those indisputable rights take their rise and find their origin? Who shall determine them? and how? A pioneer at the head-waters builds an out-house that discharges into the stream. The right of a community to build its privies in that manner, if it elects such improvident way, is surely as imprescriptible and fixed as the concession, or 'privilege,' of a solitary individual, here or there, to dam that stream, check its flow, stifle its current, and stagnate its water."

On p. 26 he says, —

"If experiments are to be tried, let Millbury and her neighbors reverse the Blackstone, Singletary, Ramshorn, and Quinsigamond, over their intervales, and pocket the profits! *Worcester does enough when she wastes her substance in the effluent stream;* for that it is waste is obvious, though not susceptible of prevention or remedy."

Now, I say, for a document of that kind to be put forth by a public commission, and to be presented by the authorized representative of the city of Worcester to this Committee as an answer to the representation of these people in Millbury, and that neighborhood, that they are suffering from this dreadful curse of the sewage of Worcester being turned upon them, is as cruel and unprovoked an insult as could possibly be given. At least, we might have had the *sympathy* of the authorities of Worcester; at least, it might have been said to us, "We are very sorry for the condition in which you are, but it is not in our power to remedy it." But we are taunted by the fact that "Worcester does enough when she wastes her substance in the effluent stream," and that we poor people of Millbury and vicinity ought to be thankful to Worcester that we have the benefit of receiving the excreta of her citizens! I say, Mr. Chairman and gentlemen, that that document, put in here as a representation of the ground taken by the city of Worcester, a concentrated sneer at the position of the petitioners, is as inhuman, as unfair, as any thing could possibly be. As I said before, if I had seen it in a newspaper, and had quoted it, I should have expected to have been told by some member of the Committee that it was not worthy of repetition as representing the views of a great city; but, when the counsel of the city brings it before the Committee with the weight of his authority and indorsement, it is important as indicating the position of one party to this controversy.

That is not all, gentlemen. I have undertaken to show that the position that has been taken here on behalf of the city is a very extraordinary one; although, as I have said, there are individual citizens of Worcester who have frankly admitted the obligations of the city as strongly as I could state them. I want now to call your attention to an incident which happened in the course of this hearing. I refer to the testimony of the mayor. The mayor of Worcester is a gentleman of honorable instincts and high intelligence, and a man whom, individually, I have no doubt, the people of Millbury, or the people of Worcester, would be perfectly justified in trusting to determine any matter that might be presented to him. He came before you in his representative capacity as mayor of the city; and he stated fairly, and, as far as I could see, with strict regard for the interests of the city of Worcester, the position which the city took in this controversy. In answer, however, to some questions that were put him, mainly by the Chairman of the Committee, he stated his individual views as to what ought to be done, or might be done, in the way of remedying this evil, suggesting a plan which he did not feel sure was adequate to the case, but which, nevertheless, he, with considerable reluctance, expressed his willingness to agree to; and he intimated that if the Committee should report a bill which should give permission to the city of Worcester to use such instrumentalities as it saw fit for remedying this evil, such a bill would be acceptable. He took occasion, however, at every stage of his testimony, to qualify it with the statement that he spoke merely as an individual, and not as a representative of the city. That fact indicated, I think, to the Committee — it certainly did to me — the sort of terror under which an official of the city of Worcester must speak, that he cannot express a humane instinct without danger of being called to account by somebody; that he cannot suggest a possible remedy, at a very slight expense, for a great evil, without its being regarded as treachery to the city; and pretty soon, gentlemen, the echoes came back from Worcester of public opinion as to the dreadful heresy committed by the mayor. "The Worcester Spy," under date of March 18, had an article on the subject, entitled "The Sewage Problem:" —

"Mayor Stoddard's testimony yesterday, before the Committee investigating the pollution of the Blackstone River, *was somewhat disconcerting to the city solicitor and other representatives of Worcester's interests present, and in an equal degree encouraging to the petitioners for relief from the alleged grievance.* It was freely said in the committee-room, and in this city, after the substance of his testimony was known here, *that the mayor had given away the city's case.* But that seems to be an extravagant statement, prompted by an exaggerated estimate of the importance of Col. Stoddard's admissions. There is no doubt that the stream is polluted; that the pollution tends to increase with the increase of population and the growth of manufactures in the Blackstone valley; that the pollution is an evil and a growing evil. So much Col. Stoddard admitted; and, if the case of the city of Worcester rests upon the disproof of these propositions, it is a hopeless case. It is not conceded, however, by Col. Stoddard, or anybody else authorized to speak for the city of Worcester, that the pollution of the stream has been, or can be proved to be, injurious to the health of the people of the valley, nor that the city of Worcester, in so far as it has contributed to the impurities of the stream, has exceeded in the least its strict legal rights, nor that it is under any obligation in law, morals, courtesy, or good neighborhood to purify the stream at its own expense, nor that it could do so while the mills above and below are pouring their refuse into the stream, nor that, if it were right and practicable for the city to do this, either of the plans submitted on behalf of the petitioners is reasonable, or would be effectual without causing inconveniences at least as great as those it was expected to remove."

There follows a general discussion of the subject which it is not material to read. The particular point of the article is a criticism upon Mayor Stoddard; because, in answer to perfectly proper questions addressed to him on the part of the Committee, he gave certain opinions, which he qualified in every instance by saying that they were *his* individual opinions. I do not wonder, however, that a city government which is controlled by the spirit which dictated the "Report of the Commission of Public Grounds" should call Mayor Stoddard to account for merely expressing the natural desire of an honorable gentleman to alleviate the miseries which his city was causing.

That is not all, gentlemen. I am considering the attitude of the authorities of the city of Worcester, which is a very important question when you come to deal with the form of law to be recommended.

You would make one bill for a city that was ready and anxious to do its full duty, and which appreciated the extent of the evil to be remedied, and quite another bill for a city which was resolutely and obstinately set against doing any thing for the people whom it had injured, and which claimed that they were not hurt at all.

I refer now to the treatment which has been given here to the experts who have been called. Why, gentlemen, if this case had been tried in a criminal court, you could hardly have had bitterer and more sarcastic abuse upon the experts employed.

Now, gentlemen, who are these men who are thus ridiculed and attacked? What is their position in the case? What right has the city of Worcester to authorize or direct its representative to treat them in this way? Consider, gentlemen, that the State Board of Health has held but one position on this question from the time when the first report was written, to the last. Last year, in an argument addressed to this Committee, I called attention to the names of the gentlemen who constituted the State Board of Health that made the first extensive investigation of this subject. They were Dr. Henry I. Bowditch, Warren Sawyer, Richard Frothingham, R. T. Davis, Dr. George Derby, P. Emory Aldrich (now Judge of the Superior Court, and then and now one of the foremost citizens of Worcester), and G. B. Fox of Lowell. That was the Board of Health (some of them distinguished physicians, as you all know) which made the report of 1873; and in that report they said, —

"We do not hesitate to say that a scheme for the treatment of the sewage is practicable. At the request of the secretary of the board, Mr. Phineas Ball, civil engineer of Worcester, has prepared a statement of a plan which he has in his own mind, by which the sewage may be collected, and conducted to a point below the city, where there is abundance of land suitable for arranging a filter on a large scale, or where the sewage may be applied for purposes of irrigation."

That was the report of 1873. You will search the records of the board from that time to this in vain to find any thing inconsistent with this position. They have always spoken of it as an evil; they have spoken of it as an increasing evil; they have spoken of it as one that might be prevented, and one that the city of Worcester ought to prevent. They have pointed out a method which could be adopted. When, therefore, my friend comes in and ridicules our position, and says that we have not been hurt, that there is no danger of our being hurt, and that there is no need of remedying it, if we have been hurt, — I say, here is the State Board of Health, an impartial tribunal, that for years and years has been calling the attention of the Legislature and the public to this trouble. With this steady action of the Board of Health in mind, consider the position of the experts. Take Dr. Walcott. He was not selected by us; we had nothing to do with his appointment on this commission; he was requested to serve by the Board of Health, because of his peculiar ability and experience. He has been the health-officer of that board. In his report, he speaks of this pollution of the Blackstone as a nuisance, and describes a remedy, which, in his judgment, would be efficient. Now, what occasion is there for attempting to depreciate the testimony of so experienced, intelligent, and impartial a medical man as is Dr. Walcott? Why should he not be listened to with respect and confidence when he says, "In my judgment there is a nuisance there, and this nuisance can be abated and ought to be abated"?

Then, take Dr. Folsom. My learned friend had a good many comments to make on Dr. Folsom, and referred depreciatingly to his testimony. We had no more to do with the selection of Dr. Folsom to make these investigations than had the city of Worcester.

Mr. GOULDING. In what terms did I refer depreciatingly, either to Dr. Folsom or Dr. Walcott? Anybody who heard my argument knows that I made no such statement with reference to either of those gentlemen. I referred to them; I did not depreciate them; I discussed their report, and showed what it actually said.

Mr. MORSE. I accept the gentleman's statement; and when we come to read the report of his argument, if what I have said is an incorrect representation of his remarks, I shall regret to have made it; but my recollection is most distinct that the gentleman argued to this Committee that the position of these petitioners was wrong, in so far as it claimed that any nuisance existed; that it was ridiculous, and that no serious trouble existed there; and, further, that he did undertake to depreciate the testimony of Dr. Folsom and Dr. Walcott, inasmuch as they both state that in their opinion there is now trouble serious enough to demand action. I understand that to be their position, and I do not understand it to be acceptable to my friend. I understand that his argument was to satisfy you that these gentlemen had exaggerated the troubles, and that their judgment in regard to the remedy was not to be relied upon.

Mr. GOULDING. If failure to accept their position is depreciating their evidence, then it is so; but in no other sense did I depreciate their evidence.

Mr. MORSE. Well, I will not spend time in discussing that. If I have inadvertently misrepresented my friend, of course the argument will speak for itself when it is printed.

I come now to another gentleman, about whom I suppose my friend will not have any question. I suppose that there is no doubt that he *did* depreciate Col. Waring, and intended to depreciate him. I suppose there is no doubt that he held him up to ridicule, and intended to do so; and I may fairly claim that he did undertake to make fun of his statements, and to treat his recommendations without any respect whatever. Now, gentlemen, the position of Col. Waring is too well known to require any statement from me. I happen to have in my hand a report of a committee of the town of Westborough on a system of sewerage, which was submitted on the 8th of August, 1881; and I refer to it to notice incidentally the broad way in which they deal with the question, but principally to show the weight that they give to the opinion of the gentleman of whom the city solicitor thinks so lightly.

They say, —

"Your Committee from the outset have been deeply impressed with a sense of the great importance of the subject intrusted to their investigation, not only in regard to its sanitary bearing, but as involving a large pecuniary outlay. For, while most men regard no outlay too large that promises speedy returns in dollars and cents, they hesitate, and doubt the necessity of putting out money for what to them is an uncertain return in improved sanitary conditions, — in a possible deliverance from many of the ills that afflict communities, and in bequeathing, not only increased length of days to the individual, but, as cleanliness is said to be next to godliness, in thereby improving the whole moral condition of society. Yet, next to the supply of a plenty of good water, — the convenience and blessing of which none are so low as not to recognize, — comes the question of how shall the waste and filth, which its greatly enlarged use has gathered up in its countless errands of mercy, be disposed of so as not to entail a curse and a scourge in exchange for the blessing of an abundant water-supply? The subject is new, not only to a large proportion of individuals in any community, but to a great majority of communities throughout the world."

They then speak of the steps which they took to get information upon this important subject; and after referring to the fact that they employed Mr. Ball of Worcester, and Mr. Heald, as civil engineers, and obtained certain recommendations from them, they go on to say, —

"Your Committee, after duly considering this report, felt that the outlay of so large a sum for little more than a bare beginning of a system of sewerage would be greater than the town would feel justified in entering upon at the present time: yet feeling that something ought to be and could be done that would meet the present and all reasonably prospective needs of the town, at a much less expenditure of money, and having also become somewhat acquainted, through Mr. Ball, and his assistant Mr. Heald, with what is known as '*The Memphis System of Sewerage*,' they determined to call to their aid *Col. George E. Waring, jun.*, of Newport, R.I., *the acknowledged foremost sanitary engineer in the United States, if not in Europe also.*"

They then give at length the recommendations which he makes, and upon which they report favorably to the town. Of course, it is not necessary to refer to that report as evidence of what you, gentlemen, particularly the medical men upon your board, know perfectly well, that Col. Waring is the highest authority in this country upon sanitary engineering; and when the town of Millbury, in its desire to aid the city of Worcester, in response to the request of the State Board of Health, in solving this difficult problem, sought to find a man whose views would be of peculiar value, it went to one of national, not to say international, reputation. Now, why should that gentleman, who has simply come here and given expression to his honest views, be attacked? Apparently for two reasons: first, because he mentioned the fact that sewers in ancient times were not used for the purpose of carrying away the contents of privies; and, secondly, because he said that Worcester seemed not to be averse to giving to others a scourge which it was unwilling to concentrate upon itself. Those are the two expressions which particularly called forth the animadversions of my friend. Now, what did Col. Waring say on this subject?

"Under all ancient practices, a sewer is only a drain and channel for the removal of waters which the proper enjoyment of territory requires to be removed. Until well into the present century this was probably the only meaning of the term; and, up to that time, the office of a sewer was simply to furnish a safe outlet for rain-water, for soil-water, for the overflow of the backing-up of streams, etc. The use of these sewers for the removal of excrementitious and other refuse matters is very recent. In Boston, according to Mr. Eliot C. Clarke, as late as 1833, and in England much later, the admission of foul matters was prohibited."

I do not find in the poetical quotations with which the learned city solicitor entertained us as an answer to this statement a single thing that is inconsistent with it. That surface drainage in ancient times may have carried with it many substances which were regarded as foul then, and would be regarded as foul now, I have no doubt; but that sewers were established in the sense in which the term is used to-day, for the purpose of removing the contents of privies, is not proved, even by "Troilus and Cressida." On the contrary, on a subject of that kind, I should prefer to take the testimony of scientific men like Col. Waring, and the researches of a practical engineer like Mr. Clarke, rather than a line from Shakspeare or even old Stow.

Then, gentlemen, as to the other expression of Col. Waring, that Worcester was not averse to turning this sewage into the stream, and thereby inflicting on others evils which it was unwilling to concentrate upon itself. Is not that a true statement of the case? *Is* Worcester averse to doing it? I wish she was. If Worcester was averse to doing it, we should have no occasion to come before you. It is because she is not, that we are under the necessity of applying for legislation.

Now, gentlemen, to sum up this part of the case, what I have endeavored to show is, that notwithstanding all that has been said and written on this subject by men of science who have been appointed, in the discharge of a public duty, to investigate this matter, and notwithstanding all that has been shown in the course of this hearing, the city of Worcester, so far as its official representatives are concerned, makes no response to the representations of Millbury and her neighbors, but to discredit them and turn them aside. "It is none of our business," they say; "we only do what others do, and you must save yourselves." That is substantially their position, their *official* position. I beg to be distinctly understood as not referring to individuals who have expressed a different opinion. But you must deal with them as a corporation, and that is their position before this Legislature. They contend that the evils upon the Blackstone River do not exist or are greatly exaggerated, and they take a determined stand against doing any thing to remedy them.

My attention is called to a paragraph in "The Worcester Gazette," of the 20th of March, 1882, which seems to illustrate what I have said: —

"In view of the unexpected character of the testimony of Mayor Stoddard, at the sewage investigation at Boston on Friday, a meeting of the Sewer Committee was called Saturday afternoon, at the request of City-Solicitor Goulding. All the members were present; and on consultation it appeared that they and the solicitor had a common understanding of the position of the city in the case, and his course was fully approved. He has conducted the case in conformity to the views of the Committee, and no occasion was found to change his instructions or to make any new suggestions; but, in view of the use that may be made of the mayor's position, it was thought best to put the views of the Committee in form.

"It was therefore unanimously voted, by a yea and nay vote, that the Committee have intended to conduct, and propose to continue to conduct, the defence of the city on the grounds that the city is using Mill Brook as a sewer by virtue of legislative authority which has not been violated; that there is no such nuisance as is claimed by the petitioners; that no plan based on theory, yet suggested, nor which may be suggested, can justly be forced upon the city, while other cities and towns are equally interested; and the city solicitor is directed to defend on these grounds, and to take entire charge and direction of the case on behalf of the city.

"The mayor was present, and made some explanations designed to convince the Committee that his position did not tend to compromise the city's case, and did not conflict with the position of the city. In view of his position, — that he is not a member of joint standing committees, — his name was not called when the vote was taken."

The CHAIRMAN. I want to say one thing right here. I am very much surprised at the criticisms upon the mayor which appear by the articles which have been read. The answers of the mayor he was led into by questions put by the Committee, and mainly by myself as I remember, having been prompted by a line of thought that was in my mind at the time. He did not compromise the city of

Worcester; and his answers were proper, and apparently those of an honest man. I hold that when a man comes upon the stand to testify, whether he is a mayor or a private citizen, he is bound either to answer truthfully or to decline to answer at all. In this case the mayor gave certain answers in which there was a distinction between what he said as representing the city and what he said as a citizen. He answered, no doubt, honestly; and I am surprised that he should have been criticised for any thing he said, and think it injudicious that these criticisms should have appeared in print.

Mr. GOULDING. I desire to say, if you will allow me, that the appearance of those articles in the papers was without my knowledge and against my desire.

The CHAIRMAN. I had no reason to think otherwise, but I felt that it was proper to take this occasion to say what I have said. The natural conclusion to be drawn from these criticisms is, that the mayor ought either to have refused to answer, or to have uttered something besides the truth when he did answer. Such criticisms tend more in the direction of harm to the city of Worcester than any thing which the mayor said in his testimony.

Mr. MORSE. I am much obliged to the Chairman for his remarks. They are a proper comment on the conduct of the Worcester Committee on Sewerage in undertaking to pass this vote, which I submit was intended to be covertly a rebuke to the mayor for the position he had taken. All this discussion has proved that a most extraordinary pressure has been brought to bear from some source in Worcester, and that a man is not permitted to breathe even the ordinary instincts of humanity upon a subject of this kind without being at once publicly and officially rebuked.

Having thus called your attention to the attitude of these parties, I come to consider the evidence as to the nuisance. I shall not discuss merely the chemical analyses. We all understand the value of chemical analysis. For the purpose of ascertaining the component parts of a given solid or liquid there is no method so exact or satisfactory. But we also know that the value of the analyses as a basis for action depends very much upon who takes the samples, when the samples are taken, and for what object they are taken. It is possible to prove any thing by figures. My learned friend, for example, succeeded in demonstrating to the Committee, yesterday, by chemical analysis, that the water of the Blackstone was purer than the Cochituate, and that the people of Millbury, therefore, could drink the water of the Blackstone with more safety than the citizens of Boston drink the Cochituate. If his argument is good for any thing, it proves that; it is established beyond controversy; the citizens of Millbury can rest upon it. He has proved it by the comparative table of chemical analyses that he produced. Now, gentlemen, I am not going to attempt to argue to this Committee as to the effect produced upon this water by turning this enormous quantity of sewage into it. Here is Mill Brook, which receives 3,000,000 gallons of sewage per day. That sewage comprises every thing that is foul and disgusting in the refuse of 40,000 of the inhabitants of Worcester. The stream which receives it has a minimum dry-weather flow of not more than 1,000,000 gallons. Its waters, thus polluted, are turned into the Blackstone River, in immediate proximity to and within two or three miles, certainly, of a considerable town, where there are dams which were in existence long before the Act under which the city thus discharges its sewage. These dams necessarily arrest a very large part of the solid matter that comes through the sewers. It is not necessary, at this day, to discuss the injurious effects of sewage and sewer gases upon the health. Are they not established beyond all controversy? Does not every practising physician on this Committee, in investigating the causes of disease properly attributable to any trouble of this sort, look first to ascertain whether there is any defect in drainage by which sewer gas can possibly have escaped into the house? Is there any case of disease so potent, or any one to which physicians have directed their attention so much, during the last few years?

Let me read, in this connection, a single sentence from Dr. Alfred Carpenter's lectures on "Preventive Medicine in Relation to the Public Health," which sums up the facts in relation to the effect of sewer gases. On p. 97, he says, —

"The influence of sewer-air in setting up disease, in permitting and spreading epidemics, and keeping disease among us in the endemic form, need not now be insisted on. You may take from me that it is an undoubted scientific fact."

Now, Mr. Chairman, my proposition is this: We may differ among us as to the extent to which this evil has now grown. Some of you gentlemen may believe it is more serious, others may think it is less serious; but we shall agree, I think, in this: that there is an evil, and that its tendency is to increase, and that the only tendency of the carrying of this sewage matter down the Blackstone River is to create disease; it is not a health-giving operation. This mixed mass of filth must leave somewhere a certain amount of solid residue, which will constantly be giving off gases that are noxious. It will pollute the water to a greater or less extent. It is true that a running stream tends to purify itself; and if the stream is large enough, and runs far enough, it may eventually dissolve the impurities that are in it. But the State Board of Health, in an article to which I called attention in the examination of Mr. Worthen, distinctly stated that the popular notion in reference to the amount of oxidation that is produced in a running stream is erroneous; and they quote high authority to the effect that there must be in every stream which carries polluted matter a certain amount, greater or less, of noxious substances and gases. Mr. Worthen undertook in some way to criticise that opinion; but his statements were very vague: and so far as he differed from the State Board of Health, and the authorities which they cite, I doubt very much whether you would be disposed to follow him as against them. But he differed from them only as to the degree of pollution. Neither he nor any other intelligent man will doubt the proposition that it pollutes a stream to put sewage into it. And upon the testimony of Mr. Worthen himself, there can be no question but that this evil, within a comparatively short time, will be serious enough to demand legislation. Mr. Worthen said that if the entire population of Worcester were now to drain into Blackstone River, he would consider the evil so serious that something should be done about it. Only he thought, that, so long as the sewers were used by forty thousand out of the fifty-eight thousand people, it might be safe to wait; that no very great risk would be run. That is the extreme proposition, stated by their own expert, — that it may be safe to let this thing stand where it is to-day; but if the present entire population of Worcester should use the same, and, still more, if Worcester grows as it must grow in the future, there will be an amount of pollution that will make purification a necessity.

I shall not trouble the Committee with details of the evidence, except in a very few instances; but I must notice the testimony of three or four gentlemen whose position in Millbury or the other towns is such that they are specially entitled to a candid consideration.

Take first the testimony of the Rev. Philip Y. Smith, which is found upon p. 60 of the report. He was asked, —

"*Q.* — Are you familiar with the schools? *A.* — Yes, sir: I am a member of the board.

"*Q.* — In your opinion, does the pollution of the river affect the salubrity of the air about those schools? *A.* — In the Saundersville school —

"*Q.* — How far is Saundersville from Wilkinsonville? *A.* — A little less than half a mile, in a straight line, from the Sutton depot.

"*Q.* — Now, will you tell about the effect of the river upon the air at those schools? *A.* — In the months of April and May last, during the latter part of April, and two weeks in May, the schools in Saundersville were very much depleted, so that in one school, for at least ten days, there were only six scholars out of an average attendance of upwards of fifty-four; and in the upper school I think they were reduced to nine, out of an average of forty-five. The prevailing troubles there were measles and scarlatina, with diphtheria. There were two cases of diphtheria near my house. The children who were sick attended that school. Their names were Annie and Susie Redpath. They were attended by Dr. Thomas T. Griggs of Grafton.

"Q. — What was the state of health among the children in the other schools in Grafton; that is, away from the river? A. — In the Centre, the number of scholars was not as small from similar causes as in the schools nearer the river.

"Q. — That is, I understand, the sickness was not as great in the other schools as in those by the river? A. — That is my understanding, sir."

There is the statement of a gentleman entitled to belief, who occupies a responsible position in connection with the schools of Grafton, that, in one school the number of children had been reduced by sickness from an average attendance of fifty-four to six, and in another, from an average of forty-five to nine. This happened last summer. A year ago the city of Worcester opposed any attempt at a remedy; and, since that action, these troubles have come. How many cases of illness of children are to happen before something is done? Are we to go on another year, and then undertake to show that more children have been sick, and more schools depleted, only to be met again with the intimation that our dangers are imaginary? Have you any reason, gentlemen, to doubt the testimony of Mr. Smith? He tells you that the prevailing troubles in those schools were measles, scarlatina, and diphtheria. Now, it may be true, that in an individual case here or there, there may be a mistake. I will agree with my friends that it is entirely possible for diseases to be misunderstood and miscalled, and, in some cases of local excitement, for troubles to be exaggerated. But, making allowance for all those things, I ask you, gentlemen, whether you doubt that serious illness among these children was due to the polluted atmosphere in which they lived.

Take, then, the testimony of Dr. Wilmot, which you will find on p. 65 of the report. He is a practising physician living in Farnumsville, which is a village of Grafton. He is questioned as follows: —

"Q. — From your experience in Farnumsville, what do you say as to the effect upon the general health of the people of the present pollution of the river? A. — I should say it was decidedly injurious.

"Q. — Your practice is not confined to Farnumsville, but extends, does it not, to Saundersville. Wilkinsonville, and other villages? A. — Saundersville, Wilkinsonville, Sutton, and down as far as Whitins' and North Uxbridge.

"Q. — Now, will you state to the Committee any particular facts that you have noticed in regard to the effect of the river upon health? A. — I have noticed that, at low water, when the shores were exposed to the rays of the sun, the emanations were still more disagreeable and cogent, and also that the river was of a disgusting appearance, black and nasty, and at all seasons of the year had a certain amount of smell.

"Q. — What sicknesses have you noticed during your practice there? A. — There is a prevailing sickness, which is scarcely worthy the full name of typhoid fever. It is more like an intermittent fever. There is no distinct medical name for it. It assumes all the appearance of a mild typhoid, without going into the extreme stage of it, *purpuræ;* without having the purple spots, which are symptomatic of the true typhoid fever, but producing lassitude and debility for some five or six weeks. It goes under the common name in the country of "slow fever."

"Q. — Do you ascribe the cause of this disease to the river, wholly or in part? A. — To a great extent, I think it is, sir, particularly at low water. There are two cases in particular that I can state to you. I refer to two sisters in the village of Rochdale.

"Q. — In what town is Rochdale? A. — I cannot say.

"Q. — Is it not in Northbridge? A. — I think it is. In this village the pond was drained very low. It was drained down lower than the average of the ponds along the river, while they were making some repairs or alterations on the dam. That was none of my business, and I did not inquire what they were. The smell from the pond there was frightful. There is no modification of the word required, — it was perfectly frightful. It was worse than the wards of a hospital.

"Q. — What did it smell like? A. — It smelt exactly like a water-closet, — 'sulphuretted hydrogen' is the scientific term, — and continued for some length of time. The repairs were extensive that they were making."

There was an attempt, in the cross-examination of this witness, or some witness, to show that all rivers smell badly when the water is low. That is true undoubtedly; but when an educated physician like Dr. Wilmot comes here and tells you that "the smell from the pond there was frightful, there is no modification of the word required, — it was perfectly frightful; it was worse than the wards of a hospital; it smelt exactly like a water-closet, — 'sulphuretted hydrogen,'" — the Committee will not suppose that he is describing an ordinary river odor. Is it to be said that he has exaggerated and misrepresented? Can it be that there is some peculiar quality in the sewage of Worcester that it does not produce the results which other sewage produces? and that when this filth and excrement comes down from Worcester and settles in these ponds, gases are not given off such as privies ordinarily discharge? What is it, gentlemen, but making of those ponds the privies of the city of Worcester?

I call your attention to the testimony of Dr. Lincoln on pp. 94 and 95. Dr. Lincoln is a practising physician in Millbury, and has been there for sixteen years. He is asked, —

"Q. — Have you noticed any thing which would enable you to say that there was a change in the general health of Millbury during that time? A. — Yes, sir.

"Q. — Will you tell what you have noticed? A. — If the Committee will allow me, and the counsel do not object, I will make a simple statement, which perhaps will make it clearer than answering questions. I came to Millbury sixteen years ago last May. The population of Millbury in 1870 was 4,397, I think: what the census was the ten years previous I have forgotten, but, if my memory serves me, it was 3,900 and something; but I won't be positive as to that. When I came there, there were two physicians in town; and they thought there was no more than they could attend to well, that there was no place for a new man, — that they had nothing more than they cared to do. There are now six physicians there, five of them in active practice; and perhaps it is safe to say that any one of the five is doing as much business as either of the two that were there before. I think that answers the question of the gentleman whether there is more sickness there now than formerly.

"Q. — In other words, your answer is that there is? A. — There is.

"Q. — What have been some of the kinds of sickness which you would think might be attributed, either in whole or in part, to the foulness of the river? A. — Well, I should say that the common sicknesses had been mostly of the zymotic type, — what we call the filth diseases; perhaps scarlet fever, diphtheria, diarrhœal troubles, dysentery, and diseases of that character. The increase would be largely of that kind."

Then on p. 95 : —

"Q. — In general, what do you think the effect of the foulness of the river has been on the health of the people of Millbury, and the towns along the river, good or bad? A. — Bad. That is the idea; but the Committee will understand me, I know of no other reason to which to attribute the amount of disease more than previously."

Dr. Webber, a practising physician of Millbury for eleven years, says on p. 98 as follows : —

"Q. (By Mr. FLAGG.) — As to your practice there, in what way have you noticed the effect of the river upon the health of people, or what can you say as to that? A. — I should say its effect was bad.

"Q. — That is stating the matter generally, — now, have you any particular cases that you would speak of? A. — I would state first, if I may be allowed, generally; and I will then go into some particular cases. The foulness of the stream, and its offensive odor, are generally acknowledged. Such a stream emits such exhalations as are conceded by all sanitary authorities to be the producing causes, often, of zymotic diseases. That in a general way. I will say further, before alluding to specific cases, that I think it not right to consider entirely and exclusively the death-rate; that there are injurious influences which the figures of death-rates do not show."

He then reads an extract from Dr. John Simon's book, which, as it is printed in the report, I will not trouble the Committee by re-reading. He then goes on to state specific cases, and alludes to some which he thinks may be referred to this foulness of the river. It is at the close of his testimony on p. 99 that he makes a statement in reference to Mr. Benjamin Flagg, whose interest in the river Mr. Lincoln, in his report for the Commission of Public Grounds, thought it a good thing to joke about, and to sneer at ; and he states that Mr. Flagg died of diarrhœa, which, in his opinion, may be referred to the condition of the river. So that, although it was considered decent to ridicule Mr. Flagg for his efforts last year to do something to prevent this trouble in the river, it seems that finally he fell himself a victim to the scourge.

Now, gentlemen, when we bring before you these cases, and many more which I have not referred to ; when you are told that men could not row over this stream or dip an oar into it, because the disturbance of the water would raise such an offensive odor as to cause illness from which they did not recover for weeks; when you find that the hands of the mills were unable to perform their customary work, so that, as Mr. C. D. Morse showed you, the time lost in the last year in his establishment greatly exceeded any thing known in his experience, — the only response that we get is, "Oh, well! you are not very unhealthy down there; you are getting along pretty well; the tables of mortality show that you are not dying very fast, or as fast as people die in some other places; we do not kill you; we make you sick for a few weeks only; we shut up your schools only for a little while; your children's diseases will not probably cause permanent trouble; we keep the mill-hands away for a short time only from their work; you ought to be thankful that you are allowed to receive this great waste of our substance which we send down to you, and, in view of that great benefit, what matters it if there are a few zymotic diseases, more or less, in Millbury?" That is the argument.

Before I pass to a consideration of the tables of mortality upon which our friends rely, I must read a few more extracts from the medical evidence; first the testimony of Dr. Booth, also a resident physician of Millbury, on p. 148. He is asked, —

"*Q.* — Whether, or not, it is your opinion, that if the unpurified sewage of Worcester, as it increases, continues to be poured into the Blackstone River, there will be a cause there capable of producing epidemics throughout the valley? *A.* — I have no doubt of it whatever."

Dr. Folsom, on p. 154, says, —

"*Q.* — There is no doubt in your mind that there is a nuisance, to a greater or less extent, existing in consequence of the emptying of the sewage of the city of Worcester? *A.* — Oh! I should think one might state that fact beforehand without seeing the conditions. You have there the sewage of a city of over fifty thousand inhabitants emptying into a stream, the greatest flow of which in the dryest weather is seven hundred and fifty thousand gallons a day. Of course, I should say beforehand that that amount of sewage, coming into a stream of that sort, would necessarily be a nuisance. Of course, the degree and extent of the nuisance would be determined by the number of people living in the vicinity of the stream, and their nearness to the stream.

"*Q.* — Have you personally observed the smells as far down as Millbury? *A.* — One day when I was at Millbury it was quite offensive: the other days that I happened to be there I did not happen to notice very much smell as far down as that.

"*Q.* — Was it your opinion that that was a sewage smell? *A.* — I think that there is no question about that."

And on p. 155: —

"*Q.* — Have you made such examination as to satisfy you that the public health of Millbury and the region around there may be impaired in consequence of this? *A.*—My opinion would be that it is to a certain extent. I could not say how far without more thorough examination. I should want to go about there pretty minutely; and, in fact, I should want to have lived there during a season to be able to judge on that point.

"*Q.*—Suppose the death-rate of Millbury should be shown to have rather improved on the whole for a period of ten years: what should you say that indicated? *A.*—I should not think it necessarily indicated any thing."

I have already said that the death-rate is not a conclusive indication of the health of any community. It is, of course, one of the facts to be taken into account; but its weight in a particular case is to be determined with reference to a great many other facts. If, for instance, the character of the prevailing diseases was such that they were not ordinarily fatal, it is clear that the death-rate would not determine the condition of the public health. Again, the birth-rate is an important factor to be ascertained and borne in mind. But, while thus denying the conclusiveness of the tables of mortality, I have been furnished an analysis which indicates that these tables fail to prove that Millbury is specially healthy, and do show the reverse.

Mr. GOULDING. We do not claim that it is specially healthy.

Mr. MORSE. I thought the claim was that we are particularly healthy. That is the argument of Mr. Lincoln in his address, which I understand you to adopt.

Mr. GOULDING. As long as you have alluded to Mr. Lincoln, let me say in this connection what I stated to the Committee when I handed them the report, that it was the opinion of one citizen of Worcester, which he had formulated; and we put it in just as you had put in the testimony of Mr. Hoar. We did not adopt it in any other sense.

Mr. MORSE. Well, it seems to be so closely in line with your argument, that I supposed you had adopted it.

The analysis shows that in 1880 the number of deaths in Millbury from filth-diseases was 35, out of a total number of deaths, from all causes, of 89. This was an increase of such diseases of 69 per cent over the average for the preceding ten years. The death-rate from filth-diseases in the State of Massachusetts in 1880 was 4.27 in a thousand; in Worcester County it was 4.46 in a thousand; and in Millbury it was 8.16 in a thousand, or nearly double the rate in the county or state; and while in Worcester County the increase of deaths from filth-diseases in 1880 was only .55 per cent over the average for the last five years, in Millbury such increase was 4.23 per cent. That is to say, the increase of deaths from filth-diseases in Millbury was almost eight times as great as it was in Worcester County, over the average, for the last ten years. The tables of mortality, therefore, show that there has been a very great and alarming increase of fatal filth-diseases in Millbury.

But, as I said before, gentlemen, I do not think that it is necessary to go to any tables to determine exactly how much injury has been caused by this pollution of the stream. The necessary tendency is to cause injury. The extent of it probably no two men would agree upon. If you gentlemen were all to go to Millbury to-day, and devote yourselves for weeks to ascertaining the facts, it is not probable that any two of you would come to exactly the same conclusion as to the amount of disease that had been caused by this pollution of the stream. You might be equally honest and equally intelligent, but you would look at the matter from different points of view; you would have different medical theories, and you would vary in your estimates of the weight to be given different witnesses. All that I seek to maintain is, that you would agree, that we must all agree, that all intelligent men must say, that, if you precipitate the entire sewage of the city of Worcester into that small stream, you must necessarily deposit in the mill-ponds an enormous amount of noxious substances, which will emit offensive and injurious gases. The incident testified to by one of the witnesses, that he applied a lighted match to this gas, and that it ignited, was only an ocular proof of what would be inferred from the other testimony, that this was probably sulphuretted hydrogen that escaped from the water.

Now, gentlemen, here is the fact that the injury is done, that these sicknesses have been caused, and that the opinion of competent medical men is that the injury must increase. We then come to the practical question, What is to be done about it? The first answer that is made is, that we can remedy the trouble by removing the dams. This remedy, however, is neither a proper nor a practicable one.

The city of Worcester has no right to require the people below her to remove the dams which have existed for a long period anterior to the construction of her sewerage system. The rights in these mill-privileges are as much the property of the people along the Blackstone as the land which the citizen of Worcester owns is his property. Money has been invested in them; they sustain the industries of a large population, and to remove them is to cause irreparable damage. Worcester, as we contend, has no right to require us to adopt a remedy of that sort. The city might as well require our people to abandon their homes, and places of work and business. Esek Saunders, I think the oldest witness that was called here on either side of this case, — one of the venerable men whom the sewage of Worcester has preserved in life till this period, — on p. 87 gives the following testimony: —

"*Q.* (By Mr. FLAGG.) — What would be the effect upon the industries of Millbury, Grafton, Sutton, Northbridge, Uxbridge, and Blackstone, with their twenty-five thousand inhabitants and thirty-two hundred operatives, of taking down the dams? *A.* — Well, it would depopulate that country. That is the business that they have been brought up to, the business that they are calculated to carry on: I do not see any other business that they could adapt themselves to.

"*Q.* — Were those dams in existence long before the city of Worcester turned its sewage into the Blackstone River? *A.* — Oh, yes!"

But, even if the dams should be removed, it would not be a permanent remedy. That was the opinion of Mr. Worthen, and you will easily see the reason. In the first place, gentlemen, we cannot possibly control the State of Rhode Island. This river runs through Rhode Island to the sea. There are dams on the Blackstone in that State; and, so long as a single dam remains, the river will tend to fill up. But if all the dams in Rhode Island were also removed, although some temporary relief might be obtained, it would be merely postponing the evil day. Of course, the more obstructions you remove, and the steadier flow you give to a stream, the more chance there is of its freeing itself from impurities; but the time will come when a winding stream, which in dry weather has only a slight current, and must always have its shallow pools and its natural obstructions, and into which polluted matter is thrown, of which so large a part is solid, must necessarily become foul. You cannot by any temporary expedient, like removing the dams, keep clear and pure the water of a stream into which such an immense mass of filth is thrown. It is not possible.

I claim, then, that the suggestion that the towns below Worcester can cure their troubles by removing the dams is impracticable. No Legislature would compel it. Yet it could not be accomplished without legislation, as it would be impossible to obtain the consent of the various parties interested. Nor would the plan turn out in the end to give an adequate return for its great cost.

But the city further says that it is under no obligation in this matter; that it has suggested a remedy which we can take into our own hands; but that, if we do not see fit to adopt it, it has nothing further to say.

I do not intend, gentlemen, to enter upon a lengthy discussion of the law on this matter. I referred to the legal authorities, in opening, for the purpose of indicating to the Committee, and also to the city, the views which the petitioners took on the point of the legal liabilities of Worcester. We wanted it understood, that, while we consider that we have a remedy at law, yet we do not think that this is a case where we should undertake to exercise it without first attempting the more peaceful and comprehensive method of trying to obtain a revision of the statute under which the city took Mill Brook as a sewer.

We felt that it was not the part of good neighbors, in the first instance, to sue the city of Worcester. We felt, in regard to the city of Worcester, as you would if a neighbor of yours with whom you had always lived on good terms should suddenly establish a bone-factory, or rendering establishment, or some other nuisance which annoyed you, and was dangerous to your health and that of your family. You would not bring a suit immediately. You would state to him the injury he was doing, and remonstrate against its continuance, and endeavor by every reasonable method to satisfy him of the propriety of your request that he should so manage his establishment as to remove your cause of complaint. While, therefore, my clients have refrained from any hostile action, I felt it proper to state the legal aspect of the case as I understood it; and in that connection I must read to the Committee an extract from the opinion of the Supreme Court in a case which I did not read from in the opening, but which, I added to the citations in the printed report.

This case, which, in its essential features, is very similar to the present, is that of Haskell v. New Bedford, in 108 Mass. 208. The precise point to which I cite it is this: that the court will not construe a statute as authorizing anybody to commit a nuisance, whether it be a city or a town or an individual. My proposition is, that the city of Worcester is not authorized to commit a nuisance. The Act of 1867 does not authorize it to commit a nuisance; and, if it does commit a nuisance, it is liable to damages for the consequences. It may be enjoined in equity, or it may be indicted.

Now, in this case of Haskell v. New Bedford, it appeared that the plaintiff, Haskell, was the owner of a wharf in New Bedford. He complained that the city had "constructed a sewer outside of the dock, but opening into and upon the same, into which sewer many tenements and houses have entered their drains; that the number of the same has largely increased; and that now very large quantities of foul and disgusting substances are continually brought down and conducted by the city, by means of the sewer, into the dock, and have already greatly and illegally obstructed the same, and now cause a great and pestilential stench at the dock, and disturb and destroy the plaintiff's privilege of maintaining the dock and wharf; and that the city has done and is continuing to do this without any color or process of law, and to the great injury and common nuisance of all citizens, as well as to the private injury and nuisance of the plaintiff."

The city undertook to justify its acts on the ground that it was authorized by law to construct the sewer, and it cited various statutes to that effect. And, at the trial, the presiding judge was so impressed by the argument in behalf of the city, that he ruled that the plaintiff could not recover; that the city had done all these things by virtue of various acts of the Legislature, and that therefore it could not be held. But, on the hearing before the full court, the chief justice, Gray, announcing the opinion of the court, says in reference to this, —

"The right conferred upon the city of New Bedford to lay out common sewers 'through any streets or private lands' does not include the right to create a nuisance, public or private, upon the property of the Commonwealth, or of an individual, within tide-water."

He then goes on to discuss the various grounds upon which the plaintiff is entitled to remedy as against that nuisance, and the various remedies he may have.

This case, as the Committee will see, is in harmony with the case upon which I commented in opening, Badger *v.* Boston, in which it was held that if a city so constructed and managed a public urinal as that it became a nuisance, it was liable in damages, although it was authorized by statute to erect it. There is no question, gentlemen, about that proposition of law; and if the city of Worcester should succeed in defeating our application for legislation, they must meet claims based upon those decisions of the court.

But, if you believe that the city of Worcester has created and now maintains this nuisance in the honest belief that it was authorized under the Act of 1867 to do so, you will, it seems to me, see an occasion for a statute that shall amend the original Act rather than to compel us to take the slow, uncertain, vexatious, and expensive processes to which we shall otherwise be compelled to resort. Suppose, gentlemen, that, when this Act of 1867 was passed, you had been members of the Legislature, and had been appealed to to grant the city of Worcester authority to turn its sewage into Mill Brook; suppose you had then had the knowledge which the experience of the last fifteen years has given, — would you not say to the city of Worcester that, while it might turn its sewage into Mill Brook, it should not do it until after it had been properly purified? Is there any doubt that you would have put that into the Act? Suppose a town now applies to the Legislature (as I believe that one or two towns have applied this year) for permission to turn its sewage into a running stream — I think ordinarily that permission is refused; but, supposing it is granted, have you any doubt that you would affix to it a condition that the sewage should be purified before it is discharged into the stream? Have you any doubt that, in legislating hereafter, you ought to insert that provision, and would insert it? Now I ask, if you would insist upon that principle in reference to any new application, why should you not now insert that qualification in the Act of 1867? Does it make any difference that for fifteen years we have suffered from the consequences of its not being there? Is that a reason why nothing should now be done? One would say it is the strongest reason why something should be done, and that speedily.

I come, then, finally, to consider what ought to be done in this case. It is clear that the nuisance exists in a greater or less degree, and I do not care which for the purposes of the argument; that it is caused by the city of Worcester, which is undertaking to act under a statute that cannot properly be construed as authorizing it to create a nuisance. The question then is, What shall be done? I have already pointed out that the answer that the city of Worcester makes, that we are to apply the remedy by removing the dams, is no answer at all. There is no remedy that we can apply that would be effective, and there is no remedy that you would compel us to apply. But are we therefore helpless? Can nothing be done? Worcester says nothing can be done; that experiments for the purification of sewage have been made in other places, but that nothing satisfactory has been accomplished; that nothing has been proved, and that science and engineering skill have failed to prevent or alleviate the devastating effects of sewage turned into a running stream. Now, there never was a more extraordinary misstatement of the fact. Science has found practical methods which have been applied with success. Not only do the eminent experts whom we have called before you testify to this, but the accomplished engineer employed by the city of Worcester to criticise the plans of others, but not to suggest a practical remedy, makes no dissent to the statements of Dr. Folsom and Col. Waring as to the successful results of various systems for purifying sewage.

In considering the weight to be given to Mr. Worthen's opinion, it should be remembered that, though he is undoubtedly an able engineer, he has not made sanitary engineering a special pursuit, that he has made no personal examination of the Blackstone Valley, and that he was not asked to recommend a plan for remedying the trouble.

The State Board of Health suggested last summer that the city of Worcester should employ a scientific man to advise as to what could be done. It did not then employ any one; but last December or January, after the examination by the Board of Health, and when the city found that this question was coming up before the Legislature, upon the report of the board, and that there was to be an attempt for some measure of relief, it then employed Mr. Worthen, not to find a remedy, but to argue what could not be done. His mission seems to be to discourage all effort to overcome the present and threatened difficulties. You would conclude, that though other places have found it necessary to purify sewage, and have succeeded in doing so, yet that for some reason such action is neither necessary nor practicable here. But, aside from such qualifications as you may consider are made by Mr. Worthen's statements, the evidence is plenary that various systems of purification under conditions similar to those of Worcester have worked successfully. Experiments must necessarily continue to be tried hereafter, as all plans are capable of improvement; but the results thus far obtained are very gratifying. Col. Waring says, —

"The entire sewage of Dantzic, on the Baltic Sea, where the climate is quite as severe as anywhere in New England, has the entire effluent of its very complete system of sewerage well purified, winter and summer, by surface irrigation. About one-eighth of the sewage of Paris, made very foul by the removal of street-dirt in a putrid condition by the sewers, and by the very considerable contamination coming from public urinals and other sources, is perfectly purified by agricultural processes on the plain of Gennevilliers. A large portion of Berlin now sends all of its sewage to the irrigation-fields at Osdorf, where it is completely purified. Croydon, in England, which is a larger city than Worcester, has most successful purification-works close to its border. The great health-resort, Malvern, purifies its sewage by intermittent filtration. So does Kendal in the north. Leamington and Rugby use broad irrigation. Over fifty other towns in England purify their sewage in a similar manner. The places named I have visited personally, and I have made a careful examination of their purification-works. I might cite other towns where satisfactory purification is effected by chemical processes; but these seem to me so unsuited to the conditions we are considering, that the discussion of chemical purification is hardly worth while. Suitable works on either of the plans submitted can with entire safety be adopted for Worcester.

. . . .

"That there is any peculiarity in the climate or in the soil of Worcester which indicates a special difficulty in the adoption of the processes of purification by agricultural treatment is clearly disproved, by the long and satisfactory experience in this very manner in connection with the insane hospital located there."

I will ask my associate, Mr. Flagg, to read some extracts which he has prepared from a recent French work upon this subject which I have not had the opportunity to examine.

Mr. GOULDING. Are those extracts to come in without our having had an opportunity to read them or refer to them in the closing argument?

Mr. MORSE. If my friend has any objection to our reading them on the ground that he has not had an opportunity to examine them or comment upon them, I have no desire to press them. I do not suppose there are any very strict rules in hearing matters of this sort; but, if he insists upon the point, I will not ask the Committee to hear these extracts.

Mr. GOULDING. I only say I have had no opportunity to answer them because I knew nothing about them.

Mr. MORSE. That is perfectly true. Still, they are only additional authority in support of our proposition.

Mr. GOULDING. You may put them in, if you deem it important for your case. I will not object.

Mr. FLAGG. The extracts are from a work entitled, "Les Travaux d'Assainissement de Danzig, Berlin, Breslau. Par M. A. Durand-Claye. Extrait de la Revue d'Hygiene. Janvier et Fevrier, 1881."

On p. 3 the author speaks of the occasion of writing his article. As the work has not yet appeared in English, I shall translate literally as I read. He says, —

"The city of Odessa, Russia, having consulted me as to its sanitary matters, and especially as to questions relating to the location of its sewers, and the purification of its sewage-water, I have taken the opportunity, upon my return from Russia, to visit certain German cities, where I have examined the recent and extensive works undertaken by those municipalities.

"At the request of 'M. le directeur des travaux' of Paris, and of the 'commission of inspectors-general of bridges and streets,' — charged with the supervision of the public works of the city of Paris, — I have undertaken to sum up the result of the observations which I have been able to make.

"Three cities came principally under my observation, — Dantzic, Berlin, Breslau. Each of these had a special interest.

"Dantzic is subject to a climate quite wintry, which would seem to present peculiar difficulties in a sanitary point of view, as well within as without.

"Berlin is one of the largest cities of Europe. Its population is more than a million. Its works should be on a scale such as are necessary for such great capitals as Paris, London, and Vienna.

"Breslau has long been notoriously bad, in a sanitary point of view. It has now completed the last of its works. It ought to profit from the experience of the two other cities of Prussia cited."

Referring, first, to Dantzic, he describes its situation on p. 5, and says, —

"The climate of Dantzic, in spite of its nearness to the sea" [i.e., the Baltic], "is cold."

A table of monthly averages of temperature follows, which, the notation being changed from Centigrade to Fahrenheit, shows the climate to be quite equal to that of New England. On p. 8 he says, —

"At the same time that it has adopted, without hesitation, what is an excellent solution of the question of water-supply, it has not hesitated, even taking into consideration the great quantity of water to be disposed of, and although but a short distance from the Baltic, to reject the evil and BARBAROUS system of discharging its filth into the sea. Thanks to the studies of engineers" [whom he quotes], "the city of Dantzic has adopted the system of sewer-discharge of filth, and purification of its sewage by the soil and by vegetation."

After describing the works, upon p. 16, he says, —

"In a sanitary view, the results are complete. The sewage-water is wholly filtered by the earth, purified, and taken up by vegetable growth. The subterranean water is limpid, without the least odor, sometimes colored by natural constituents of the soil."

Upon p. 19, under the head of " Operation in Winter," he says, —

" The successful treatment continues without interruption all winter, in spite of the severity of the season."

On p. 20, under the head of " Berlin," he says, —

"The sanitary works of the city of Berlin are the same in principle as those of Dantzic, and, for that matter, of a great number of English cities; viz., —

"1st, Suppression of cesspools ; 2d, Network of sewers, with plenty of water, and receiving street and house sewage; 3d, Purification of sewage-water by irrigation."

On p. 22 : —

"The winter is severe, and like that of Dantzic."

He describes the works of Berlin, and, on p. 36, says, as to the results of purification, —

"The results obtained, in a sanitary point of view, are remarkable."

On p. 37 : —

" I have drunk of the effluent drainage: it is as clear, as pure, and as fresh as that obtained at Gennevilliers. It is impossible to recognize in the least any influence caused by previous sewage contamination."

On p. 39, under the head of " Breslau : " —

"This city has imitated Berlin in its interior works, its sewers, etc. For sanitary effects outside, that is to say, in its plan for purification of its sewage, it has imitated the works of Dantzic."

On p. 47 he states his " conclusions : " —

" I may sum up the results of my investigations in few words. In Germany it is admitted to-day, by all, without dispute, that municipal health depends upon three principles: —

"1st, Total drainage of filth by sewers. 2d, Abundant water in dwellings, and frequent flushing of sewers. 3d, Purification of sewage by the earth and vegetable growth.

"When I have conversed with our German colleagues upon questions affecting Paris, and especially as to any hesitation as to suppressing cesspools or purification of sewage, I have universally been met with expressions of profound astonishment. These points are considered settled abroad. Sixty-eight English cities purify their sewage by the soil. For twelve years the city of Paris has had in practice a like system at Gennevilliers."

Mr. MORSE. This authority, gentlemen, confirms the statement of Col. Waring. You have also the opinions of Dr. Walcott and Dr. Folsom to the same effect.

Now, all that it is necessary for the petitioners in this case to do is to show that some remedy can be applied. We are not bound to prove that the remedy would prevent all trouble. We are not bound to satisfy the Committee that the first experiment could be made without possibility of mistake, nor are we bound to prescribe the method, nor is this Committee called upon to prescribe the method, that shall be adopted. If we show that an evil exists, for which Worcester is responsible, then we are entitled to relief, if relief can be had. I agree that if no system can be adopted that will be of any benefit, the city can do nothing; and the only result must be what I have already indicated,— the depopulation of the region: but if relief can be given by any plan, it is immaterial, so far as this hearing is concerned, precisely what form of relief shall be adopted. The experts that the State Board of Health selected had peculiar qualifications for their work. In addition to Dr. Folsom and Dr. Walcott, the commission comprised Mr. Davis, whom many of this Committee know to have been one of the most eminent and valuable engineers ever in the employ of the city of Boston, — the one under whom the great system of sewerage now in process of construction was planned. He has considered this matter carefully as an engineer, with a view of determining the cost. You have his judgment in addition to that of Drs. Folsom and Walcott. The commission reports that, with an expenditure of four hundred and eight thousand dollars, it is possible for Worcester to establish a system of purification which will remove, for all time to come, all cause of complaint on this river. Col. Waring suggests a plan that would be less expensive, which would cost about half that amount. Those are the only plans which have been carefully examined and figured upon. It is rather remarkable — I call it to your attention in passing — that an able city engineer like Mr. Allen, who was present here last winter, and heard all the discussions on this subject, and whose attention must have been called to this matter more or less, has not been asked to consider, and has not, in fact, considered the question of remedy, and has made no figures and submitted no plan. The plan recommended by the commission appointed by the State Board of Health, which is more expensive than that advised by Col. Waring, does not, nevertheless, involve a cost disproportionate to the outlay upon the water and sewerage systems of Worcester, or beyond the reasonable ability of the city to pay. This plan involves, as I stated in the opening, and appears more fully in the report, the construction of new lateral sewers to conduct the sewage of the city, leaving Mill Brook to carry its natural flow only, the pumping of the sewage upon lands specially prepared to act as filters, different sections being used alternately, and the discharge of the water into the Blackstone only after it is thus freed from noxious substances.

I have been furnished a statement which shows the extent of the burden that would be imposed upon the city of Worcester, assuming that it should adopt the plan recommended by the State Board of Health, and that the cost of the plan would be $408,000 as estimated

If you assume that that amount is borrowed at three and a half per cent, the interest will be	$14,280
Add expenses for pumping, in accordance with the estimate of the State Board of Health	3,500
Add expenses for superintending and extra outlay for the first five years over and above the income	2,500
Then the total interest account and running expenses per year will be,	$20,280

This will amount to a tax of thirty-two cents per head for the first year, estimating the population at 63,000. Add for the sinking-fund twenty cents per head, and you will have a charge of fifty-two cents per year for the first year. These assessments will be reduced each year, so that at the end of twenty years the city will have money enough to pay the entire sinking-fund, and a handsome balance in the treasury. Again, assuming that the population of Worcester increases uniformly at the rate of twenty per cent in five years, — and that is in accordance with its past increase, — the rate of increase each year is 3.71 per cent. At this rate, the population will double in nineteen years. An annual appropriation of $10,000, invested at four per cent, would yield in twenty-five years over $400,000 (exactly $416,459.08) ; that is to say, if $10,000 were to be invested at four per cent each year for twenty-five years, at the end of that time it would pay this entire expense. An appropriation of $10,000 the first year, $10,400 the second, with an annual increase of four per cent upon the appropriation of the preceding year, to correspond with the increase of population, with the accumulated interest at four per cent per annum, would in twenty years amount to a little more than $400,000, — $400,243, counting nineteen annual payments with interest at the end of the twentieth year. In other words, by making an appropriation of $10,000 the first year, and $10,400 the second year, and so on, increasing annually four per cent upon the appropriations of the preceding year in order to correspond with the increase in population, you will have at the end of twenty years a sum sufficient to pay the amount recommended by the State Board of Health.

My object in introducing these figures is to satisfy the Committee that if the city of Worcester were to adopt the most expensive plan, — and nobody has asked that it be made obligatory upon the city of Worcester to take that plan, — it can with a very small annual appropriation meet this entire expenditure in twenty, or, at the farthest, twenty-five, years. Of course, their own financiers are better able than we are to suggest the ways in which they can deal with the question of raising the money; but these figures show that, for a great and growing city like Worcester, the expense is a comparatively small sum. It is nothing at all in comparison with the amount of injury that is being caused by the present condition of things.

But their final objection is, that granting all that can be said of the practicability of a system of purification, and of the reasonableness of the cost, yet the city of Worcester should not be compelled to adopt it, and that the utmost that should be passed is a bill permitting the city to adopt those measures. Well, gentlemen, it is hardly worth while seriously to argue that proposition. What chance is there of the city of Worcester doing this voluntarily? We know how difficult it is in ordinary cities to get an appropriation for many matters which look merely to the health and pleasure of their own citizens. Appropriations for sewerage, and for public parks, which are necessary or desirable for public health and the general benefit, are ordinarily very difficult to obtain. As a rule, towns oppose the adoption of a system of sewerage because of its expense. The city of Boston even, which has to-day a very large and admirable system under way for a part of the city, has come to it with great difficulty and after much opposition. And, as you are aware, the system of sewerage that is now talked of with reference to the Mystic Valley is opposed by most of the towns interested on account of its expense.

Now, I undertake to say that, when it is so difficult to obtain appropriations for the construction of a system for the protection of the health of the inhabitants of the place which is to have the immediate benefit, it is impossible to suppose that the city of Worcester will voluntarily assume an expense for the benefit of its neighbors, particularly when it believes that neither legally nor morally it is bound to do it. It will not be done, gentlemen. If you report a permissive bill, it is a purely harmless statute. It goes on the statute-book, nobody objects to it; and there it will lie, and be of no sort of value whatever. *Authorize* the city of Worcester to remedy this evil! Why, gentlemen, they have had ample opportunity all these years to do it, if they wanted to do something. They do not come here asking you for any act. They will accept it as a mild dispensation, and no doubt as a very happy relief. But the idea that the city of Worcester, whose committee met in brother Goulding's office, and adopted the grim resolution which I have read, after they thought that their cause had been given away here, — the idea of their voluntarily doing any thing, under a permissive bill, for the relief of their neighbors below them, is preposterous. They do not intend it, and they say they do not intend it. They are perfectly frank about it. Under those circumstances, what use is there in reporting a permissive bill?

But what we are entitled to, and what the justice of the State ought to give us, is an obligatory bill, requiring the city of Worcester to do something. We are not particular as to the limitation of time within which they shall adopt some plan. The period of four months was put into the bill presented by us, without any special consideration as to its sufficiency. I said the other day, and I repeat it now, that we are perfectly willing to extend that time to a year, so that another Legislature will sit before the city of Worcester is committed to any serious expense. The city will thus have ample time to consult proper authorities, and determine upon the best course to pursue. We do not desire a controversy with our neighbors. We have no desire to appeal to the court hastily. We simply ask this Legislature to so amend or construe the Act of 1867 as to prevent a continuance of this nuisance, and to establish some reasonable limit of time within which it shall be done. But, when you have fixed that time, we ask you to make the act efficient, and provide that within that time the city shall do something. It may consult any body it pleases; it may adopt any system; it may incur only such expense as shall be actually necessary, and the less expense that will enable them to carry out a proper plan will be the most satisfactory to us. We only ask that the city shall be required to do what a court of equity hereafter shall say is equitable, — what it can do reasonably to prevent and redress this wrong. The city of Worcester is a very powerful corporation. We admit it. We know perfectly well, that if that city oppose us with all its strength in this Legislature, through their senator and representatives, and through all the instrumentalities that it can bring to bear, try to defeat this measure, it will be very difficult to carry it through, even with the powerful help of a favorable report from this Committee. We understand that perfectly well. In comparison with Worcester, the towns which are injured are small, and their industries unimportant. Their people are of modest means. You may blot them all out of existence, and you will do nothing like the injury to the State, or its material interests, or to its greatness, that would be caused by destroying or even seriously injuring the city of Worcester. If it is the determination of the city of Worcester that nothing shall be done, that resistance shall be made to every application that these people make for relief from the great wrong which it is doing them, we appreciate the power of their opposition. And yet, gentlemen, remembering that the policy and proud tradition of this State have been justice to all, whether high or low, rich or poor, strong or weak, and considering that we have a cause that appeals to the sense of equity and justice, we believe, that if you shall grant this petition, and set forth the plain facts in your report, not only will the Legislature indorse your action, but the people of Worcester themselves, after they shall have forgotten the heat and irritation of this controversy, will be ready to admit frankly the justice of our claim, and the reasonableness of the obligation imposed upon them.

The following is the bill presented by Mr. Morse: —

AN ACT FOR THE PRESERVATION OF THE PUBLIC HEALTH IN THE TOWNS BORDERING UPON THE BLACKSTONE RIVER, AND OF THE PURITY OF THE WATERS OF SAID RIVER.

Be it enacted by the Senate and House of Representatives in General Court assembled, and by the authority of the same, as follows:

SECTION 1. The city of Worcester is hereby directed, within four months after the passage of this act, to provide for purifying from all offensive, contaminating, noxious, and polluting properties the waters or substances that may thereafter be discharged from its sewers into Blackstone River, so that said waters and substances shall not of themselves, or in connection with other matter, create a nuisance or endanger the public health; and said city thereafter shall cease to empty from its sewers into Blackstone River any waters or substances containing said properties until the same shall have been first so purified.

SECT. 2. Said city is hereby authorized to take and hold such lands, on or near Blackstone River, and to construct such works, as it may deem necessary, to enable said city to treat its sewage and free the same from all offensive, contaminating, noxious, and polluting properties and substances. Said city shall make compensation to the owners for such lands as it shall take under this act; and if said city and said owners do not agree, any person aggrieved shall be entitled to have his damages ascertained in the manner provided by law for the recovery of damages in the taking of lands for highways.

SECT. 3. The city of Worcester is hereby authorized to raise and appropriate, in such manner as its city government shall determine, such sums of money as shall be required by said city to carry out the provisions of this act.

SECT. 4. The supreme judicial court, or any justice thereof, in term time or vacation, sitting in equity for either of the counties of Suffolk or Worcester, shall have jurisdiction in equity to enforce the provisions of this act, by injunction or by any other appropriate equitable remedy, on complaint of the selectmen of any town in the county of Worcester situate on the Blackstone River.

INDEX.

	Page
ARGUMENT FOR PETITIONERS,—	
Opening	5-23
Closing	328-361
Bill	361
Map	opp. 364

WITNESSES CALLED BY COMMITTEE.

Folsom, Dr. Charles F.	153-165
Walcott, Dr. Henry P.	174-177

PETITIONERS' WITNESSES.

Bancroft, H. L.	136-140
Bates, Dr. J. N.	281
Booth, Dr. Robert	145-153
Chase, G. F.	29-37
Durand-Claye (quoted)	355
Durand, Nehemiah	71-74
Ewell, Rev. J. L.	74-77
Fisher, George W.	92-94
Gegenheimer, John	123-128
Greenwood, N. H.	24-29
Heap, Thomas	120-123
Hull, S. E.	44-48
Harrington, William H.	104, 105, 108-111
Hoar, Hon. George F. (quoted)	19
Johnston, B. J.	180
Lincoln, Dr. William H.	94-98
Morse, C. D.	140-144
Pratt, Herbert A.	129-133
Saunders, Esek	77-88
Simon, Dr. John (quotation)	98
Simpson, Peter	111-120
Smith, Joel	88-92
Smith, Rev. P. Y.	59-65
Thomson, Elijah	181
Verry, George F. (quoted)	10
Waring, George E., jun.	165-174
Webber, Dr. G. C.	98-104
Wheelock, Thomas	49-59
Whitin, Charles E.	177-180
Whitworth, Charles	133-136
Whitney, L. L.	37-44
Wilmot, Dr. Thomas	65-71

For City of Worcester.

	Page
Argument, Closing	281–327

Witnesses for Worcester.

	Page
Adams, Charles F.	185–199
Allen, Charles A.	265–270, 273, 274
Barnard, George A.	279, 280
Chamberlin, Robert H.	274, 275
Coes, Loring	276, 277
Dyson, Joseph M.	277, 278
Harrington, Stephen	213–225
Lovell, A. B.	229–234, 278, 279
McClellan, John	209–213
Martin, Dr. Oramel	199–204
Perry, Joseph S.	225–228
Pratt, Charles D.	207–209
Rice, Dr. J. M.	204–207
Stoddard, E. B., Mayor	234–253
Taylor, Lucien A.	271–273
Walker, Benjamin	275, 276
Worthen, William E.	253–265

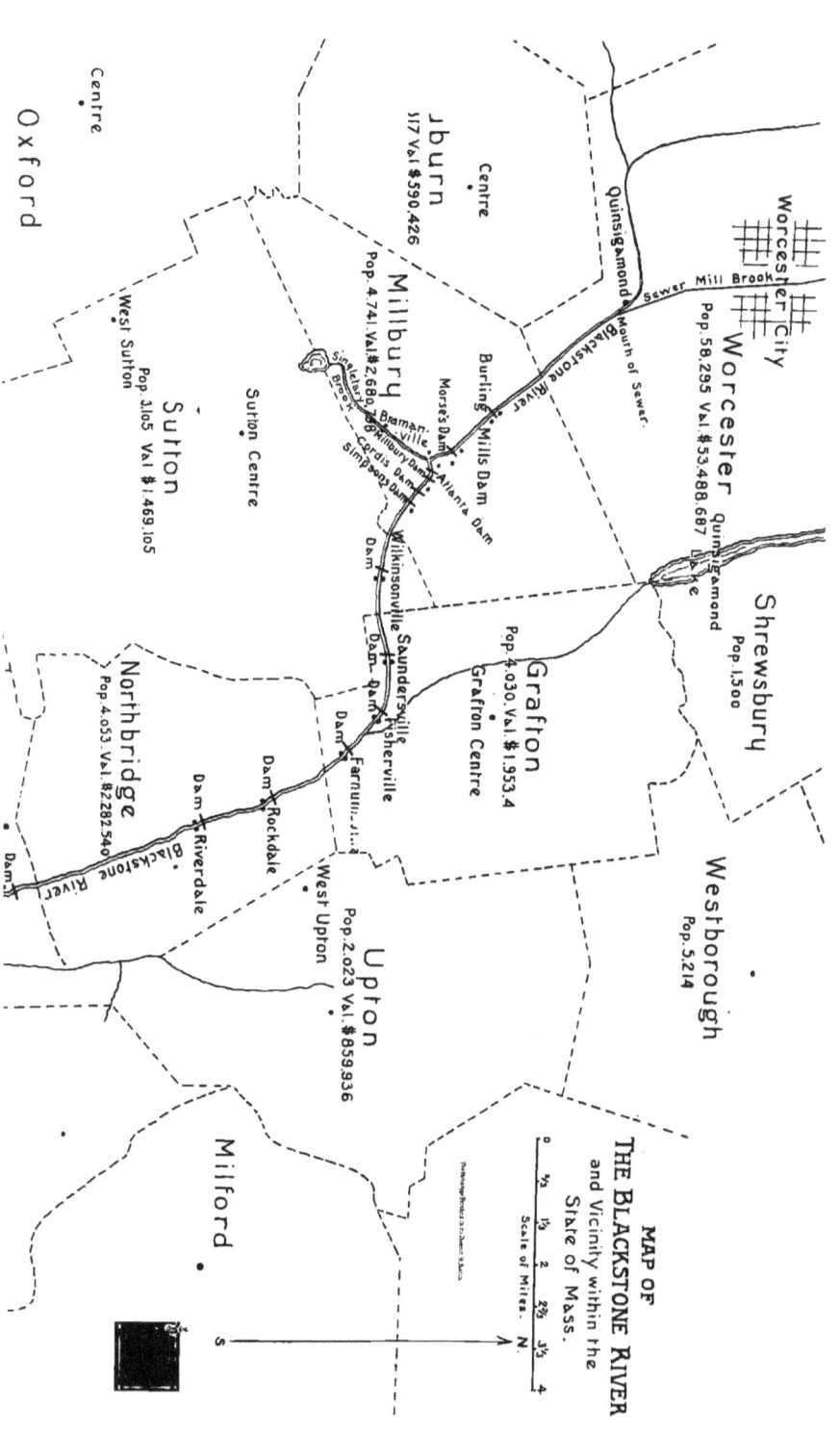

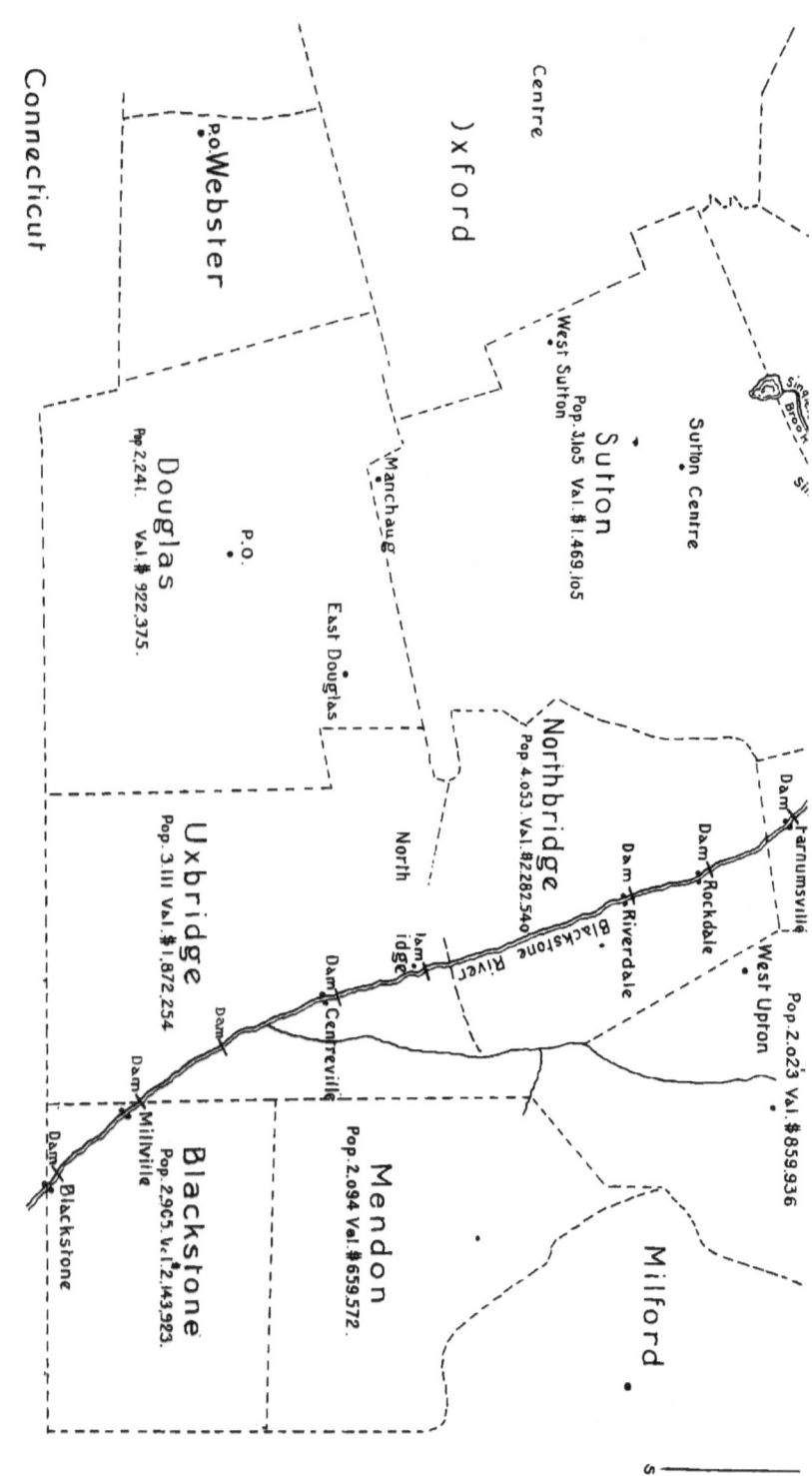

www.ingramcontent.com/pod-product-compliance
Lightning Source LLC
Chambersburg PA
CBHW020318240426
43673CB00039B/845